Communications and Control Engineering

For other titles published in this series, go to
www.springer.com/series/61

Series Editors

A. Isidori • J.H. van Schuppen • E.D. Sontag • M. Thoma • M. Krstic

Published titles include:

Stability and Stabilization of Infinite Dimensional Systems with Applications
Zheng-Hua Luo, Bao-Zhu Guo and Omer Morgul

Nonsmooth Mechanics (Second edition)
Bernard Brogliato

Nonlinear Control Systems II
Alberto Isidori

L_2-Gain and Passivity Techniques in Nonlinear Control
Arjan van der Schaft

Control of Linear Systems with Regulation and Input Constraints
Ali Saberi, Anton A. Stoorvogel and Peddapullaiah Sannuti

Robust and H_∞ Control
Ben M. Chen

Computer Controlled Systems
Efim N. Rosenwasser and Bernhard P. Lampe

Control of Complex and Uncertain Systems
Stanislav V. Emelyanov and Sergey K. Korovin

Robust Control Design Using H_∞ Methods
Ian R. Petersen, Valery A. Ugrinovski and Andrey V. Savkin

Model Reduction for Control System Design
Goro Obinata and Brian D.O. Anderson

Control Theory for Linear Systems
Harry L. Trentelman, Anton Stoorvogel and Malo Hautus

Functional Adaptive Control
Simon G. Fabri and Visakan Kadirkamanathan

Positive 1D and 2D Systems
Tadeusz Kaczorek

Identification and Control Using Volterra Models
Francis J. Doyle III, Ronald K. Pearson and Babatunde A. Ogunnaike

Non-linear Control for Underactuated Mechanical Systems
Isabelle Fantoni and Rogelio Lozano

Robust Control (Second edition)
Jürgen Ackermann

Flow Control by Feedback
Ole Morten Aamo and Miroslav Krstic

Learning and Generalization (Second edition)
Mathukumalli Vidyasagar

Constrained Control and Estimation
Graham C. Goodwin, Maria M. Seron and José A. De Doná

Randomized Algorithms for Analysis and Control of Uncertain Systems
Roberto Tempo, Giuseppe Calafiore and Fabrizio Dabbene

Switched Linear Systems
Zhendong Sun and Shuzhi S. Ge

Subspace Methods for System Identification
Tohru Katayama

Digital Control Systems
Ioan D. Landau and Gianluca Zito

Multivariable Computer-controlled Systems
Efim N. Rosenwasser and Bernhard P. Lampe

Dissipative Systems Analysis and Control (Second edition)
Bernard Brogliato, Rogelio Lozano, Bernhard Maschke and Olav Egeland

Algebraic Methods for Nonlinear Control Systems
Giuseppe Conte, Claude H. Moog and Anna M. Perdon

Polynomial and Rational Matrices
Tadeusz Kaczorek

Simulation-based Algorithms for Markov Decision Processes
Hyeong Soo Chang, Michael C. Fu, Jiaqiao Hu and Steven I. Marcus

Iterative Learning Control
Hyo-Sung Ahn, Kevin L. Moore and YangQuan Chen

Distributed Consensus in Multi-vehicle Cooperative Control
Wei Ren and Randal W. Beard

Control of Singular Systems with Random Abrupt Changes
El-Kébir Boukas

Nonlinear and Adaptive Control with Applications
Alessandro Astolfi, Dimitrios Karagiannis and Romeo Ortega

Stabilization, Optimal and Robust Control
Aziz Belmiloudi

Control of Nonlinear Dynamical Systems
Felix L. Chernous'ko, Igor M. Ananievski and Sergey A. Reshmin

Periodic Systems
Sergio Bittanti and Patrizio Colaneri

Discontinuous Systems
Yury V. Orlov

Constructions of Strict Lyapunov Functions
Michael Malisoff and Frédéric Mazenc

Controlling Chaos
Huaguang Zhang, Derong Liu and Zhiliang Wang

Stabilization of Navier-Stokes Flows
Viorel Barbu

Distributed Control of Multi-agent Networks
Wei Ren and Yongcan Cao

Bongsob Song · J. Karl Hedrick

Dynamic Surface Control of Uncertain Nonlinear Systems

An LMI Approach

Bongsob Song
Department of Mechanical Engineering
Ajou University
San 5, Wonchon-dong, Yeongtong-gu
443-749 Suwon
Korea, Republic of (South Korea)
bsong@ajou.ac.kr

J. Karl Hedrick
Department of Mechanical Engineering
University of California at Berkeley
5104 Etcheverry Hall
Mailstop 1740
94720 Berkeley
USA
khedrick@me.berkeley.edu

ISSN 0178-5354
ISBN 978-0-85729-631-3 e-ISBN 978-0-85729-632-0
DOI 10.1007/978-0-85729-632-0
Springer London Dordrecht Heidelberg New York

British Library Cataloguing in Publication Data
A catalogue record for this book is available from the British Library

Library of Congress Control Number: 2011929130

Cover design: VTeX UAB, Lithuania

Printed on acid-free paper

Springer is part of Springer Science+Business Media (www.springer.com)

Preface

This monograph provides a detailed practical and theoretical introduction to nonlinear control system design for the use of engineers and scientists. It will present systematic design methods, such as those that have been developed for linear systems, to address nonlinear problems. Linear control theory has a limited ability to tackle highly nonlinear systems. Our purpose here is to extend the recent developments in convex optimization of linear systems to nonlinear systems. A new nonlinear control algorithm known as "dynamic surface control (DSC)" using convex optimization is presented. It provides an effective design methodology for designing robust controllers for uncertain nonlinear systems. Sample applications will be provided to demonstrate how DSC can be effectively used to solve design problems in both the automotive and robotic fields.

This book is primarily intended for graduate students in nonlinear control theory, but can also serve as a source of applications for researchers in control design in the area of mechatronic systems such as automotive and robotic control. A wide variety of problems ranging from the design of DSC to extensions to output feedback, input saturation, multi-input multi-output, and fault tolerant control are considered. The results are shown to apply to a class of nonlinear and interconnected systems, in particular to automated vehicle control and biped robot control.

This book is divided in two parts. The first part addresses theoretical results for nonlinear control system design. In Chap. 2 a new method of analyzing the stability of a class of nonlinear systems by using the DSC design approach is presented. Based on quadratic stability theory, feasibility of the fixed controller gains for quadratic stabilization and tracking can be tested by solving a convex optimization problem. This approach is extended to problems with consideration of the following constraints, as we advance from chapter to chapter:

- a class of uncertainties: Chap. 3
- output feedback: Chap. 4
- input saturation: Chap. 5
- multi-input multi-output: Chap. 6.

The second part of the book introduces applications of theoretical results to vehicle and robot control. The relation between chapters and the results of the first part of the book is summarized as follows:

- Fault classification for vehicle control in Chap. 7: extension of the results in Chaps. 2 and 3 to switched nonlinear systems
- Fault tolerant control for an automated vehicle in Chap. 8: extension of the results in Chap. 4 to switched nonlinear systems
- Biped robot control in Chap. 9: application of the results in Chaps. 2 and 6 to interconnected mechanical systems.

The authors are particularly indebted to former graduate students in the Vehicle Dynamics and Control Lab at UC Berkeley: J. Green and M. Won contributed to the idea of multiple sliding surface control; S. Choi and D. McMahon conducted engine and vehicle control using multiple sliding surface control; D. Swaroop and C. Gerdes triggered the use of dynamic surface control for uncertain nonlinear systems and applied the idea to vehicle control; P. Yip extended the result to adaptive dynamic surface control; S. Raghavan and R. Rajamani provided a systematic procedure for nonlinear observer design. There are probably more names we should acknowledge for their contributions in a long line of simulations and applications. We would like to thank each one of them.

We also want to acknowledge the collaboration with California PATH at UC Berkeley. Especially more than 15 researchers including the first author developed the automated transit bus and demonstrated its feasibility in 2003, and some of results and pictures are included in this book. The implementation of the longitudinal control described in this book would not have succeeded without A. Howell, S. Dickey, and many other researchers at the California PATH research program.

The first author wishes to thank his former students at the Automatic Control Lab at Ajou University; J. Choi worked on simulations of the biped robot control. He also wants to thank his family for their endless support: his wife Moonjeung and sons Ryan and Kyle. Furthermore, he is grateful to Ajou University and UC Berkeley for providing an environment to write this monograph at Berkeley.

The second author would like to thank his wife Carlyle and daughters Ashley, Tristan and Ryan for their interest and support over the years. He would also like to thank his many, many students, who have taught him so much.

Suwon, Korea Bongsob Song
Berkeley, California, USA J. Karl Hedrick

Contents

Abbreviations[1]

Notation

$:=$	Defined as
$\approx$	Approximately equal
$\forall$	For all
$\subset$	Subset of
$\in$	Belongs to
$\rightarrow$	Tends to
$\Rightarrow$	Implies
$\Leftrightarrow$	Equivalent to, if and only if
max	Maximum
min	Minimum
sup	Supremum, the least upper bound
inf	Infimum, the greatest lower bound
$\Re$, $\Re^k$, $\Re^{m\times n}$	The set of all real numbers, real k-vectors, real $m \times n$ matrices
$\Re_+$	The set of all non-negative real numbers, $\Re_+ = [0, +\infty)$
$\mathrm{sgn}(\cdot) : \Re \rightarrow \Re$	The sign function: $\mathrm{sgn}(\sigma) = \lvert\sigma\rvert\sigma^{-1}$ for $\sigma \neq 0$, $\mathrm{sgn}(0) = 0$
[,] and (,)	Denotes a closed and an open interval, respectively
$\lVert x\rVert$ or $\lvert x\rvert$	Norm of an element x in a vector space
$\lVert x\rVert_2$	Euclidean norm of vector x
x_i	The ith element of vector x in $\Re^n$, respectively
$x \leq y$	$x_i \leq y_i$ for all $1 \leq i \leq n$ For vectors $x, y \in \Re^n$
$0_n \in \Re^n$	Denotes a zero vector in $\Re^n$
$\mathbf{0}_{mn} \in \Re^{m\times n}$	A rectangular zero matrix in $\Re^{m\times n}$
$\mathbf{0}_n$	A square zero matrix in $\Re^{n\times n}$
$\mathbf{I}_{mn}, \mathbf{I}_{m\times n} \in \Re^{m\times n}$	An identity matrix in the sense that all diagonal elements are 1 whatever the dimension of the matrix is

[1]Throughout the monograph, the following notation is used and other non-standard notation is defined when introduced in the text.

$\mathbf{I}_n$	A square identity matrix in $\Re^{n\times n}$
a_{ij}	The ith row and jth column element of matrix A in $\Re^{m\times n}$
A_i	The ith column vector of matrix A
$\bar{A}_i$	The ith row vector of matrix A
A^T	Transpose of matrix A, i.e., $(A^T)_{ij}=a_{ji}$
$\mathrm{Tr}(A)$	Trace of matrix $A\in\Re^{n\times n}$, i.e., $\sum_{i=1}^n a_{ii}$
$\lambda(A)$, $\lambda_i(A)$	All eigenvalues of matrix A and the ith eigenvalue
$\mathrm{Re}[\lambda_i(A)]$	The real value of the ith eigenvalue of A
$\lambda_{\max}(A)$, $\lambda_{\min}(A)$	The maximum and the minimum eigenvalues of matrix A, respectively
$A>(\text{or}\geq)\ 0$	Represents a positive definite (or semidefinite) matrix
$A>B$	A and B are symmetric and $A-B>0$
$\underline{A}$	A matrix where the last row of A is eliminated
$\mathrm{diag}(x,i)$	A square matrix of size $(m+i)$ with the vector $x\in\Re^m$ forming the ith super-diagonal of the matrix for positive integer i
$\mathrm{diag}(x,-i)$	A square matrix of size $(m+i)$ with the vector $x\in\Re^m$ forming the ith sub-diagonal of the matrix for positive integer i
$\mathrm{diag}(x):=\mathrm{diag}(x,0)$	A $n\times n$ diagonal matrix with the vector $x\in\Re^n$ forming the diagonal
$f:\mathscr{A}\to\mathscr{B}$	f maps the domain $\mathscr{A}$ into the codomain $\mathscr{B}$
$\dot{f}$	Time derivative of scalar (or vector) function f
$\mathscr{C}([t_0,t_1],\Re])$	Vector space of continuous functions $[t_0,t_1]\to\Re$
$\mathscr{C}^k([t_0,t_1],\Re])$	Vector space of continuous functions $[t_0,t_1]\to\Re$ with k continuous derivatives
$\mathscr{C}^k([t_0,t_1],\Re^n])$	Vector space of continuous functions $[t_0,t_1]\to\Re$ whose first k derivatives are continuous
$\mathscr{L}_2^n([t_0,t_1],\Re^n)$	The set of all piecewise continuous functions $u:[t_0,t_1]\to\Re^n$ such that $\|u\|_{\mathscr{L}_2}=\sqrt{\int_{t_0}^{t_1}u^T(t)u(t)\,dt}<\infty$
$\Psi:\mathscr{D}\to\Re_+$	A Gauge function such that $\Psi(z)>0$ for $\forall z\neq 0$; $\Psi(\alpha z)=\alpha\Psi(z)$ for $\forall\alpha\geq 0$, and $\Psi(y+z)\leq\Psi(y)+\Psi(z)$ for $\forall y,z\in\mathscr{D}$

Acronyms

ARE	Algebraic Riccati Equation
COP	Convex Optimization Problem
DIP	Double Impact Phase

DNLDI	Diagonal Norm-bounded Linear Differential Inclusion
DDSC	Decentralized Dynamic Surface Control
DSC	Dynamic Surface Control
DSP	Double-leg Support Phase
FDD	Fault Detection and Diagnosis
FMS	Fault Management System
FTC	Fault Tolerant Control
IB	Integrator Backstepping
LDI	Linea Differential Inclusion
LMI	Linear Matrix Inequality
LMIP	Linear Matrix Inequality Problem
LPF	Low-Pass Filter
LTI	Linear Time-Invariant
MIMO	Multi-Input Multi-Output
MSS	Multiple Sliding Surface control
NLDI	Norm-bounded Linear Differential Inclusion
ODSC	Observer-based Dynamic Surface Control
PID	Proportional Integral Derivative
PLDI	Polytopic Linear Differential Inclusion
PQB	Piecewise Quadratic Boundedness
PQS	Piecewise Quadratic Stability
ROA	Region Of Attraction
SQB	Simultaneous Quadratic Boundedness
SQS	Simultaneous Quadratic Stability
SISO	Single-Input Single-Output
SMC	Sliding Mode Control
SSP	Single-leg Support Phase
ZMP	Zero Moment Point

Part I
Theory

Chapter 1
Introduction

1.1 A Brief History of Dynamic Surface Control

There have been great advances in nonlinear feedback control based on differential geometric theory in the 1980s and thorough results using feedback linearization can be found in the literature [44, 64]. Due to the inability of feedback linearization to handle uncertainties, much attention has been given recently to Lyapunov-based control design techniques, such as integrator backstepping (IB) and sliding mode control [51, 52, 55, 80]. They have been studied for nonlinear systems with model uncertainties and applied to many control systems. Sliding mode control requires a heavier mathematical background than other smooth control techniques such as integrator backstepping. For instance, the theory of differential inclusion must be used to guarantee solutions in the case that differential equations have discontinuous functions of the state [4]. Also, the use of integrator backstepping is problematic because of an "explosion of terms" in the control law, while sliding mode control cannot generally be used for a system with mismatched uncertainties [107].

An alternative control design method called *multiple sliding surface* control (MSS), Slotine and Hedrick [79], Won and Hedrick [107], was developed independently of integrator backstepping but is mathematically very similar. It has been investigated in a long line of applications including engine control, Green and Hedrick [33], Choi [20], McMahon [61]. MSS, however, has the same problem as integrator backstepping in that it leads to an "explosion of terms" as will be shown in the following sections.

In order to avoid the drawback of both IB and MSS above, a robust nonlinear control technique called *dynamic surface control* (DSC) has been developed by Swaroop et al. [94], Hedrick and Yip [37], and Gerdes and Hedrick [29]. The DSC method is basically composed of MSS and a series of first-order low-pass filters. Due to the characteristics of MSS, DSC avoids the mathematical difficulties of an "explosion of terms" and was implemented for automated longitudinal control of a passenger vehicle including the throttle and brake dynamics [28, 37, 59]. Moreover, existence of DSC gains and filter time constants for semi-global stability was theoretically proved in [94] and it was extended to adaptive dynamic surface control to consider a specific class of parametric uncertainty [111].

B. Song, J.K. Hedrick, *Dynamic Surface Control of Uncertain Nonlinear Systems*,
Communications and Control Engineering,
DOI 10.1007/978-0-85729-632-0_1,

1.2 Sliding Mode Control

Great strides have been made in the past several decades in the area of controller design for nonlinear uncertain systems. Sliding mode control (or variable structure control) uses a discontinuous controller structure to guarantee perfect tracking for a class of systems satisfying the well known "matching conditions" [80, 102].

Consider the single-input second-order dynamic system with uncertainty

$$\begin{aligned} \dot{x}_1 &= x_2, \\ \dot{x}_2 &= u + f(x) + \Delta f(x) \end{aligned} \tag{1.1}$$

where $x = [x_1 \ x_2]^T \in \Re^2$ is the state, $u \in \Re$ is the control input, $f : \Re^2 \to \Re$ is a nonlinear function, and the uncertainty Δf is bounded such that $|\Delta f(x)| \le \rho(x)$. Suppose the control objective is $x_1(t) \to x_{1d}(t)$ in the presence of uncertainty. After defining the error $\tilde{x} = x_1 - x_{1d}$, we define a time-varying *sliding surface* $S(t)$ by the scalar equation $s(x,t) = 0$ as in [80], where

$$s(x,t) = \left(\frac{d}{dt} + \lambda\right)\tilde{x} = \dot{\tilde{x}} + \lambda\tilde{x}$$

and λ is a strictly positive constant. Taking a derivative of s,

$$\dot{s} = \ddot{\tilde{x}} + \lambda\dot{\tilde{x}} = \ddot{x}_1 - \ddot{x}_{1d} + \lambda\dot{\tilde{x}} = u + f + \Delta f - \ddot{x}_{1d} + \lambda\dot{\tilde{x}}.$$

Let a Lyapunov candidate function be $V = s^2/2 > 0$ for all nonzero s. Then

$$\frac{dV}{dt} = s\dot{s} = s\big(u + f + \Delta f - \ddot{x}_{1d} + \lambda\dot{\tilde{x}}\big). \tag{1.2}$$

Next, the control input is chosen as

$$u = -f + \ddot{x}_{1d} - \lambda\dot{\tilde{x}} - (\eta + \rho)\mathrm{sgn}(s)$$

to satisfy the *sliding condition* such that

$$\dot{V} = s\dot{s} = -s(\eta + \rho)\mathrm{sgn}(s) + s\Delta f \le -\eta|s| \quad \text{for } \eta > 0$$

outside of the set, $S(t)$ defined by $s(x,t) = 0$. $S(t)$ is an invariant set in the sense that once the system trajectories are on the surface, they are defined by the equation of the set itself

$$\left(\frac{d}{dt} + \lambda\right)\tilde{x} = 0.$$

This system's behavior on the surface is often called the *sliding mode*.

There have been several "smoothing" extensions to sliding mode control that still retain the concept of an "attractive" surface but eliminate the discontinuous nature of the control [80]. In general, the smoothing nature also eliminates the perfect tracking condition and results in a definable boundary layer around the desired surface. Recently, the area of robust nonlinear control has received a great deal of attention in the literature and has resulted in several new approaches. Most methods employ a Lyapunov synthesis approach. Corless and Leitmann applied this method to open-loop stable, mismatched systems [22].

1.3 Integrator Backstepping

1.3.1 Mismatched Uncertainties

When uncertainty is considered in a nonlinear system, it can be classified as either a *matched* or a *mismatched* uncertainty [55]. It is called a mismatched uncertainty when it appears in the state equation before the control input while the matched uncertainty enters the equation at the same point as the control input. Consider an example with the mismatched uncertainty as follows:

$$\begin{aligned} \dot{x}_1 &= x_2 + \Delta f_1(x_1), \\ \dot{x}_2 &= u \end{aligned} \tag{1.3}$$

where the uncertainty $|\Delta f_1| \in \mathscr{C}^1$ is an unknown function. The control objective is to stabilize the system in the presence of the mismatched uncertainty. If sliding mode control is applied to achieve robust stability, we first define the sliding surface

$$s = \dot{x}_1 + \lambda x_1.$$

After differentiating it and using (1.3), the surface error dynamics is written as

$$\begin{aligned} \dot{s} = \ddot{x}_1 + \lambda \dot{x}_1 &= \frac{d}{dt}(x_2 + \Delta f_1) + \lambda(x_2 + \Delta f_1) \\ &= u + \frac{\partial \Delta f_1}{\partial x_1}(x_2 + \Delta f_1) + \lambda(x_2 + \Delta f_1). \end{aligned}$$

However, since Δf_1 is unknown, $\frac{\partial \Delta f_1}{\partial x_1}$ cannot be computed, u cannot be found to make $dV/dt = s\dot{s} < 0$ for all x as similarly done in (1.2).

1.3.2 Design Methodology

Integrator backstepping is a recursive procedure that interlaces the choice of a Lyapunov function with the design of feedback control. The recursive design allows us to break up a design problem for a full system into a sequence of design problems for lower-order subsystems. Furthermore, its design with *nonlinear damping* guarantees boundedness in the presence of the mismatched uncertainty which is unknown but its bound is known. More detailed analysis and design tools can be found in the book by Krstić et al. [55]. Let us consider the following example to illustrate the backstepping approach:

$$\begin{aligned} \dot{x}_1 &= x_2 + f_1(x_1) + \Delta f_1(x_1), \\ \dot{x}_2 &= u \end{aligned} \tag{1.4}$$

where f_1 and $[\partial f_1/\partial x_1]$ are continuous on $\mathscr{D} \in \Re^2$, Δf_1 is locally Lipshitz on $\mathscr{D}$ to guarantee the existence and uniqueness of the solution of (1.4) and is unknown but is bounded by a known function $\rho_1 \in \mathscr{C}^1$ such that $|\Delta f_1| \le \rho_1(x_1)$. Furthermore, suppose the control objective is $x_1 \to 0$.

First, let $z_1 = x_1$ and after differentiating z_1 and using (1.4),

$$\dot{z}_1 = x_2 + f_1(x_1) + \Delta f_1(x_1). \tag{1.5}$$

To design a stabilizing function $\alpha(z_1)$ for x_2 in (1.5) with respect to $V_1(z_1) = z_1^2/2$, we examine the inequality $\dot{V}_1 < 0$, i.e.,

$$z_1\big[\alpha(z_1) + f_1(z_1) + \Delta f_1(z_1)\big] < 0, \quad \forall z_1 \neq 0. \tag{1.6}$$

To satisfy the inequality condition, let the stabilizing function $x_{2d} \triangleq \alpha(z_1)$ be

$$x_{2d} = -f_1(z_1) - z_1\frac{\rho_1^2}{2\varepsilon} - Kz_1. \tag{1.7}$$

Then, (1.6) is rewritten as

$$z_1\left[-z_1\frac{\rho_1^2}{2\varepsilon} - Kz_1 + \Delta f_1(z_1)\right] \leq -Kz_1^2 + \frac{\varepsilon}{2} \tag{1.8}$$

where the inequality comes from *Young's inequality*, which states that if the constants $p > 1$ and $q > 1$ are such that $(p-1)(q-1) = 1$, we have

$$xy \leq \frac{k^p}{p}|x|^p + \frac{1}{qk^q}|y|^q \tag{1.9}$$

for all $k > 0$ and all $(x, y) \in \Re^2$. Choosing $p = q = 2$ and $\varepsilon = k^2$, (1.9) becomes

$$xy \leq \frac{\varepsilon}{2}|x|^2 + \frac{1}{2\varepsilon}|y|^2. \tag{1.10}$$

After assigning $x = 1$ and $y = |z_1||\rho_1|$, it becomes the inequality used in (1.8):

$$\frac{z_1^2\rho_1^2}{2\varepsilon} + \frac{\varepsilon}{2} \geq |z_1|\big|\rho_1(x_1)\big| \geq z_1\Delta f_1(x_1). \tag{1.11}$$

Next, we add and subtract the stabilizing function $x_{2d} = \alpha(z_1)$ in the first row equation of (1.4) and define $z_2 := x_2 - x_{2d}$. Then, (1.4) can be written in the z coordinate as follows:

$$\begin{aligned}
\dot{z}_1 &= (x_2 - x_{2d}) + x_{2d} + f_1 + \Delta f_1(x_1) = z_2 - Kz_1 - z_1\frac{\rho_1^2}{2\varepsilon} + \Delta f_1(x_1), \\
\dot{z}_2 &= u - \dot{x}_{2d} = u + \phi_1(x_2 + f_1 + \Delta f_1)
\end{aligned} \tag{1.12}$$

where

$$\phi_1 = -\frac{\partial x_{2d}}{\partial x_1} = \frac{\partial f_1}{\partial x_1} + \frac{z_1\rho_1}{\varepsilon}\frac{\partial \rho_1}{\partial x_1} + \frac{\rho_1^2(x_1)}{2\varepsilon} + K. \tag{1.13}$$

We now select a Lyapunov function candidate by augmenting $V_1(z_1)$ with a quadratic term in the error variable z_2 as follows:

$$V(z) := \frac{z_1^2 + z_2^2}{2}.$$

The derivative of V along the solutions of (1.4) is

$$\dot{V} = z_1\left(z_2 - Kz_1 - z_1\frac{\rho_1^2}{2\varepsilon} + \Delta f_1\right) + z_2\{u + \phi_1(x_2 + f_1 + \Delta f_1)\}$$
$$= -Kz_1^2 + z_1\left(-z_1\frac{\rho_1^2}{2\varepsilon} + \Delta f_1\right) + z_2\{u + z_1 + \phi_1(x_2 + f_1 + \Delta f_1)\}.$$

As derived in (1.7), similarly u is defined recursively using the idea of nonlinear damping [55] as

$$u = -z_1 - \phi_1(x_2 + f_1) - \frac{\phi_1^2\rho_1^2 z_2}{2\varepsilon} - Kz_2 \tag{1.14}$$

where the controller gain K in (1.7) and (1.14) is assumed to be equal for clarity of exposition. Consequently, the derivative of V is

$$\dot{V} = -Kz_1^2 + z_1\left(-z_1\frac{\rho_1^2}{2\varepsilon} + \Delta f_1\right) - Kz_2^2 + z_2\left(-\frac{\phi_1^2\rho_1^2 z_2}{2\varepsilon} + \phi_1\Delta f_1\right)$$
$$\leq -2KV + \varepsilon$$

where the inequality comes from Young's inequality such as (1.11) and

$$\frac{\phi_1^2\rho_1^2 z_2^2}{2\varepsilon} + \frac{\varepsilon}{2} \geq |z_2||\phi_1||\rho_1| \geq z_2\phi_1\Delta f_1(x_1) \tag{1.15}$$

by choosing $x = 1$ and $y = |z_2||\phi_1||\rho_1|$ in (1.10). Therefore, $\dot{V}(t) \leq -2KV(t) + \varepsilon$ results in ultimately uniformly bounded stabilization of the state. Since ε is arbitrary, the ultimate error bound in stabilization can be made arbitrarily small. It is remarked that (1.13) and (1.14) begin to show the property that we have called the "explosion of terms" that occurs in integrator backstepping and will also occur in multiple sliding surface control.

1.4 Multiple Sliding Surface Control

A procedure similar to integrator backstepping, called multiple sliding surface control (MSS) was developed to simplify the controller design of systems with mismatched uncertainty at about the same time as integrator backstepping. A "multiple-surface" method was suggested by Green and Hedrick when sliding mode control was applied to a speed tracking controller of an automobile engine whose model does not satisfy the matching condition [33]. This approach was implemented successfully and showed an advantage in implementation over conventional sliding surface control methodology. To illustrate the principle of multiple sliding surface control, consider the same nonlinear system in (1.4) with the control objective $x_1 \to 0$. However, the model uncertainty is assumed to be Lipschitz for clarity of exposition as follows:

$$|\Delta f_1| \leq \gamma|x_1| \tag{1.16}$$

where $\gamma > 0$ is a Lipschitz constant.

Applying MSS control to the system, the first sliding surface is defined as $S_1 = x_1$. After differentiating S_1 and using (1.4),

$$\dot{S}_1 = x_2 + f_1 + \Delta f_1.$$

Then, the second sliding surface is defined as

$$S_2 = x_2 - x_{2d}$$

where x_{2d} called the synthetic input will be designed to drive $S_1 \to 0$. The derivative of S_2 is

$$\dot{S}_2 = u - \dot{x}_{2d}.$$

Then, (1.4) is written in the term of S_1, S_2, and x_{2d} as follows:

$$\begin{aligned} \dot{S}_1 &= S_2 + x_{2d} + f_1 + \Delta f_1, \\ \dot{S}_2 &= u - \dot{x}_{2d} \end{aligned} \tag{1.17}$$

where u will be designed to drive $S_2 \to 0$.

Assuming that $x_2 \to x_{2d}$ and thus $S_2 \to 0$, a reasonable choice for x_{2d} would be to set

$$x_{2d} = -f_1 - K_1 S_1 \tag{1.18}$$

where K_1 is a controller gain and is determined later. Then, the control input u is chosen to drive $s_2 \to 0$ such as

$$u = \dot{x}_{2d} - K_2 S_2.$$

However, there is uncertainty in $\dot{x}_{2d}$:

$$\dot{x}_{2d} = -\left(\frac{\partial f_1}{\partial S_1} + K_1\right)\dot{S}_1 = -\left(\frac{\partial f_1}{\partial S_1} + K_1\right)(S_2 + x_{2d} + f_1 + \Delta f_1).$$

Using (1.18),

$$\begin{aligned} \dot{x}_{2d} &= -\left(\frac{\partial f_1}{\partial S_1} + K_1\right)(S_2 - K_1 S_1 + \Delta f_1) \\ &= -\left(\frac{\partial f_1}{\partial S_1} + K_1\right)(S_2 - K_1 S_1) - \left(\frac{\partial f_1}{\partial S_1} + K_1\right)\Delta f_1. \end{aligned}$$

Ignoring Δf_1 in $\dot{x}_{2d}$, we let u be

$$u = -\left(\frac{\partial f_1}{\partial S_1} + K_1\right)(S_2 - K_1 S_1) - K_2 S_2. \tag{1.19}$$

If both (1.18) and (1.19) are put into (1.17), the closed-loop error dynamics becomes

$$\begin{aligned} \dot{S}_1 &= S_2 - K_1 S_1 + \Delta f_1, \\ \dot{S}_2 &= -K_2 S_2 + \left(\frac{\partial f_1}{\partial S_1} + K_1\right)\Delta f_1. \end{aligned} \tag{1.20}$$

To determine stability, let the Lyapunov function candidate be

$$V = \frac{S_1^2 + S_2^2}{2}$$

and its derivative along the solution of (1.20) is

$$\dot{V} = S_1\dot{S}_1 + S_2\dot{S}_2 = S_1(S_2 - K_1S_1 + \Delta f_1) + S_2\left\{-K_2S_2 + \left(\frac{\partial f_1}{\partial S_1} + K_1\right)\Delta f_1\right\}.$$

Since f_1 is a $\mathscr{C}^1$ and Lipschitz function, there is a positive constant M such that

$$\left|\frac{\partial f_1(S_1)}{\partial S_1}\right| \le M.$$

Using (1.16) and the inequality above, the derivative of V is bounded as

$$\begin{aligned}\dot{V} &\le -K_1S_1^2 - K_2S_2^2 + S_1S_2 + \gamma S_1^2 + \gamma(M+K_1)|S_1||S_2|\\ &\le -K_1S_1^2 - K_2S_2^2 + \frac{S_1^2+S_2^2}{2} + \gamma S_1^2 + \frac{\gamma(M+K_1)}{2\varepsilon}S_1^2 + \frac{\varepsilon\gamma(M+K_1)}{2}S_2^2\end{aligned}$$

where the last inequality comes from Young's inequality as

$$S_1S_2 \le \frac{S_1^2+S_2^2}{2}, \qquad |S_1||S_2| \le \frac{S_1^2}{2\varepsilon} + \frac{\varepsilon S_2^2}{2} \tag{1.21}$$

for any positive number ε. If ε is chosen as $\varepsilon := \gamma(M+K_1)$, the upper bound of $\dot{V}$ is

$$\dot{V} \le -(K_1 - \gamma - 1)S_1^2 - \left(K_2 - \frac{1}{2} - \frac{\gamma^2(M+K_1)^2}{2}\right)S_2^2.$$

Let $K_1 := K + \gamma + 1$ and $K_2 := K + \frac{1+\gamma^2(M+K_1)^2}{2}$ where $K > 0$. Then

$$\dot{V} \le -2KV.$$

Therefore, it results in $V(S) \le V(S(0))e^{-2Kt}$ for all error trajectories $S(t) = [S_1\ S_2]^T \in \Re^2$ and it is shown that the overall system with MSS becomes exponentially stable.

In the non-Lipschitz case of (1.4), we use same multiple sliding surfaces, S_1 and S_2. A choice for x_{2d} to make $S_1\dot{S}_1 < 0$ is modified as

$$x_{2d} = -f_1 - KS_1 - \rho_1\operatorname{sgn}(S_1) \tag{1.22}$$

where sgn is the signum function. The control input u is chosen to drive $s_2 \to 0$ such as

$$u = \dot{x}_{2d} - KS_2.$$

As a result, the closed-loop error dynamics is described by

$$\begin{aligned}\dot{S}_1 &= S_2 - KS_1 + \Delta f_1 - \rho_1\operatorname{sgn}(S_1),\\ \dot{S}_2 &= -KS_2.\end{aligned}$$

When the Lyapunov function candidate is supposed to be $V = (S_1^2 + S_2^2)/2$, its derivative is

$$\begin{aligned}\dot{V} &= S_1\dot{S}_1 + S_2\dot{S}_2 = S_1\{S_2 - KS_1 + \Delta f_1 - \rho_1\operatorname{sgn}(S_1)\} - KS_2^2\\ &\le -K\left(S_1^2 + S_2^2\right) + S_1S_2\end{aligned}$$

which can be made negative definite for a choice of $K > 1/2$. The difficulty of this approach is also to compute $\dot{x}_{2d}$ since $\dot{S}_1$ involves Δf_1, which is unknown, and $\mathrm{sgn}(S_1)$ is a discontinuous function. The earlier work on this problem used an ad hoc approach by numerical differentiation, i.e.,

$$\dot{x}_{2d}(n) \approx \frac{x_{2d}(n) - x_{2d}(n-1)}{\Delta T} \tag{1.23}$$

where ΔT is the sample time. This MSS control approach has been implemented in many applications ranging from active suspension and fuel-injection control to automated vehicles [33, 36, 37].

Instead of a discontinuous function in (1.22), x_{2d} can be redesigned using the idea of nonlinear damping as follows:

$$x_{2d} = -f_1 - KS_1 - \frac{\rho_1^2 S_1}{2\varepsilon_1} \tag{1.24}$$

where ε_1 is an arbitrary positive constant. Then, the derivative of x_{2d} is

$$\dot{x}_{2d} = -\eta_1 \dot{S}_1 = -\eta_1(S_2 + x_{2d} + f_1) - \eta_1 \Delta f_1$$

where

$$\eta_1 = \frac{\partial f_1}{\partial S_1} + K_1 + \frac{\rho_1^2}{2\varepsilon_1} + \frac{S_1}{\varepsilon_1}\frac{\partial \rho_1}{\partial S_1}.$$

As in (1.19), similarly the control input is

$$u = -\eta_1(S_2 + x_{2d} + f_1) - K_2 S_2 - \frac{\rho_1^2 S_2}{2\varepsilon_2} \tag{1.25}$$

where ε_2 is an arbitrary positive constant. Then, the closed-loop error dynamics is

$$\begin{aligned} \dot{S}_1 &= S_2 - K_1 S_1 + \Delta f_1 - \frac{\rho_1^2 S_1}{2\varepsilon_1}, \\ \dot{S}_2 &= -K_2 S_2 - \frac{\rho_1^2 S_2}{2\varepsilon_2} + \eta_1 \Delta f_1 \end{aligned} \tag{1.26}$$

and the derivative of $V = \frac{S_1^2 + S_2^2}{2}$ along the trajectories of (1.26) is

$$\begin{aligned} \dot{V} &= S_1\left(S_2 - K_1 S_1 + \Delta f_1 - \frac{\rho_1^2 S_1}{2\varepsilon_1}\right) + S_2\left(-K_2 S_2 - \frac{\rho_1^2 S_2}{2\varepsilon_2} + \eta_1 \Delta f_1\right) \\ &= -K_1 S_1^2 - K_2 S_2^2 + S_1 S_2 + S_1\left(\Delta f_1 - \frac{\rho_1^2 S_1}{2\varepsilon_1}\right) + S_2\left(\eta_1 \Delta f_1 - \frac{\rho_1^2 S_2}{2\varepsilon_2}\right). \end{aligned}$$

Since there is a maximum constant M such that $|\eta_1| \le M$ for all S in a convex and compact set $\mathscr{D}$ by continuity of η_1 and $|\Delta f_1(x_1)|$ is bounded by $\rho_1(x_1)$, using Young's inequality in (1.21)

$$\begin{aligned} S_1 \Delta f_1 &\le |S_1||\Delta f_1| \le |S_1|\rho_1 \le \frac{\varepsilon_1}{2} + \frac{\rho_1^2 S_1^2}{2\varepsilon_1}, \\ S_2 \eta_1 \Delta f_1 &\le |S_2||\eta_1||\Delta f_1| \le M\rho_1|S_2| \le \frac{\varepsilon_2 M^2}{2} + \frac{\rho_1^2 S_2^2}{2\varepsilon_2}. \end{aligned}$$

Therefore,

$$\dot{V} \leq -K_1 S_1^2 - K_2 S_2^2 + S_1 S_2 + \frac{\varepsilon_1}{2} + \frac{\varepsilon_2 M^2}{2}.$$

Let $K_1 = K_2 = K + 1/2$ and $\varepsilon_2 = \varepsilon_1 / M^2$. Then

$$\dot{V} \leq -2KV + \varepsilon_1.$$

Thus, the error eventually resides in a ball of radius ε_1 which is chosen arbitrarily. In spite of the semi-global regulation, MSS also has the property of "explosion of terms" due to repeated differentiation of nonlinear functions and the need to bound the uncertainties, e.g., calculation of η_1 and the addition of $\frac{\rho_1^2 S_2}{2\varepsilon_2}$ in (1.25).

1.5 Dynamic Surface Control

1.5.1 Motivating Example

A dynamic extension to MSS control that overcomes the problem of explosion of terms associated with the integrator backstepping technique as well as the MSS method has been developed. The first structured approach to the use of dynamic filters in the framework of MSS is found in [28] and called "dynamic surface control (DSC)" while the use of a low-pass filter to smooth the signal produced by (1.23) for calculating $\dot{x}_{2d}$ is discussed in [33]. For better understanding, DSC is applied to the example discussed earlier. Let

$$S_1 = x_1,$$
$$\dot{S}_1 = x_2 + f_1 + \Delta f_1.$$

Then, the synthetic input x_{2d} in (1.24) is renamed as

$$\bar{x}_2 = -f_1 - K S_1 - S_1 \frac{\rho_1^2}{2\varepsilon}$$

and $\bar{x}_2$ is passed through a first-order filter, i.e.,

$$\tau \dot{x}_{2d} + x_{2d} = \bar{x}_2, \quad x_{2d}(0) = \bar{x}_2(0) \tag{1.27}$$

where τ is the filter time constant. We now define the second sliding surface

$$S_2 = x_2 - x_{2d}$$

using the filtered signal (x_{2d}) instead of $\bar{x}_2$. Then, the control input is obtained by

$$u = \dot{x}_{2d} - K S_2 = \frac{\bar{x}_2 - x_{2d}}{\tau} - K S_2. \tag{1.28}$$

If $\bar{x}_2$ and x_{2d} are added and subtracted in the first row of (1.4), and u in (1.28) is applied to the second row of (1.4), (1.4) is written as

$$\begin{aligned} \dot{S}_1 &= (x_2 - x_{2d}) + (x_{2d} - \bar{x}_2) + \bar{x}_2 + f_1 + \Delta f_1, \\ \dot{S}_2 &= -K S_2. \end{aligned} \tag{1.29}$$

After defining the filter error, $\xi_2 = x_{2d} - \bar{x}_2$, and including the low-pass filter dynamics in (1.27), the augmented closed-loop dynamics are

$$\begin{cases} \dot{S}_1 = S_2 + \xi - KS_1 + \Delta f_1 - S_1 \frac{\rho_1^2}{2\varepsilon}, \\ \dot{S}_2 = -KS_2, \\ \dot{\xi}_2 = -\frac{\xi_2}{\tau} + \frac{d}{dt}(f_1 + S_1 \frac{\rho_1^2}{2\varepsilon} + KS_1) := -\xi_2/\tau + \eta(S_1, S_2, \xi_2, K, \varepsilon), \end{cases} \tag{1.30}$$

where η is a nonlinear function of S_1, S_2, and ξ_2. Suppose a Lyapunov function candidate is as follows:

$$V(z) := \frac{S_1^2 + S_2^2 + \xi^2}{2}. \tag{1.31}$$

The derivative of V is

$$\begin{aligned} \dot{V} &= S_1\dot{S}_1 + S_2\dot{S}_2 + \xi_2\dot{\xi}_2 \\ &= S_1\left(S_2 + \xi_2 - KS_1 + \Delta f_1 - S_1\frac{\rho_1^2}{2\varepsilon}\right) - KS_2^2 + \xi_2(-\xi_2/\tau + \eta) \\ &\le \frac{2S_1^2 + S_2^2 + \xi_2^2}{2} - K\left(S_1^2 + S_2^2\right) + \varepsilon - \frac{\xi_2^2}{\tau} + \frac{\xi_2^2\eta^2}{2\varepsilon} \end{aligned} \tag{1.32}$$

where the last inequality comes from Young's inequality

$$\begin{aligned} &\frac{S_1^2 + S_2^2}{2} \ge S_1S_2, \qquad &&\frac{S_1^2 + \xi_2^2}{2} \ge S_1\xi_2, \\ &\frac{S_1^2\rho^2}{2\varepsilon} + \frac{\varepsilon}{2} \ge |S_1|\rho \ge S_1\Delta f_1, \qquad &&\frac{\xi_2^2\eta^2}{2\varepsilon} + \frac{\varepsilon}{2} \ge \xi_2\eta. \end{aligned} \tag{1.33}$$

Consider the set $\mathscr{B} = \{z \in \Re^3 | S_1^2 + S_2^2 + \xi_2^2 \le 2p,\ p > 0,\ z = [S_1\ S_2\ \xi_2]^T\}$, which is compact and convex. Then, η has a maximum, $M > 0$, on $\mathscr{B}$ for the surface gain $K := 2 + K_0$ where $K_0 > \varepsilon/p$, and choose the time constant such that [94]

$$\frac{1}{\tau} = 1 + \frac{M^2}{2\varepsilon} + K_0. \tag{1.34}$$

Then the inequality (1.32) is written as

$$\begin{aligned} \dot{V} &\le \frac{2S_1^2 + S_2^2 + \xi_2^2}{2} - (2 + K_0)\left(S_1^2 + S_2^2\right) + \varepsilon - \xi_2^2\left(1 + \frac{M^2}{2\varepsilon} + K_0\right) + \frac{M^2\xi_2^2}{2\varepsilon}\frac{\eta^2}{M^2} \\ &\le -2K_0V + \varepsilon - \left(1 - \frac{\eta^2}{M^2}\right)\frac{M^2\xi_2^2}{2\varepsilon}. \end{aligned}$$

Therefore, $\dot{V} \le -2K_0V + \varepsilon$ on $\mathscr{B}$ and a positive constant M such that $\eta \le M$ on $\mathscr{B}$. As discussed in the integrator backstepping, similarly this inequality results in ultimately uniformly bounded stabilization of the state. Furthermore, since ε is arbitrary, the ultimate error bound can be made arbitrarily small.

Based on this example, it is summarized that the design procedure of DSC becomes simpler and sequential due to inclusion of first-order low-pass filters, while integrator backstepping is designed recursively and is complex due to the explosion of terms. However, the design of DSC increases the overall system dimension due to the addition of $(n-1)$th-order filter dynamics to the nth-order nonlinear system, thus requiring a more rigorous analysis for appropriate assignment of controller gains. In the above example, calculation of M, which is the upper bound of η on $\mathscr{B}$, is necessary to design a set of controller gains but it also depends on the controller gains as well as the set $\mathscr{B}$. In order to develop a methodology for analysis and design of DSC, we introduce an LMI approach.

1.5.2 An LMI Approach

The key idea of the LMI approach is that the closed-loop error dynamics with DSC are considered as a class of linear differential inclusions (LDIs) [12]. Then, their stability and performance analysis is formulated in the form of a linear matrix inequality (LMI) and solved numerically in the framework of convex optimization [13]. For the system given in (1.4), the closed-loop error dynamics in (1.30) can be decoupled into linear and nonlinear terms as follows:

$$\begin{cases} \dot{S}_1 = S_2 + \xi_2 - KS_1 + d := c_z z + d, \\ \dot{S}_2 = -KS_2, \\ -K\dot{S}_1 + \dot{\xi}_2 = -\frac{\xi_2}{\tau} + \frac{\partial}{\partial S_1}(f_1 + S_1\frac{\rho_1^2}{2\varepsilon})\dot{S}_1 := -\xi_2/\tau + \phi c_z z + \phi d \end{cases}$$

where $z = [S_1\ S_2\ \xi_2]^T \in \Re^3$, $c_z = [-K\ 1\ 1] \in \Re^{1\times 3}$,

$$d = \Delta f_1 - S_1\frac{\rho_1^2}{2\varepsilon}, \quad \text{and} \quad \phi = \frac{\partial f_1}{\partial S_1} + \frac{\rho_1^2}{2\varepsilon} + \frac{S_1\rho_1}{\varepsilon}\frac{\partial \rho_1}{\partial S_1}.$$

Then, they can be rewritten in matrix form as

$$\begin{bmatrix} 1 & 0 & 0 \\ 0 & 1 & 0 \\ -K & 0 & 1 \end{bmatrix} \dot{z} = \begin{bmatrix} -K & 1 & 1 \\ 0 & -K & 0 \\ 0 & 0 & -1/\tau \end{bmatrix} z + \begin{bmatrix} 0 \\ 0 \\ 1 \end{bmatrix} \phi c_z z + \begin{bmatrix} 1 & 0 \\ 0 & 0 \\ 0 & 1 \end{bmatrix} \begin{bmatrix} d \\ \phi d \end{bmatrix}. \tag{1.35}$$

Furthermore, the first matrix in (1.35) is always invertible such that

$$\begin{bmatrix} 1 & 0 & 0 \\ 0 & 1 & 0 \\ -K & 0 & 1 \end{bmatrix}^{-1} = \begin{bmatrix} 1 & 0 & 0 \\ 0 & 1 & 0 \\ K & 0 & 1 \end{bmatrix}.$$

After multiplying the inverse matrix to both sides in (1.35), the augmented error dynamics are rewritten as

$$\dot{z} = \begin{bmatrix} -K & 1 & 1 \\ 0 & -K & 0 \\ -K^2 & K & K - 1/\tau \end{bmatrix} z + \begin{bmatrix} 0 \\ 0 \\ 1 \end{bmatrix} \phi c_z z + \begin{bmatrix} 1 & 0 \\ 0 & 0 \\ K & 1 \end{bmatrix} \begin{bmatrix} d \\ \phi d \end{bmatrix}$$

$$\Longrightarrow \quad \dot{z} = A_{cl} z + B_w w + B_n n \tag{1.36}$$

where $w = \phi c_z z \in \Re$ and $n = [d\ \phi d]^T \in \Re^2$. Thus, if DSC is applied to the example, the closed-loop error dynamics in (1.36) are regarded as a linear system subject to a vanishing perturbation w and a nonvanishing perturbation n. The n is called the nonvanishing perturbation in the sense that $n \neq 0$ for $z = 0$ while the w is the vanishing one, i.e., $w = 0$ for $z = 0$.

Furthermore, an inequality constraint for w can be imposed on a convex set $\mathscr{B}$. That is, since f_1 and ρ_1 are in $\mathscr{C}^1$, there exists $\gamma > 0$ such that $|\phi| \leq \gamma$. Then,

$$|w| = |\phi c_z z| \leq \gamma |c_z z| := |C_z z|.$$

Therefore, the augmented error dynamics are written as

$$\begin{aligned} \dot{z} &= A_{cl} z + B_w w + B_n n, \\ v &= C_z z, \\ w^T w &\leq v^T v. \end{aligned} \tag{1.37}$$

For the given controller gains, K and τ, $\dot{z} = A_{cl} z$ can be regarded as a linear time-invariant system and the nonvanishing perturbation n can be considered as the exogenous input. Then, they are in a special class of linear differential inclusions(LDIs) called norm-bounded LDIs (NLDI) [4, 12]. Using this fact, the ultimate goal of this monograph is to extend the recent developments in convex optimization of linear systems to nonlinear systems.

1.6 Book Organization

Chapter 2 presents a new method to analyze the stability of a class of nonlinear systems controlled via DSC. Since the augmented closed-loop error dynamics of the DSC system can be described by linear error dynamics with bounded perturbation terms as a function of the error, the desired eigenvalues can be assigned to the linear part of the error dynamics. Furthermore, based on quadratic stability theory, the feasibility of the fixed controller gains for quadratic stabilization and tracking can be tested by solving a convex optimization problem. It provides a systematic procedure for choosing the controller gains and filter time constants. Moreover, the smallest ellipsoidal approximation of the upper bound on the errors for the tracking problem can be obtained via convex optimization. Moreover, Chap. 3 extends the proposed method to uncertain nonlinear systems in the presence of modeling uncertainty. Using nonlinear damping, robust DSC is designed based on an LMI approach and a quadratic ultimate error bound is computed through convex optimization. Furthermore, input–output stability is considered in terms of the $\mathscr{L}_2$ gain.

Chapter 4 discusses the analysis method to design nonlinear control systems whose full state information is not available via sensors, due to either cost considerations or sensor fidelity. Extensions of the Luenberger observer design are considered to estimate the state for a class of Lipschitz nonlinear systems. Combining the nonlinear observer with DSC, we will develop a nonlinear compensator design technique within the framework of convex optimization. Although the separation principle from linear system theory does not generally hold for nonlinear systems, a separation principle for the nonlinear compensator system will be shown for a class of nonlinear systems, thus enabling the independent design of the observer and DSC. Furthermore, using the fact that model uncertainty in the system does not directly affect the DSC error dynamics, we only need to design the robust observer to compensate for the uncertainty.

Chapter 5 focuses on a stabilization problem for constrained input nonlinear systems. In general, the DSC is a "high-gain" controller whose gains are required to be sufficiently large to compensate for model uncertainty. Thus, if the control input is bounded, an actuator could be saturated for an inappropriately assigned initial condition. Moreover, the DSC-based control input relies on the choice of control gains and filter time constants. Therefore, a method of estimating the ellipsoidal initial condition set which guarantees quadratic stability will be developed for a fixed set of controller gains and filter time constants. Finally, this initial condition set will be enlarged to allow some degree of control input saturation, while preserving quadratic stability. Consequently, it will provide more flexibility for designing the DSC.

Chapter 6 addresses a synthesis method and a decentralized approach of DSC for a multi-input multi-output interconnected mechanical system. While a centralized design approach of DSC is developed in Chap. 2, the decentralized approach to deal with large-scale interconnected systems is proposed under the assumption that interconnected functions among subsystems are unknown but bounded. Furthermore, the proposed analysis and design methodology of decentralized DSC is extended to a Lagrangian system of N particles. This is the outcome of a modeling technique based on a variational method whose starting point is the definition of the energy function in terms of sets of generalized variables. This result is extended and applied to a holonomic system of N interconnected rigid bodies, i.e., biped walking of a 5-link biped robot in Chap. 9.

In Chap. 7, all previous results and methodologies are generalized for nonlinear systems including faults as well as model uncertainties. A new method of analyzing the performance loss caused by faults in the nonlinear systems and classifying the faults will be presented. First, it will be shown that DSC is a *passive* fault tolerant approach in the sense that it gives robust stabilization and tracking in the presence of model uncertainties and even a specific class of faults. Then, based on the amount of performance loss measured by a quadratic Lyapunov function and information from the fault detection and diagnosis (FDD) system, a fault classification technique will be developed for both single and multiple faults. Finally, this technique will be applied to the automated longitudinal control of a vehicle implemented and demonstrated by the California PATH Program. Due to the given architecture for coordinated throttle and brake control, the augmented error dynamics is described by a hybrid system with a switching condition. Extensions of the control design methodology in Chap. 2 and the fault classification in this context will be discussed for the hybrid error dynamics.

Chapter 8 takes the ideas developed in Chap. 7 and develops subsystems necessary for a hierarchical active fault tolerant control architecture proposed for AHS.

Once a fault is classified to be intolerable and isolatable, controller reconfiguration will be needed to accommodate the fault in the framework of a multi-controller structure including a series of controllers. Among all possible faults, observer-based DSC and trajectory reconfiguration will be focused to make up for the performance loss due to either sensor or actuator faults. Finally, both fault classification and controller reconfiguration will be designed and simulated for a longitudinal fault tolerant control of a transit bus.

Chapter 9 presents the analysis and design of a biped walking with a variable step size as another DSC application. Under the assumption that the biped model consists of a single-leg support, a double impact, and a double-leg support phase model, DSC is applied to the model which is a piecewise multi-input multi-output nonlinear system. Once the system becomes closed-loop with DSC, piecewise augmented error dynamics with provable stability properties are derived in the form of a piecewise linear system subject to exogenous inputs. Based on the error dynamics, a convex optimization problem is formulated to estimate the ellipsoidal error bound to guarantee the piecewise quadratic boundedness.

1.7 Origin of the Book

This LMI approach of DSC was originally suggested by Song et al. [88] and its extensions to output feedback systems and fault tolerant systems were developed by Song and Hedrick [83–85, 89]. This series of developments using the new analysis and design approach motivated the authors to write this book for further extensions and applications. This monograph makes several contributions to the areas of robust nonlinear control as well as to the study of mechatronic systems such as automated highway systems and biped robots. These contributions are discussed below.

1. *Dynamic Surface Control* (*DSC*): a systematic method to choose appropriate gains and filter time constants for DSC and to analyze both stability and tracking performance is developed. Although existence of the gains and filter time constants for semi-global stability had been proved in earlier work [94], typically these values were determined through heuristic methods and the stability was not guaranteed. In this context, the closed-loop error dynamics of the DSC systems including filter dynamics are described by linear error dynamics with bounded vanishing and nonvanishing perturbation terms. This fact offers the reformulation of the stability analysis method into convex optimization problems. Based on quadratic stability theory, sufficient conditions to provide quadratic stabilization and tracking are discussed in the framework of convex optimization problems. Thus, the stability and performance for the given controller are tested numerically using a convex optimization tool.
2. *Output Feedback DSC*: a separation principle for observer-based DSC systems is developed to separate the observer design from the controller design. As done above, similarly the augmented closed-loop error dynamics via DSC and a nonlinear observer is presented in the form of a *diagonal norm-bounded linear dif-*

ferential inclusion (DNLDI) [12], and we show that the eigenvalues of the observer and DSC can be assigned independently. Furthermore, using the fact that model uncertainties in the error dynamics are related with only the observer error dynamics, we are able to use DSC design methodology in Chap. 2 without considering the model uncertainty as long as the observer is robust enough to compensate for the uncertainty. Finally, sufficient conditions for stability and tracking performance of the nonlinear compensator are formulated within the framework of convex optimization problems.

3. *Constrained Stabilization*: for single-input constrained Lipschitz nonlinear systems, an estimation method is developed to determine an initial condition set which guarantees local quadratic stabilization via DSC without control input saturation. Also, an algorithm to obtain a region of attraction with a specific decay rate is designed via a Linear Matrix Inequality (LMI) approach. Furthermore, we allow some degree of control input saturation to get an enlarged initial condition set, while preserving the local quadratic stability of the system in the domain.
4. *Fault Tolerant Control*:

   - *Fault classification*: for a hybrid structure using a combination of both the passive and active FTC approaches, a switching logic based on fault classification is developed. This architecture leads quite naturally to a fault classification scheme which can be used in a controller reconfiguration strategy. To minimize false alarms in FDD due to model uncertainty and disturbance, the fault is classified by tolerable performance of DSC as well as isolatability in FDD. That is, the tolerable performance is represented by the ellipsoidal error bound discussed in Chap. 2. Moreover, it is shown that the DSC provides robust performance with respect to model uncertainties as well as pre-defined faults consistent with the passive approach.
   - *Switched control for controller reconfiguration*: based on a multiple controller structure proposed as the active FTC approach in Chap. 8, a nonlinear compensator discussed in Chap. 4 and a trajectory reconfiguration scheme via convex optimization are developed for a faulty system in the framework of controller reconfiguration. They are focused on reducing performance loss as well as maintaining quadratic stability by accommodation of either sensor or actuator faults.

Chapter 2
Dynamic Surface Control

2.1 Motivation

As mentioned earlier, dynamic surface control (DSC) has been developed to overcome an "explosion of terms", which is generally problematic in integrator backstepping control, through the use of dynamic filters [28]. Its design procedure can applied to both a Lipschitz and a non-Lipschitz nonlinear system, and its existence for semi-global stability was shown in [94]. However, a systematic method to choose appropriate gains and filter time constants for a dynamic surface controller has not been fully addressed yet in the literature. Typically, their values may be determined through somewhat heuristic methods, e.g. high gain assignment, and these heuristic methods do not always guarantee stability. Moreover, there is no analytic way to check the stability for the given gains and filter time constants without numerous simulations.

In order to illustrate the DSC design procedure, as well as the choice of surface gains and filter time constants, consider the following example:

$$\begin{aligned} \dot{x}_1 &= x_2 - x_1^2 := x_2 + f_1(x_1), \\ \dot{x}_2 &= x_3, \\ \dot{x}_3 &= u. \end{aligned} \tag{2.1}$$

The goal is to make x_1 track the desired value, $x_{1d}(t) = \sin t$. Let

$$S_1 = x_1 - x_{1d}(t).$$

Then, after taking a derivative and using (2.1),

$$\dot{S}_1 = \dot{x}_1 - \dot{x}_{1d} = x_2 + f_1 - \dot{x}_{1d}.$$

Using the idea of a low-pass filter as in the DSC design procedure [37], consider the following choice of $\bar{x}_2$ and introduce a first-order low-pass filter as follows:

$$\bar{x}_2 := \dot{x}_{1d}(t) - f_1 - K_1 S_1,$$
$$\tau_2 \dot{x}_{2d} + x_{2d} = \bar{x}_2, \quad x_{2d}(0) := \bar{x}_2(0).$$

B. Song, J.K. Hedrick, *Dynamic Surface Control of Uncertain Nonlinear Systems*, Communications and Control Engineering,
DOI 10.1007/978-0-85729-632-0_2, © Springer-Verlag London Limited 2011

Then, after defining the second sliding surface as

$$S_2 = x_2 - x_{2d},$$

similarly we can choose $\bar{x}_3$ and the second low-pass filter as follows:

$$\begin{aligned} \bar{x}_3 &:= \dot{x}_{2d} - K_2 S_2, \\ \tau_3 \dot{x}_{3d} + x_{3d} &= \bar{x}_3, \quad x_{3d}(0) = \bar{x}_3(0). \end{aligned} \tag{2.2}$$

Finally, the control input u is calculated with the definition of the third sliding surface

$$\begin{aligned} S_3 &= x_3 - x_{3d}, \\ \dot{S}_3 &= u - \dot{x}_{3d}, \\ u &= \dot{x}_{3d} - K_3 S_3 = \frac{\bar{x}_3 - x_{3d}}{\tau_3} - K_3 S_3 \end{aligned}$$

where the last equality comes from (2.2).

Next, we need to assign the set of controller gains, K_i and τ_i, to guarantee stability and/or ultimate error boundedness. As discussed for the second-order nonlinear system in Sect. 1.5, similarly it is guaranteed that there exists a set of controller gains to make the errors of the closed-loop system ultimately bounded. However, we need to find the specific controller gains to guarantee stability. Suppose $K_i = 50$ for $i = 1, 2, 3$, and $\tau_i = 0.021$ for $i = 2, 3$ for simplicity. Then, the time responses of x_1, S_1, and u are shown in Fig. 2.1. As noticed in the plot of S_1, the first surface error is diverging, thus the system becomes unstable. However, when either $\tau_i = 0.019$ or $K_i = 47$, which is slightly smaller than the original gain set, the system becomes stable and the error is bounded on a small boundary, as shown in Fig. 2.2. These results show that design of the controller gains itself is an important issue in the framework of dynamic surface control even though the design procedure is well structured.

In general, instability of the closed-loop system with DSC may happen by two main causes: one is due to inappropriate gain assignment and the other is when the given controller cannot compensate for model uncertainty or disturbances. In this chapter, we will focus on the development of a systematic procedure of stability and performance analysis for design of the set of controller gains without considering the uncertainty. Model uncertainty and disturbances are considered in Chap. 3. One of challenging problems regarding design of the controller gains is to determine the filter time constants (τ_i). If the gain assignment scheme (1.34) introduced in the example of Sect. 1.5 is used, we need to know M, which is an upper bound of η, and η is in general highly nonlinear with respect to error subspaces, S_i and ξ_i and depends on K_i and τ_i as well. One ad hoc approach to design τ_i is to choose the value as small as possible. However, the filter time constant cannot be reduced arbitrarily in most applications due to hardware performance limitations, such as the control sampling frequency in a real-time implementation. Furthermore, it is desired in some cases to increase the surface gains, K_i, to compensate for the uncertainty. Thus it results in an increase of M and thus requires smaller filter time constants.

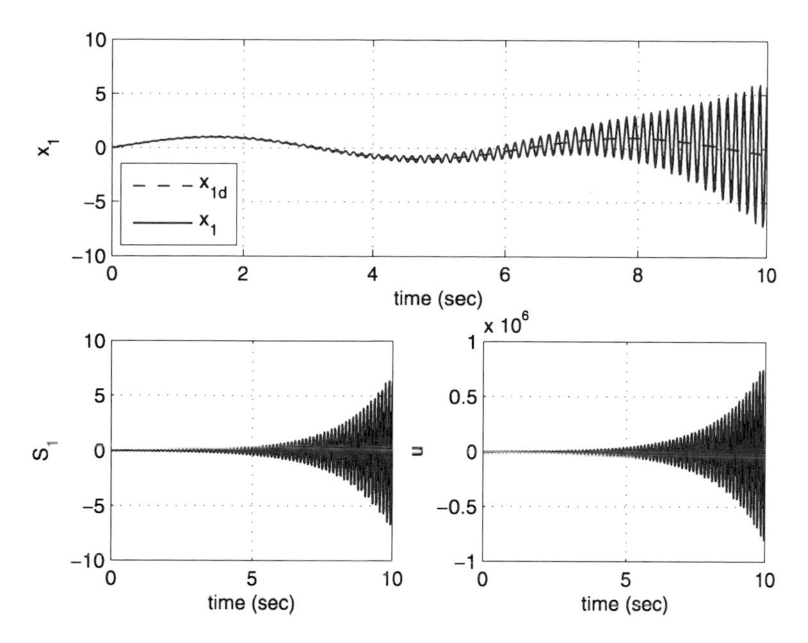

Fig. 2.1 Tracking performance and control input of DSC with $K_i = 50$ and $\tau_i = 0.021$

Therefore, we need a compromise for choosing the surface gains (K_i) and filter time constants (τ_i) to balance the goals of stability and robustness.

The remainder of this chapter is divided as follows: Sect. 2.3 outlines the development of the DSC controller structure for the stabilization problem to the system given in Sect. 2.2. Then, it is shown in Section 2.4 that the closed-loop error dynamics is derived and regarded as a linear system subject to a perturbation. Once the error dynamics is derived with the given bound assumption of the perturbation, the analysis and design method for the stabilization problem is proposed in the framework of an LMI approach in Sect. 2.5. Finally, Sect. 2.6 will discuss the performance analysis of the closed-loop nonlinear system for the tracking problem in the term of quadratic ultimate boundedness.

2.2 Problem Statement

Consider the class of nonlinear systems

$$
\begin{aligned}
\dot{x}_1 &= x_2 + f_1(x_1), \\
\dot{x}_2 &= x_3 + f_2(x_1, x_2), \\
&\vdots \\
\dot{x}_{n-1} &= x_n + f_{n-1}(x_1, \ldots, x_{n-1}), \\
\dot{x}_n &= u + f_n(x_1, \ldots, x_n)
\end{aligned}
\tag{2.3}
$$

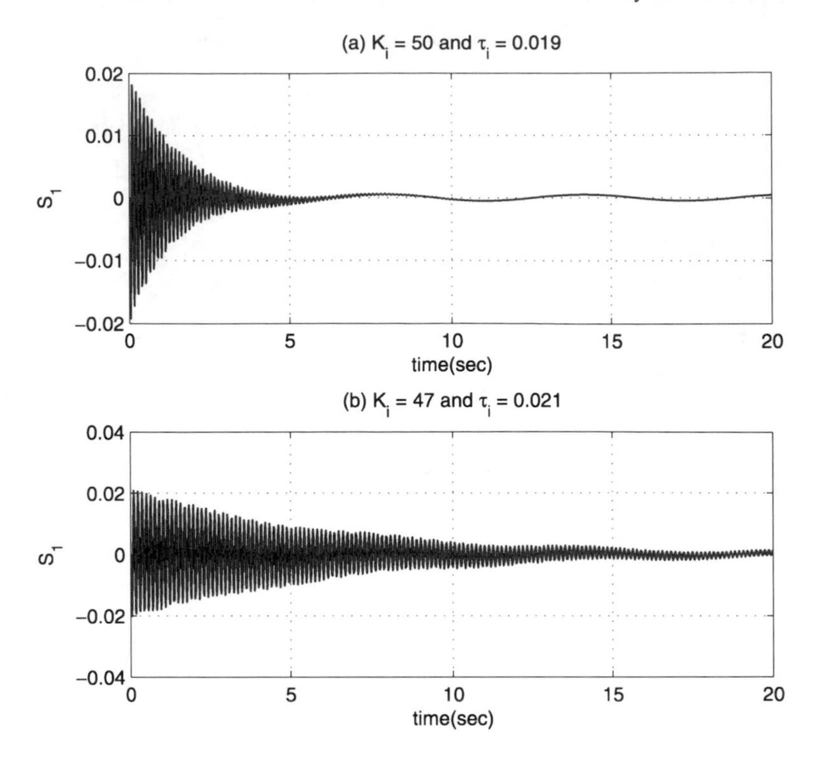

Fig. 2.2 Tracking performance of DSC with different gain sets

where $f(x)$ and $[\partial f(x)/\partial x]$ are continuous on $\mathscr{D} \in \Re^n$, and $f_i : \mathscr{D} \to \Re$ is in *strict-feedback* form in the sense that the f_i depend only on $x_1, \ldots, x_i$ [55]. Then, it is implied that f is locally Lipschitz in x on $\mathscr{D}$ and the existence and uniqueness of the solution of (2.3) are guaranteed [51]. Furthermore, by continuity, $[\partial f(x)/\partial x]$ is bounded on $\mathscr{D}_i$ which is convex and compact, and contained in $\mathscr{D}$. Therefore, there exists a constant $\gamma > 0$ such that

$$\left\| \frac{\partial f(x)}{\partial x} \right\| = \left\| J(x) \right\| \leq \gamma \tag{2.4}$$

for all x in a convex subset $\mathscr{D}_i \subset \mathscr{D}$ where $J(x)$ is a Jacobian matrix of f in the form of a lower triangular matrix, i.e.,

$$J(x) = \begin{bmatrix} \frac{\partial f_1}{\partial x_1} & 0 & 0 & \cdots & 0 \\ \frac{\partial f_2}{\partial x_1} & \frac{\partial f_2}{\partial x_2} & 0 & \cdots & 0 \\ \vdots & \vdots & \vdots & \ddots & \vdots \\ \frac{\partial f_n}{\partial x_1} & \frac{\partial f_n}{\partial x_2} & \frac{\partial f_n}{\partial x_3} & \cdots & \frac{\partial f_n}{\partial x_n} \end{bmatrix}.$$

Without loss of generality, we will assume that $f_i(0, \ldots, 0) = 0$. If $f_i(0, \ldots, 0) = f_{i0} \neq 0$, any equilibrium point can be shifted to the origin via a change of variables [51]. That is, by defining new variables $y_i = x_i - f_{i0}$, the derivatives of y_i is given by

$$\dot{y}_i = x_{i+1} + f_i(x_1 + f_{10}, \ldots, x_i + f_{i0}) \triangleq y_{i+1} + g_i(y_1, \ldots, y_i)$$

where $g_i(0, \ldots, 0) = 0$. Furthermore, it is assumed that f_i is a nonlinear function. If the f_i includes linear combinations of x_i, the linear part can be isolated from f_i and the class of systems can be then rewritten as

$$\dot{x} = Ax + Bu + \bar{f}(x) \tag{2.5}$$

where the matrix $A = U + \bar{A}$ is a *lower-Hessenberg* matrix, i.e., a matrix whose elements above the first super-diagonal are all zero, the matrix $U = \text{diag}([1, \ldots, 1], 1)$ is a square matrix whose first super-diagonal elements are one and elsewhere zero, the matrix $\bar{A} = f(x) - \bar{f}(x)$ is the linear part of $f(x)$ in the strict-feedback form, and $B = [0 \cdots 0\ 1]^T \in \Re^n$.

2.3 Design Procedure

The standard design procedure for the DSC which stabilizes the Lipschitz nonlinear system is described in [94]. An outline of this procedure is as follows: Define the first error surface as $S_1 := x_1 - x_{1d}$ where x_{1d} is the desired value as the control objective, e.g., $x_{1d} = \dot{x}_{1d} = 0$ for the stabilization problem. After taking the time derivative of S_1 and using (2.3),

$$\dot{S}_1 = x_2 + f_1(x_1) - \dot{x}_{1d}. \tag{2.6}$$

The surface error S_1 will converge to zero if $S_1\dot{S}_1 < 0$, however there is no direct control over the surface dynamics. But if x_2 is considered as the forcing term for the surface dynamics, then the sliding condition outside some boundary layer is satisfied if $x_2 = \bar{x}_2$ where

$$\bar{x}_2 = \dot{x}_{1d} - f_1(x_1) - K_1 S_1. \tag{2.7}$$

Consequently, the next step is to force $x_2 \to \bar{x}_2$, so define $S_2 := x_2 - x_{2d}$ where x_{2d} equals $\bar{x}_2$ passed through a first-order low-pass filter, i.e.,

$$\tau_2 \dot{x}_{2d} + x_{2d} = \bar{x}_2, \quad x_{2d}(0) := \bar{x}_2(0). \tag{2.8}$$

Similarly, if we choose $\bar{x}_3$ as

$$\bar{x}_3 = \dot{x}_{2d} - f_2(x_1, x_2) - K_2 S_2 \tag{2.9}$$

and force $x_3 \to \bar{x}_3$. Continuing this process for each consecutive state, define the $(i-1)$th error surface as $S_{i-1} = x_{i-1} - x_{(i-1)d}$ and $\bar{x}_i$ is

$$\bar{x}_i = \dot{x}_{(i-1)d} - f_{i-1}(x_1, \ldots, x_{i-1}) - K_{i-1} S_{i-1}. \tag{2.10}$$

x_{id} is obtained by filtering $\bar{x}_i$, i.e.,

$$\tau_i \dot{x}_{id} + x_{id} = \bar{x}_i, \quad x_{id}(0) := \bar{x}_i(0). \tag{2.11}$$

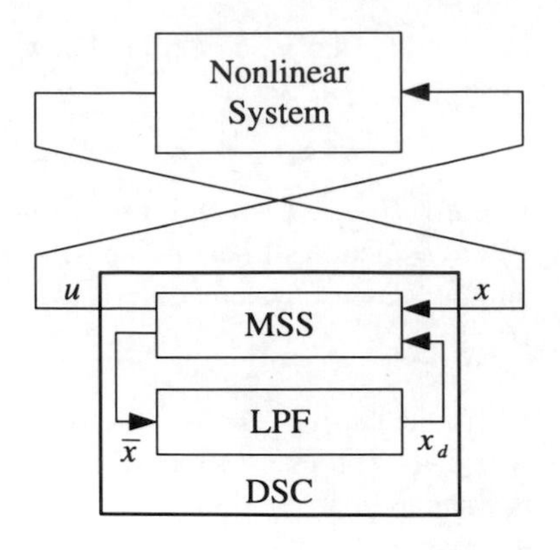

Fig. 2.3 Schematic structure of DSC systems

After continuing this procedure for $2 \le i \le n-1$, define $S_n := x_n - x_{nd}$. Finally, the control input is chosen as

$$\begin{aligned} u &= \dot{x}_{nd} - f_n(x_1, \ldots, x_n) - K_n S_n \\ &= \frac{\bar{x}_n - x_{nd}}{\tau_n} - f_n(x_1, \ldots, x_n) - K_n S_n. \end{aligned} \tag{2.12}$$

It is noted that $\dot{x}_{nd}$ is replaced by $(\bar{x}_n - x_{nd})/\tau_n$ based on (2.11).

The structural design procedure of DSC can be summarized using the graphical representation as shown in Fig. 2.3. In words, the DSC is composed of two blocks: multiple sliding surface (MSS) and low-pass filter (LPF) blocks. As introduced in Sect. 1.4, the MSS block defines multiple sliding surfaces (S_i) and calculates the forcing state values ($\bar{x}_i$), based on the state information as well as the filtered signal (x_{id}). Then, all the forcing state values pass through the LPF block and the corresponding filtered values go back into the MSS. After n iterations, the control input(u) is calculated and fed into the system.

Remark 2.1 The design procedure of DSC can be applied to a more general class of nonlinear system such as [29]

$$\begin{aligned} \dot{x}_1 &= g_1(x_1)x_2 + f_1(x_1), \\ \dot{x}_2 &= g_2(x_1, x_2)x_3 + f_2(x_1, x_2), \\ &\vdots \\ \dot{x}_n &= g_n(x_1, \ldots, x_n)u + f_n(x_1, \ldots, x_n) \end{aligned} \tag{2.13}$$

where $f_i : \mathscr{D}_i \to \Re$, $g_i : \mathscr{D}_i \to \Re$. Then, (2.10) is rewritten as

$$\bar{x}_i = [\dot{x}_{(i-1)d} - f_{i-1} - K_{i-1}S_{i-1}]/g_{i-1}.$$

Then the control input is rewritten as

$$u = [\dot{x}_{nd} - f_n - K_n S_n]/g_n = \left[\frac{\bar{x}_n - x_{nd}}{\tau_n} - f_n - K_n S_n\right] \Big/ g_n.$$

2.4 Augmented Error Dynamics

Once the design procedure is applied, the closed-loop error dynamics needs to be derived for stability analysis. By subtracting and adding $\bar{x}_{i+1}$ and $x_{(i+1)d}$ in x_{i+1}, and $\dot{x}_{nd}$ in u, respectively, they can be rewritten as $x_{i+1} = \bar{x}_{i+1} + [x_{i+1} - x_{(i+1)d}] + [x_{(i+1)d} - \bar{x}_{i+1}]$ and $u = \dot{x}_{nd} + (u - \dot{x}_{nd})$. Then, the system equation in the right hand side terms of (2.3) can be written as

$$\dot{x}_i = \bar{x}_{i+1} + [x_{i+1} - x_{(i+1)d}] + [x_{(i+1)d} - \bar{x}_{i+1}] + f_i \quad \text{for } i = 1, \dots, n-1,$$
$$\dot{x}_n = \dot{x}_{nd} + (u - \dot{x}_{nd}) + f_n.$$

By use of (2.10) and the definition of error surfaces, the above equations can be described as the error equations of DSC as follows:

$$\begin{aligned} \dot{S}_1 &= -K_1 S_1 + S_2 + (x_{2d} - x_2), \\ &\vdots \\ \dot{S}_{n-1} &= -K_{n-1} S_{n-1} + S_n + (x_{nd} - \bar{x}_n), \\ \dot{S}_n &= -K_n S_n. \end{aligned} \tag{2.14}$$

In addition, we need to consider the augmented error dynamics due to inclusion of a set of the first-order low-pass filters. Let us define the filter error as $\xi_i := x_{id} - \bar{x}_i$ for $2 \le i \le n$. Then, the filter dynamics is

$$\dot{\xi}_i = \dot{x}_{id} - \dot{\bar{x}}_i = -\xi_i/\tau_i - \dot{\bar{x}}_i \tag{2.15}$$

where the last equality comes from (2.11). By differentiating (2.10), we can write $\dot{\bar{x}}_i$ as

$$\begin{aligned} \dot{\bar{x}}_2 &= -\dot{f}_1 + \ddot{x}_{1d} - K_1 \dot{S}_1, \\ \dot{\bar{x}}_i &= -\dot{f}_{i-1} + \ddot{x}_{(i-1)d} - K_{i-1} \dot{S}_{i-1} \\ &= -\dot{f}_{i-1} - \dot{\xi}_{i-1}/\tau_{i-1} - K_{i-1} \dot{S}_{i-1} \end{aligned}$$

for $i = 3, \dots, n$. Combining (2.15) with (2.16), we have the filter error dynamics,

$$\begin{aligned} \dot{\xi}_2 - K_1 \dot{S}_1 &= -\frac{\xi_2}{\tau_2} + \dot{f}_1 - \ddot{x}_{1d}, \\ \dot{\xi}_3 - \frac{\dot{\xi}_2}{\tau_2} - K_2 \dot{S}_2 &= -\frac{\xi_3}{\tau_3} + \dot{f}_2, \\ &\vdots \\ \dot{\xi}_n - \frac{\dot{\xi}_{n-1}}{\tau_{n-1}} - K_{n-1} \dot{S}_{n-1} &= -\frac{\xi_n}{\tau_n} + \dot{f}_{n-1}. \end{aligned} \tag{2.16}$$

Therefore, the overall error dynamics including both the nth-order closed-loop nonlinear system in (2.14) and $(n-1)$th-order low-pass filter error equations in (2.16) can be given as

$$T\dot{z} = A_z z + \bar{B}_w \dot{f} + \bar{B}_e \ddot{x}_{1d} \tag{2.17}$$

where the error state $z \in \Re^{n_z}$ and $w \in \Re^{n_w}$ are

$$z = [S\ \xi]^T = [S_1\ \cdots\ S_n\ \xi_2\ \cdots\ \xi_n]^T \in \Re^{2n-1},$$
$$S = [S_1\ \cdots\ S_n], \qquad \xi = [\xi_2\ \cdots\ \xi_n],$$
$$\dot{f} = [\dot{f}_1\ \cdots\ \dot{f}_{n-1}]^T \in \Re^{n-1},$$

the system matrices in the above equation are

$$T = \begin{bmatrix} \mathbf{I}_n & \mathbf{0} \\ -K & T_\xi \end{bmatrix}, \qquad A_z = \begin{bmatrix} A_{11} & A_{12} \\ \mathbf{0} & A_{22} \end{bmatrix},$$
$$\bar{B}_w = \begin{bmatrix} \mathbf{0} \\ \mathbf{I}_{n_w} \end{bmatrix}, \qquad \bar{B}_e = \begin{bmatrix} 0_n \\ -b_r \end{bmatrix},$$

and the corresponding sub-block matrices are the following:

$$T_\xi = \begin{bmatrix} 1 & 0 & \cdots & 0 & 0 \\ -\frac{1}{\tau_2} & 1 & \cdots & 0 & 0 \\ 0 & -\frac{1}{\tau_3} & \cdots & 0 & 0 \\ \vdots & \vdots & \ddots & \vdots & \vdots \\ 0 & 0 & \cdots & -\frac{1}{\tau_{n-1}} & 1 \end{bmatrix} = \mathbf{I}_{n-1} + \text{diag}\left(\left[-\frac{1}{\tau_2}, \ldots, -\frac{1}{\tau_{n-1}}\right], -1\right),$$

$$K = \left[\text{diag}(K_1, \ldots, K_{n-1})\ 0_{n-1}\right] \in \Re^{(n-1)\times n},$$

$$A_{11} = \begin{bmatrix} -K_1 & 1 & \cdots & 0 \\ 0 & -K_2 & \ddots & \vdots \\ \vdots & \vdots & \ddots & 1 \\ 0 & 0 & \cdots & -K_n \end{bmatrix} \in \Re^{n\times n}, \qquad A_{12} = \begin{bmatrix} \mathbf{I}_{n-1} \\ 0_{n-1}^T \end{bmatrix} \in \Re^{n\times(n-1)},$$

$$A_{22} = -\,\text{diag}\left(\frac{1}{\tau_2}, \ldots, \frac{1}{\tau_n}\right) \in \Re^{(n-1)\times(n-1)}, \qquad b_r = [1\ 0\ \cdots\ 0]^T \in \Re^{n_w}.$$

It is noted that $\dot{f}$ is defined as $\dot{f} = [\dot{f}_1\ \cdots\ \dot{f}_{n-1}]^T \in \Re^{n-1}$, not $\dot{f} = [\dot{f}_1\ \cdots\ \dot{f}_n]^T \in \Re^n$ since $\dot{f}_n$ does not affect the filter error dynamics as seen in (2.16). Therefore, the reduced order of Jacobian matrix J is redefined as

$$J(x) = \begin{bmatrix} \frac{\partial f_1}{\partial x_1} & 0 & 0 & \cdots & 0 \\ \frac{\partial f_2}{\partial x_1} & \frac{\partial f_2}{\partial x_2} & 0 & \cdots & 0 \\ \vdots & \vdots & \vdots & \ddots & \vdots \\ \frac{\partial f_{n-1}}{\partial x_1} & \frac{\partial f_{n-1}}{\partial x_2} & \frac{\partial f_{n-1}}{\partial x_3} & \cdots & \frac{\partial f_{n-1}}{\partial x_{n-1}} \end{bmatrix} \in \Re^{(n-1)\times(n-1)}$$

and without loss of generality, there exists a constant $\gamma > 0$ satisfying the inequality (2.4) for all x in a convex subset $\mathcal{D}_i \subset \mathcal{D}$.

Furthermore, since T_ξ is full rank of $n-1$, both T and T_ξ are invertible with inverses given by

$$T_\xi^{-1} = \begin{bmatrix} 1 & 0 & \cdots & 0 & 0 \\ \frac{1}{\tau_2} & 1 & \cdots & 0 & 0 \\ \frac{1}{\tau_2\tau_3} & \frac{1}{\tau_3} & \cdots & 0 & 0 \\ \vdots & \vdots & \ddots & \vdots & \vdots \\ \frac{1}{\tau_2\cdots\tau_{n-1}} & \frac{1}{\tau_3\cdots\tau_{n-1}} & \cdots & \frac{1}{\tau_{n-1}} & 1 \end{bmatrix},$$

we can compute the inverse matrix of T using the following property in linear algebra: if the matrices X and Z are square and invertible,

$$\begin{bmatrix} X & 0 \\ Y & Z \end{bmatrix}^{-1} = \begin{bmatrix} X^{-1} & 0 \\ -Z^{-1}YX^{-1} & Z^{-1} \end{bmatrix}$$

so that

$$T^{-1} = \begin{bmatrix} I_n & 0 \\ T_\xi^{-1}K & T_\xi^{-1} \end{bmatrix}.$$

Then, after multiplying T^{-1} to both sides in (2.17), the augmented closed-loop error dynamics is rewritten as

$$\dot{z} = A_{cl}z + B_w\dot{f} + B_e\ddot{x}_{1d} \tag{2.18}$$

where

$$A_{cl} = T^{-1}A_z = \begin{bmatrix} A_{11} & A_{12} \\ T_\xi^{-1}KA_{11} & T_\xi^{-1}(KA_{12} + A_{22}) \end{bmatrix} \in \Re^{n_z\times n_z},$$

$$B_w = T^{-1}\bar{B}_w = \begin{bmatrix} \mathbf{0}_{n\times n_w} \\ T_\xi^{-1} \end{bmatrix} \in \Re^{n_z\times n_w}, \qquad B_e = T^{-1}\bar{B}_e = \begin{bmatrix} 0_n \\ -T_\xi^{-1}b_r \end{bmatrix} \in \Re^{n_z}.$$

Finally, after decoupling $\dot{f}$ in (2.18) into a vanishing and a nonvanishing term, i.e., the nonvanishing term may not be zero for $z = 0$, while the vanishing term becomes zero when $z = 0$, the augmented error dynamics is summarized as follows:

Lemma 2.1 *For the given class of nonlinear system* (2.3), *the augmented closed-loop error dynamics with DSC is*

$$\begin{aligned} \dot{z} &= A_{cl}z + B_w w + B_r r, \\ \|w\| &\le \gamma\|C_z z\| \end{aligned} \tag{2.19}$$

where

$$\dot{f} = [\dot{f}_1 \ \cdots \ \dot{f}_{n-1}]^T = \frac{\partial f}{\partial x}\dot{x} = J(x)C_z z + J_1\dot{x}_{1d} = w + J_1\dot{x}_{1d} \in \Re^{n_w},$$

$w = J(x)C_z z$, $\quad J = \dfrac{\partial f}{\partial x}$, *$J_1$ is the first column of J,*

$$r = \begin{bmatrix} J_1 \dot{x}_{1d} \\ \ddot{x}_{1d} \end{bmatrix} \in \Re^{n_w+1} := \Re^{n_r},$$

$$B_r = \begin{bmatrix} \mathbf{0}_{n \times n_w} & 0_n \\ T_\xi^{-1} & -T_\xi^{-1} b_r \end{bmatrix} = [B_w \; B_e] \in \Re^{n_z \times n_r},$$

and $C_z = [\underline{A}_{11} \; T_\xi] \in \Re^{n_w \times n_z}$ where the notation $\underline{A}_{11}$ is the matrix where the last row of the matrix A_{11} is eliminated.

The proof is given in Appendix A.1. It is worth noting that the perturbation term, w, only affects filter error dynamics since the first block matrix in the matrix B_w in (2.18) is a zero matrix. That is, the augmented error dynamics is also rewritten as

$$\begin{aligned} \dot{z} &= T^{-1}(A_z + \bar{B}_w J C_z) z + T^{-1} \bar{B}_r r \\ &= A(x) z + B_r r \end{aligned} \tag{2.20}$$

where

$$A(x) = T^{-1}(A_z + \bar{B}_w J C_z) = \begin{bmatrix} \mathbf{I}_n & \mathbf{0}_{n(n-1)} \\ T_\xi^{-1} K & T_\xi^{-1} \end{bmatrix} \begin{bmatrix} A_{11} & A_{12} \\ J(x)\underline{A}_{11} & A_{22} + J(x) T_\xi \end{bmatrix}$$

$$= \begin{bmatrix} A_{11} & A_{12} \\ T_\xi^{-1}(K A_{11} + J \underline{A}_{11}) & T_\xi^{-1}(K A_{12} + A_{22} + J T_\xi) \end{bmatrix}.$$

Example 2.1 (Derivation of closed-loop error dynamics) Let us consider the third-order nonlinear system as follows:

$$\begin{aligned} \dot{x}_1 &= x_2 + x_1^2 := x_2 + f_1(x_1), \\ \dot{x}_2 &= x_3 - x_1 x_2^2 := x_3 + f_2(x_1, x_2), \\ \dot{x}_3 &= u \end{aligned} \tag{2.21}$$

where f_i are locally Lipschitz on $\Re^3$ since it is continuously differentiable on $\Re^3$. However, it is not globally Lipschitz, since $[\partial f / \partial x]$ is not bounded on $\Re^3$. On any compact subset of $\Re^3$, f_i is Lipschitz. The Jacobian matrix is given by

$$J(x) := \begin{bmatrix} \frac{\partial f}{\partial x} \end{bmatrix} = \begin{bmatrix} 2x_1 & 0 \\ x_2^2 & 2x_1 x_2 \end{bmatrix}.$$

Suppose that the control objective is $x_1 \to x_{1d}(t)$. If the design procedure of DSC in Sect. 2.3 is applied to the system, the first error surface, synthetic input $\bar{x}_2$, and first-order low-pass filter are

$$\begin{aligned} S_1 &:= x_1 - x_{1d}, \\ \dot{S}_1 &= \dot{x}_1 - \dot{x}_{1d} = x_2 + f_1 - \dot{x}_{1d}, \\ \bar{x}_2 &:= \dot{x}_{1d} - f_1 - K_1 S_1, \\ \tau_2 \dot{x}_{2d} + x_{2d} &= \bar{x}_2, \quad x_{2d}(0) = \bar{x}_2(0). \end{aligned}$$

Similarly, the synthetic input $\bar{x}_3$ and control input u are

$$\begin{aligned}
S_2 &:= x_2 - x_{2d}, \\
\dot{S}_2 &= \dot{x}_2 - \dot{x}_{2d} = x_3 + f_2 - \dot{x}_{2d}, \\
\bar{x}_3 &:= \dot{x}_{2d} - f_2 - K_2 S_2, \\
\tau_3 \dot{x}_{3d} + x_{3d} &= \bar{x}_3, \quad x_{3d}(0) = \bar{x}_3(0), \\
S_3 &:= x_3 - x_{3d}, \\
\dot{S}_3 &= u - \dot{x}_{3d}, \\
u &:= \dot{x}_{3d} - K_2 S_3.
\end{aligned}$$

Then, the augmented error dynamics is

$$\dot{z} = A_{cl} z + B_w \dot{f} + B_d \ddot{x}_{1d} \tag{2.22}$$

where $z = [S_1\ S_2\ S_3\ \xi_2\ \xi_3]^T \in \Re^5$, $\dot{f} = [\dot{f}_1\ \dot{f}_2]^T \in \Re^2$, and the matrices are

$$A_{cl} = \left[\begin{array}{ccc|cc}
-K_1 & 1 & 0 & 1 & 0 \\
0 & -K_2 & 1 & 0 & 1 \\
0 & 0 & -K_3 & 0 & 0 \\
\hline
-K_1^2 & K_1 & 0 & K_1 - \frac{1}{\tau_2} & 0 \\
-\frac{K_1^2}{\tau_2} & \frac{K_1}{\tau_2} - K_2^2 & K_2 & \frac{K_1}{\tau_2} - \frac{1}{\tau_2^2} & K_2 - \frac{1}{\tau_3}
\end{array}\right],$$

$$B_w = \left[\begin{array}{cc}
0 & 0 \\
0 & 0 \\
0 & 0 \\
\hline
1 & 0 \\
\frac{1}{\tau_2} & 1
\end{array}\right], \qquad B_d = \left[\begin{array}{c}
0 \\
0 \\
0 \\
\hline
-1 \\
-\frac{1}{\tau_2}
\end{array}\right].$$

Furthermore, $\dot{f}$ is

$$\begin{aligned}
\dot{f}_1 &= J_{11}\{S_2 + \xi_2 - K_1 S_1\} + J_{11}\dot{x}_{1d} = J_{11} c_{z1} z + J_{11}\dot{x}_{1d} = \bar{J}_1 C_z z + J_{11}\dot{x}_{1d}, \\
\dot{f}_2 &= J_{21}(c_{z1} z + \dot{x}_{1d}) + J_{22} c_{z2} z = [J_{21}\ J_{22}]\begin{bmatrix} c_{z1} \\ c_{z2} \end{bmatrix} z + J_{21}\dot{x}_{1d} = \bar{J}_2 C_z z + J_{21}\dot{x}_{1d}
\end{aligned}$$

where

$$C_z := \begin{bmatrix} c_{z1} \\ c_{z2} \end{bmatrix} = \begin{bmatrix} -K_1 & 1 & 0 & 1 & 0 \\ 0 & -K_2 & 1 & -\frac{1}{\tau_2} & 1 \end{bmatrix}.$$

Therefore,

$$\dot{f} = J C_z z + J_1 \dot{x}_{1d}$$

where $J_1 = [J_{11}\ J_{21}]^T \in \Re^2$.

Finally, by use of Lemma 2.1, the closed-loop error dynamics can be written as

$$\begin{aligned}
\dot{z} &= A_{cl} z + B_w w + B_r r, \\
w &= J C_z z
\end{aligned}$$

where $r = [J_{11}\dot{x}_{1d} \; J_{21}\dot{x}_{1d} \; \ddot{x}_{1d}]^T \in \Re^3$ and

$$B_r = \begin{bmatrix} 0 & 0 & 0 \\ 0 & 0 & 0 \\ 0 & 0 & 0 \\ \hline 1 & 0 & -1 \\ \frac{1}{\tau_2} & 1 & -\frac{1}{\tau_2} \end{bmatrix}.$$

If $\|\cdot\|_\infty$ for vectors in $\Re^3$ and the induced matrix norm for matrices are used, we have

$$\|w\|_\infty = \|JC_z z\|_\infty \le \big\|J(x)\big\|_\infty \|C_z z\|_\infty$$

where

$$\big\|J(x)\big\|_\infty = \max\big\{2|x_1|, \big|x_2^2\big| + 2|x_1 x_2|\big\}.$$

If we are interested in the tracking problem over the convex set $\mathscr{D}_\delta = \{x \in \Re^3 \,|\, |x_1| \le \delta_1, |x_2| \le \delta_2\}$, all points in $\mathscr{D}_\delta$ satisfy

$$2|x_1| \le 2\delta_1, \qquad \big|x_2^2\big| + 2|x_1 x_2| \le \delta_2^2 + 2\delta_1\delta_2.$$

Hence,

$$\|w\|_\infty \le \gamma \|C_z z\|_\infty$$

where $\gamma := 2\delta_1 + \delta_2^2 + 2\delta_1\delta_2$.

2.5 Quadratic Stabilization

If either stabilization (i.e., $x_{1d} = 0$) or regulation (i.e., $x_{1d} = c \neq 0$) problem is considered, the r in (2.19) is zero since $\dot{x}_{1d} = \ddot{x}_{1d} = 0$. Based on Lemma 2.1, the augmented closed-loop error dynamics becomes

$$\dot{z} = A_{cl} z + B_w w, \quad w = JC_z z. \tag{2.23}$$

Moreover, $J(x)$ can be written as a nonlinear function of z, i.e., $J(x) = G(z)$ because x can be expressed as a function of z as follows [94]:

$$x_1 = S_1 := \eta_1(S_1),$$
$$x_2 = S_2 + \xi_2 + \bar{x}_2 \;=\; S_2 + \xi_2 - K_1 S_1 - f_1\big\{\eta_1(S_1)\big\} := \eta_2(S_1, S_2, \xi_2).$$

By induction, for $3 \le i \le n$,

$$x_i = S_i + \xi_i + \bar{x}_i = S_i + \xi_i - \frac{\xi_{i-1}}{\tau_{i-1}} - K_{i-1}S_{i-1} - f_{i-1}(\eta_1, \ldots, \eta_{i-1})$$
$$:= \eta_i(S_1, \ldots, S_i, \xi_2, \ldots, \xi_i)$$

where η_i are continuous nonlinear functions and $\eta_i(0, \ldots, 0) = 0$. Therefore, the error dynamics for the stabilization or regulation problem can be described graphically as shown in Fig. 2.4.

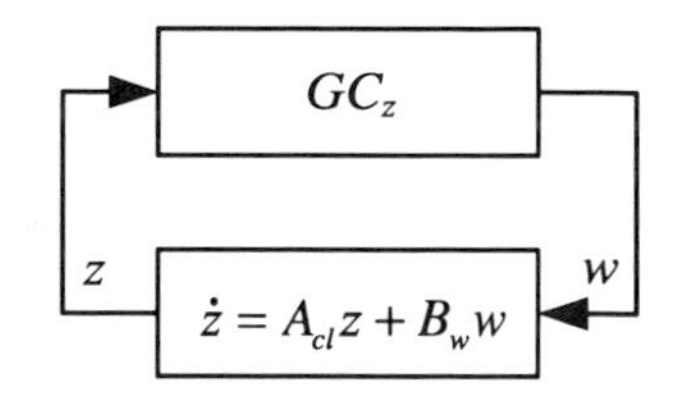

Fig. 2.4 Graphical interpretation for augmented closed-loop error dynamics

Definition 2.1 Let $z = 0$ be an exponentially stable equilibrium point of the nominal system (2.25) when A_{cl} is Hurwitz for the given set of controller gains, $\Theta = \{K_1, \ldots, K_n, \tau_2, \ldots, \tau_n\}$. Then, a nonlinear system (2.3) is *quadratically stabilizable* via DSC if there exists a positive definite matrix P such that

$$\frac{d}{dt}V(z) = \frac{d}{dt}\left(z^T P z\right) = (A_{cl}z + B_w w)^T Pz + z^T P(A_{cl}z + B_w w) < 0. \quad (2.24)$$

It is important to note that we are interested in finding a quadratic Lyapunov function which will be calculated in the framework of linear matrix inequality (LMI) and convex optimization although there may exist a different type of Lyapunov function to guarantee the stability. Therefore, we will investigate how the quadratic Lyapunov function satisfying LMI (2.24) can be found for different types of bounds of w, which is the vanishing perturbation. If a quadratic Lyapunov function exists for this system, then the system is said to be *quadratically stable*. The concept of quadratic stabilization was introduced for the robust stabilization of uncertain linear systems by linear feedback [6, 70]. These results can be applied to the error dynamics above as long as the perturbation term w is bounded by a linear function of the augmented error z.

2.5.1 Nominal Error Dynamics

For the stabilization problem, the augmented error dynamics (2.23) can be regarded as linear nominal error dynamics,

$$\dot{z} = A_{cl}z \quad (2.25)$$

subject to a nonlinear perturbation function w, which is the vanishing perturbation in the sense that $w = 0$ for $z = 0$ from $w = JC_z z$ for all x [51]. If the matrix A_{cl} is Hurwitz; that is, $\text{Re}\{\lambda_i(A_{cl})\} < 0$ for all eigenvalues of A_{cl}, the nominal system (2.25) is exponentially stable at $z = 0$. This is well known in linear system theory.

The assignment of $\lambda(A_{cl})$ in (2.25) can be considered as a generalized eigenvalue problem as follows:

$$\begin{aligned} \lambda v = A_{cl}v \quad &\Longrightarrow \quad \lambda T v = A_z v \quad \Longrightarrow \quad (\lambda T - A_z)v = 0 \\ &\Longrightarrow \quad \det(\lambda T - A_z) = 0. \end{aligned}$$

Moreover, using the definitions of T and A_z in (2.17), it can be written as

$$\lambda T - A_z = \lambda \begin{bmatrix} \mathbf{I}_n & \mathbf{0} \\ -K & T_\xi \end{bmatrix} - \begin{bmatrix} A_{11} & A_{12} \\ \mathbf{0} & A_{22} \end{bmatrix} = \begin{bmatrix} \lambda \mathbf{I}_n - A_{11} & -A_{12} \\ -\lambda K & \lambda T_\xi - A_{22} \end{bmatrix}, \tag{2.26}$$

$$\det(\lambda T - A_z) = \det\left(\begin{bmatrix} \lambda \mathbf{I}_n - A_{11} & -A_{12} \\ -\lambda K & \lambda T_\xi - A_{22} \end{bmatrix}\right) = 0.$$

Since the nth row of $\lambda T - A_z$ has only one nonzero element such as $\lambda + K_n$ in the nth column, $-K_n$ is always one of eigenvalues of A_{cl}. After using this fact and eliminating the nth row and column of $\lambda T - A_z$, (2.26) is equivalent to

$$\det\left(\begin{bmatrix} \lambda \mathbf{I}_{n-1} - \tilde{A}_{11} & -\mathbf{I}_{n-1} \\ -\lambda \tilde{K} & \lambda T_\xi - A_{22} \end{bmatrix}\right) = 0 \tag{2.27}$$

where

$$\tilde{K} = \text{diag}(K_1, \ldots, K_{n-1}) \in \Re^{(n-1)\times(n-1)},$$

$$\tilde{A}_{11} = \begin{bmatrix} -K_1 & 1 & \cdots & 0 \\ 0 & -K_2 & \ddots & \vdots \\ \vdots & \vdots & \ddots & 1 \\ 0 & 0 & \cdots & -K_{n-1} \end{bmatrix} \in \Re^{(n-1)\times(n-1)}.$$

Furthermore, using linear algebra, (2.27) can be converted into computation of a lower dimension matrix. The basic idea is as follows: if the matrices A and D are square, and D is invertible,

$$\begin{bmatrix} A & B \\ C & D \end{bmatrix} = \begin{bmatrix} A - BD^{-1}C & BD^{-1} \\ 0 & I \end{bmatrix} \begin{bmatrix} I & 0 \\ C & D \end{bmatrix},$$

$$\det\left(\begin{bmatrix} A & B \\ 0 & C \end{bmatrix}\right) = \det(A)\det(C)$$

so that

$$\det\left(\begin{bmatrix} A & B \\ C & D \end{bmatrix}\right) = \det\left(A - BD^{-1}C\right)\det(D).$$

Therefore, if $(\lambda T_\xi - A_{22})$ is invertible, i.e.

$$\det(\lambda T_\xi - A_{22}) \neq 0 \quad \text{or} \quad \det\left(\lambda I - T_\xi^{-1} A_{22}\right) \neq 0,$$

(2.27) is equivalent to

$$\det\left[\lambda \mathbf{I}_{n-1} - \tilde{A}_{11} - \lambda(\lambda T_\xi - A_{22})^{-1}\tilde{K}\right] = 0 \quad \text{or}$$
$$\det\left[\lambda \mathbf{I}_{n-1} - \tilde{A}_{11} - \lambda\left(\lambda I - T_\xi^{-1} A_{22}\right)^{-1} T_\xi^{-1}\tilde{K}\right] = 0. \tag{2.28}$$

It is noted that $T_\xi^{-1}A_{22}$ is a lower triangular matrix such that

$$T_\xi^{-1}A_{22} = -\begin{bmatrix} \frac{1}{\tau_2} & 0 & \cdots & 0 & 0 \\ \frac{1}{\tau_2^2} & \frac{1}{\tau_3} & \cdots & 0 & 0 \\ \frac{1}{\tau_2^2\tau_3} & \frac{1}{\tau_3^2} & \cdots & 0 & 0 \\ \vdots & \vdots & \ddots & \vdots & \vdots \\ \frac{1}{\tau_2^2\cdots\tau_{n-1}} & \frac{1}{\tau_3^2\cdots\tau_{n-1}} & \cdots & \frac{1}{\tau_{n-1}^2} & \frac{1}{\tau_n} \end{bmatrix}$$

so the eigenvalues of $T_\xi^{-1}A_{22}$ are diagonal elements such as

$$\lambda_i\left(T_\xi^{-1}A_{22}\right) = -\frac{1}{\tau_i}.$$

Therefore, either $(\lambda T_\xi - A_{22})$ or $(\lambda I - T_\xi^{-1}A_{22})$ is invertible if $\lambda \neq -1/\tau_i$ for $i = 2, \ldots, n$. Thus, (2.28) can be used as long as $\lambda \neq -1/\tau_i$.

Remark 2.2 If we consider a special class of the systems (2.5), i.e., linear systems, they becomes

$$\dot{x} = Ax + Bu \tag{2.29}$$

where the matrix A is a *lower-Hessenberg* matrix described in (2.5). Then, the closed-loop error dynamics is described as

$$\dot{z} = \tilde{A}_{cl}z \tag{2.30}$$

where the matrix $\tilde{A}_{cl}$ is

$$A_{cl} = \begin{bmatrix} A_{11} & A_{12} \\ T_\xi^{-1}\tilde{K}A_{11} & T_\xi^{-1}(\tilde{K}A_{12} + A_{22}) \end{bmatrix} \in \Re^{n_z \times n_z}$$

and

$$\tilde{K} = K + (A - U)$$

where $\underline{A}$ is the matrix the last row of the matrix $A - U$ is eliminated, i.e., $(A - U)_{1:n-1,1:n}$. Therefore, the linear system in (2.29) is *quadratically stabilizable* via DSC if there is a set of controller gains, Θ, such that $\tilde{A}_{cl}$ in (2.30) is Hurwitz.

Example 2.2 (Eigenvalues of nominal error dynamics) If the second-order nonlinear system is controlled by DSC with a set of gains $\Theta = \{K_1, K_2, \tau_2\}$ as was done in Sect. 1.5, A_{cl} in (1.36) for different gains K_1 and K_2 is rewritten as

$$T = \begin{bmatrix} 1 & 0 & 0 \\ 0 & 1 & 0 \\ -K_1 & 0 & 1 \end{bmatrix}, \qquad A_z = \begin{bmatrix} -K_1 & 1 & 1 \\ 0 & -K_2 & 0 \\ 0 & 0 & -1/\tau_2 \end{bmatrix}.$$

Following (2.26), the characteristic equation is

$$\det(\lambda T - A_z) = \det\left(\begin{bmatrix} \lambda + K_1 & -1 & -1 \\ 0 & \lambda + K_2 & 0 \\ -\lambda K_1 & 0 & \lambda + 1/\tau_2 \end{bmatrix}\right) = 0.$$

Since $-K_2$ is the eigenvalue of A_{cl}, the characteristic equation for the reduced-order nominal error dynamics can be written as follows: using (2.27),

$$\det\left(\begin{bmatrix} \lambda + K_1 & -1 \\ -\lambda K_1 & \lambda + \frac{1}{\tau_2} \end{bmatrix}\right) = 0,$$
$$(\lambda + K_1)(\lambda + 1/\tau_2) - K_1\lambda = 0$$

or using (2.28), if $\lambda \neq -1/\tau_2$,

$$(\lambda + K_1) - (\lambda + 1/\tau_2)^{-1}\lambda K_1 = 0,$$
$$\lambda^2 + \frac{1}{\tau_2}\lambda + \frac{K_1}{\tau_2} = 0.$$

Therefore, A_{cl} is Hurwitz as long as $\tau_2 > 0$.

Example 2.3 (Stabilization of nominal error dynamics) Consider the nonlinear system in (2.1) whose the control objective is $x_1 \to 0$. Then, if DSC is applied, the augmented error dynamics is described as follows:

$$\dot{z} = A_{cl} z$$

subject to a nonlinear perturbation w where $z = [S_1\ S_2\ S_3\ \xi_2\ \xi_3]^T \in \Re^5$ and the matrix A_{cl} is derived in (2.22). If $\lambda \neq -1/\tau_i$ for $i = 2, 3$, $\lambda = -K_2$ is one of eigenvalues of A_{cl}. Then, using (2.28), the characteristic equation for the reduced-order nominal error dynamics is

$$\det\left[\begin{pmatrix} \lambda + K_1 & -1 \\ 0 & \lambda + K_2 \end{pmatrix} - \begin{pmatrix} \lambda + \frac{1}{\tau_2} & 0 \\ \frac{1}{\tau_2^2} & \lambda + \frac{1}{\tau_3} \end{pmatrix}^{-1} \begin{pmatrix} 1 & 0 \\ \frac{1}{\tau_2} & 1 \end{pmatrix} \begin{pmatrix} \lambda K_1 & 0 \\ 0 & \lambda K_2 \end{pmatrix}\right]$$
$$= 0.$$

Although one may derive the characteristic equation algebraically, it can be also calculated using Symbolic Math Toolbox™ and MATLAB® [98] and used for higher-order systems. As shown in the MATLAB Program 2-1 below, the computed equation is

$$\frac{\tau_2\tau_3\lambda^4 + (\tau_2 + \tau_3)\lambda^3 + (K_2\tau_2 + 1)\lambda^2 + (K_1 + K_2)\lambda + K_1K_2}{(\tau_2\lambda + 1)(\tau_3\lambda + 1)} = 0.$$

Therefore, inequality conditions of Θ for A_{cl} to be Hurwitz may be derived analytically if the Routh stability criterion, which is introduced in most classical feedback control textbooks, is used.

Suppose that the given set of controller gains is $K = K_1 = K_2 = K_3$ and $\tau = \tau_2 = \tau_3$ for simplicity. Then, the above characteristic equation is rewritten as

$$\frac{\tau^2\lambda^4 + 2\tau\lambda^3 + (K\tau + 1)\lambda^2 + 2K\lambda + K^2}{(\tau\lambda + 1)^2} = 0.$$

Using the Routh stability criterion, the inequality condition for A_{cl} to be Hurwitz is derived to be

$$\tau K < 1 \tag{2.31}$$

for $\lambda \neq -1/\tau$. As used in Sect. 2.1, let $K = 50$ and $\tau = 0.021$, not satisfying (2.31). Then, $\lambda(A_{cl})$ have $\lambda(A_{cl}) = \{-50.0, -48.207 \pm i7.5522, 0.588 \pm i48.7915\}$, thus A_{cl} is not Hurwitz. Therefore, the error vector, z, is diverging due to the eigenvalues with positive real part. Consequently, the control input u is also diverging (see Fig. 2.1). However, when either the surface gains are changed to $K = 49$ or filter time constants to $\tau = 0.019$, the set of controller gains satisfy the inequality condition (2.31) and the corresponding eigenvalues are

$$\lambda(A_{cl}) = \{-51.2998, -43.6278, -47.0, -0.1553 \pm i47.3083\}$$

for $K = 47$ and

$$\lambda(A_{cl}) = \{-60.2603, -43.6702, -50.0, -0.6663 \pm i51.2946\}$$

for $\tau = 0.019$, respectively. Therefore, A_{cl} is Hurwitz for both cases and it is expected that the error vector converges to the origin for the nominal error dynamics.

MATLAB Program 2-1

```
%***** Generation of the symbolic objects *****
x = sym('x');
K1 = sym('K1');  K2 = sym('K2');
tau2 = sym('tau2');  tau3 = sym('tau3');

%***** Symbolic math computation *****
K = [x+K1  -1; 0  x+K2 ];
P = [x+1/tau2  0; 1/tau2^2  x+1/tau3];
Q = [1 0; 1/tau2 1];
R = [x*K1 0; 0 x*K2];
f = det(K - inv(P)*Q*R)
```

2.5.2 *Norm-Bounded Error Dynamics*

According to [37], it is said that semi-global stability of the resulting closed-loop system under the DSC is guaranteed by showing the existence of a set of controller gains for exponential regulation of DSC for Lipschitz nonlinear systems, and the corresponding proof is given in [110]. However, as motivated in Sect. 1.5, determination of the controller gains is not straightforward since it requires the upper bound of a highly nonlinear function of S_i, ξ_i, K_i, and τ_i, e.g., see the function η in (1.30). Although higher surface gains and lower filter time constants are preferable ideally, the filter time constant cannot be arbitrarily small in most applications due to hardware limitation and the surface gains cannot be arbitrarily large because they may result in input saturation to relatively small surface errors due to uncertainties and/or disturbances.

In this section, the LMI approach will be proposed in the framework of the augmented error dynamics which will allow us to calculate the Lyapunov function candidate $V(z) = z^T Pz$ where $P > 0$ while the existence of the Lyapunov function candidate $V(z) = \frac{1}{2}z^T z$ is shown for stability in [37]. Using Lemma 2.1, the closed-loop error dynamics for the stabilization or regulation problem is written as

$$\begin{cases} \dot{z} = A_{cl}z + B_w w, \\ w = J(x)C_z z, \\ \|w\| \leq \gamma \|C_z z\| := \|\tilde{C}_z z\| \end{cases} \tag{2.32}$$

where γ is a Lipschitz constant on the convex set $\mathscr{D}_i \in \mathscr{D}$ and $\tilde{C}_z = \gamma C_z$. It is noted that the calculation of γ is straightforward and simpler than the upper bound of the nonlinear function since $J(x)$ is a function of x in a convex domain $\mathscr{D}_i \subset \mathscr{D}$.

Quadratic stability of the augmented error dynamics under the DSC is guaranteed by the following theorem.

Theorem 2.1 *Suppose that the closed-loop error dynamics* (2.32) *is given for the given set of controller gains,* $\Theta = \{K_1, \dots, K_n, \tau_2, \dots, \tau_n\}$*, for all* x *in a domain* $\mathscr{D}$*. If there exist* $P > 0$ *and* $\sigma \geq 0$ *such that*

$$\begin{bmatrix} A_{cl}^T P + P A_{cl} + \sigma \tilde{C}_z^T \tilde{C}_z & P B_w \\ B_w^T P & -\sigma I \end{bmatrix} < 0, \tag{2.33}$$

the origin in (2.32) *is then exponentially stable in* $\mathscr{D}$*. Thus the nonlinear system* (2.3) *is quadratically stabilizable via DSC with the given* Θ *on* $\mathscr{D}$*.*

Proof Suppose the quadratic Lyapunov function $V(z) = z^T P z$ satisfies

$$\dot{V}(z) = z^T \left(A_{cl}^T P + P A_{cl}\right) z + w^T B_w^T P z + z^T P B_w w < 0 \tag{2.34}$$

for all nonzero z. Then it is claimed that this is equivalent to

$$\dot{V}(z) = \begin{bmatrix} z \\ w \end{bmatrix}^T \begin{bmatrix} A_{cl}^T P + P A_{cl} & P B_w \\ B_w^T P & 0 \end{bmatrix} \begin{bmatrix} z \\ w \end{bmatrix} < 0 \tag{2.35}$$

for all nonzero z and w satisfying $w^T w \leq z^T \tilde{C}_z^T \tilde{C}_z z$. To show the equivalence, we need to show that the set

$$\mathscr{A} = \left\{(z, w) | z \neq 0\right\}$$

equals the set

$$\mathscr{B} = \left\{(z, w) | (z, w) \neq 0, \|w\| \leq \left\|\tilde{C}_z z\right\|\right\}.$$

It suffices to show that $\{(z, w) | z = 0, w \neq 0, \|w\| \leq \|\tilde{C}_z z\|\} = \emptyset$, which is an empty set. Therefore, the inequality condition that $\dot{V}(z) < 0$ for all nonzero z is equivalent to (2.35) for any nonzero (z, w) satisfying

$$\begin{bmatrix} z \\ w \end{bmatrix}^T \begin{bmatrix} -\tilde{C}_z^T \tilde{C}_z & 0 \\ 0 & I \end{bmatrix} \begin{bmatrix} z \\ w \end{bmatrix} < 0. \tag{2.36}$$

The S-procedure can then be used to give a sufficient condition for a quadratic constraint to be satisfied given that some other quadratic constraints are also satisfied. The S-procedure can be summarized as follows [12]: Given matrices P_i for $i = 0, \dots, n$, then the condition

$$x^T P_0 x < 0 \quad \text{for all } x \neq 0 \quad \text{such that} \quad x^T P_i x < 0, \quad i = 1, \dots, n$$

holds if $\exists \sigma_i \geq 0$ for $i = 1, \ldots, n$ such that

$$x^T P_0 x - \sum_{i=1}^{n} \sigma_i x^T P_i x < 0, \quad \forall x \neq 0.$$

Using the S-procedure, the inequality condition (2.34) for the quadratic stability is equivalent to the existence of $P > 0$ and $\sigma \geq 0$ satisfying

$$z^T \left(A_{cl}^T P + P A_{cl}\right) z + 2z^T P B_w w - \sigma \left\{w^T w - (\tilde{C}_z z)^T (\tilde{C}_z z)\right\} < 0 \tag{2.37}$$

or

$$\begin{bmatrix} z \\ w \end{bmatrix}^T \begin{bmatrix} A_{cl}^T P + P A_{cl} & P B_w \\ B_w^T P & 0 \end{bmatrix} \begin{bmatrix} z \\ w \end{bmatrix} - \sigma \begin{bmatrix} z \\ w \end{bmatrix}^T \begin{bmatrix} -\tilde{C}_z^T \tilde{C}_z & 0 \\ 0 & I \end{bmatrix} \begin{bmatrix} z \\ w \end{bmatrix} < 0$$

which is equivalent to LMI (2.33). □

For a specific class of nonlinear functions f, quadratic stability is guaranteed globally as follows:

Corollary 2.1 *Suppose f is continuous and globally Lipschitz in $\Re^n$. If there exist $P > 0$ and $\sigma \geq 0$ satisfying LMI* (2.33) *for the given set of controller gains Θ, the origin in* (2.23) *is globally exponentially stable. Thus the nonlinear system* (2.3) *is globally quadratically stabilizable via DSC with the given Θ.*

Since f_i are known functions, $\dot{f}_i$ can be derived mathematically and the upper bound of $\|\dot{f}\|$ can be computed for the given convex set as done in Example 2.1. That is, we first need to define the convex and compact set $\mathscr{D}_i \subset \mathscr{D}$ to calculate γ. Then, the quadratic stability analysis can be performed using Theorem 2.1 for the given γ. However, it may be more interesting to estimate the maximum value of γ to guarantee the quadratic stability for the given set of controller gains. This allows us to estimate the region of attraction for the stabilization problem. Since the error dynamics is regarded as a norm-bounded linear differential inclusion (LDI) [12], the region of attraction can be computed in the framework of convex optimization and the computed region is defined as the convex and compact set $\mathscr{D}_i \subset \mathscr{D}$. Then we do not need to compute γ, but are able to compute a region of attraction.

Theorem 2.2 *Suppose that the closed-loop error dynamics* (2.32) *is given for the given set of controller gains, Θ. If A_{cl} is Hurwitz, i.e., there exist $P > 0$ and $Q = Q^T > 0$ such that*

$$P A_{cl} + A_{cl}^T P = -Q, \tag{2.38}$$

and $\gamma < \frac{\lambda_{\min}(Q)}{2\lambda_{\max}(P)\|B_w C_z\|_2}$ for $\mathscr{D}_i = \{x \in \Re^n \mid \|J\| \leq \gamma\} \subset \mathscr{D}$, the origin in (2.32) *is exponentially stable on $\mathscr{D}_i$. Thus a nonlinear system* (2.3) *is quadratically stabilizable via DSC with the given Θ on $\mathscr{D}_i$. Furthermore, $\mathscr{D}_i$ is the region of attraction.*

Proof Suppose the quadratic Lyapunov candidate, $V(z) = z^T P z$, satisfies (refer to [51, §9])

$$\begin{aligned}
\lambda_{\min}(P)\|z\|_2^2 \le V(z) \le \lambda_{\max}(P)\|z\|_2^2, \\
\frac{\partial V}{\partial z} A_{cl} z = -z^T Q z \le -\lambda_{\min}(Q)\|z\|_2^2, \\
\left\| \frac{\partial V}{\partial z} \right\|_2 = \left\| 2z^T P \right\|_2 \le 2\|P\|_2 \|z\|_2 = 2\lambda_{\max}(P)\|z\|_2.
\end{aligned}$$

The derivative of $V(z)$ along the trajectories of the perturbed error dynamics satisfies

$$\begin{aligned}
\dot{V}(z) &= z^T \left(A_{cl}^T P + P A_{cl}\right) z + w^T B_w^T P z + z^T P B_w w \\
&= -z^T Q z + (B_w J C_z z)^T P z + z^T P (B_w J C_z z) \\
&\le -\lambda_{\min}(Q)\|z\|_2^2 + 2\gamma \lambda_{\max}(P)\|B_w C_z\|_2 \|z\|_2^2
\end{aligned}$$

for all x in the convex set $\mathscr{D}_i = \{x \in \Re^n \mid \|J\| \le \gamma\}$. Hence, $\dot{V}(z) < 0$ if $\gamma < \frac{\lambda_{\min}(Q)}{2\lambda_{\max}(P)\|B_w C_z\|_2}$. Therefore, the origin is exponentially stable for all x in $\mathscr{D}_i \subset \mathscr{D}$. □

Remark 2.3 It is known that the maximum of γ is calculated when $Q = I$ [51, §9], that is, $\gamma < \frac{1}{2\lambda_{\max}(P)\|B_w\|\|C_z\|}$.

There is another way to calculate the maximum value of γ which is computed in the framework of convex optimization. First, consider *quadratic stability margin* which is the largest nonnegative α for which the origin in (2.32) satisfying $w^T w \le \alpha^2 z^T C_z^T C_z z$ is exponentially stable. That is, if the inequality condition (2.37) is satisfied for all nonzero z and w on $\mathscr{D}_\alpha = \{x \in \Re^n \mid \|w\| \le \alpha \|C_z z\|\}$. Then, by defining $\tilde{P} = P/\sigma$, the inequality condition (2.37) can rewritten as

$$z^T \left(A_{cl}^T \tilde{P} + \tilde{P} A_{cl}\right) z + 2 z^T \tilde{P} B_w w - \left\{ w^T w - \beta (C_z z)^T (C_z z) \right\} < 0, \tag{2.39}$$

where $\beta = \alpha^2$. This inequality is equivalent to LMI (2.40). Then, the quadratic stability margin can be computed as follows:

Algorithm 2.1 Quadratic stability margin of closed-loop error dynamics

$$\begin{aligned}
&\text{maximize} \quad \beta \\
&\text{subject to} \quad \tilde{P} > 0, \qquad \beta \ge 0, \\
&\qquad \begin{bmatrix} A_{cl}^T \tilde{P} + \tilde{P} A_{cl} + \beta C_z^T C_z & \tilde{P} B_w \\ B_w^T \tilde{P} & -I \end{bmatrix} < 0.
\end{aligned} \tag{2.40}$$

To solve the above algorithm, the convex optimization programming method called CVX can be used [32]. It allows us to solve standard problems such as linear programs (LPs), quadratic programs (QPs), second-order cone programs (SOCPs), and semidefinite programs (SDP) and simplifies the task of specifying the problem. Furthermore, it supports two core solvers, SeDuMi [93] and SDPT3 [100]. MATLAB Program 2-2 below shows an example to program the Algorithm 2.1 using CVX in the framework of MATLAB.

MATLAB Program 2-2

```
%***** Define the dimension of matrices *****
n = size(Acl, 1);
m = size(Bw, 2);

%***** cvx version *****
cvx_begin
   variable P(n,n) symmetric;
   variable beta;
   maximize(beta);
   beta >= 0;
   P == semidefinite(n);
   -[Acl'*P + P*Acl + beta*Cz'*Cz P*Bw; Bw'*P -eye(m)] == semidefinite(n+m);
cvx_end
```

Example 2.4 (Region of attraction for the second-order nonlinear system) Consider the second-order nonlinear system:

$$\begin{aligned} \dot{x}_1 &= x_2 - ax_1^2 := x_2 + f_1(x_1), \\ \dot{x}_2 &= u - x_2 \end{aligned} \tag{2.41}$$

where a is a known positive constant. The control objective is $x_1(t) \to 0$. Then, after applying DSC to the system, the augmented error dynamics is

$$\begin{aligned} \dot{z} &= A_{cl}z + B_w w, \\ w &= J(x)C_z z \end{aligned} \tag{2.42}$$

where $z = [S_1\ S_2\ \xi_2]^T \in \Re^3$, w is

$$w = -2ax_1\dot{x}_1 = -2ax_1(S_2 + \xi_2 - K_1 S_1) := J(x)C_z z \tag{2.43}$$

where $J(x) = \partial f_1/\partial x_1 = -2ax_1$, $C_z = [-K_1\ 1\ 1] \in \Re^{1\times 3}$, and the matrices A_{cl}, B_w are

$$A_{cl} = \left[\begin{array}{cc|c} -K_1 & 1 & 1 \\ 0 & -K_2 & 0 \\ \hline -K_1^2 & K_1 & K_1 - \frac{1}{\tau_2} \end{array}\right], \qquad B_w = \left[\begin{array}{c} 0 \\ 0 \\ \hline 1 \end{array}\right]. \tag{2.44}$$

Moreover,

$$|w| = 2a|x_1||C_z z| \le \gamma |C_z z| \tag{2.45}$$

for the convex set $\mathscr{D}_i = \{x \in \Re^2 | |x_1| \le \frac{\gamma}{2a}\}$.

Suppose the set of controller gains is $\{K_1, K_2, \tau_2\} = \{1, 2, 0.1\}$. Then $\lambda(A_{cl}) = \{-1.127, -2, -8.873\}$ is Hurwitz and the solution of (2.38) for $Q = I$ is

$$P = \begin{bmatrix} 0.4600 & 0.1588 & 0.0400 \\ 0.1588 & 0.3412 & 0.0235 \\ 0.0400 & 0.0235 & 0.0600 \end{bmatrix}.$$

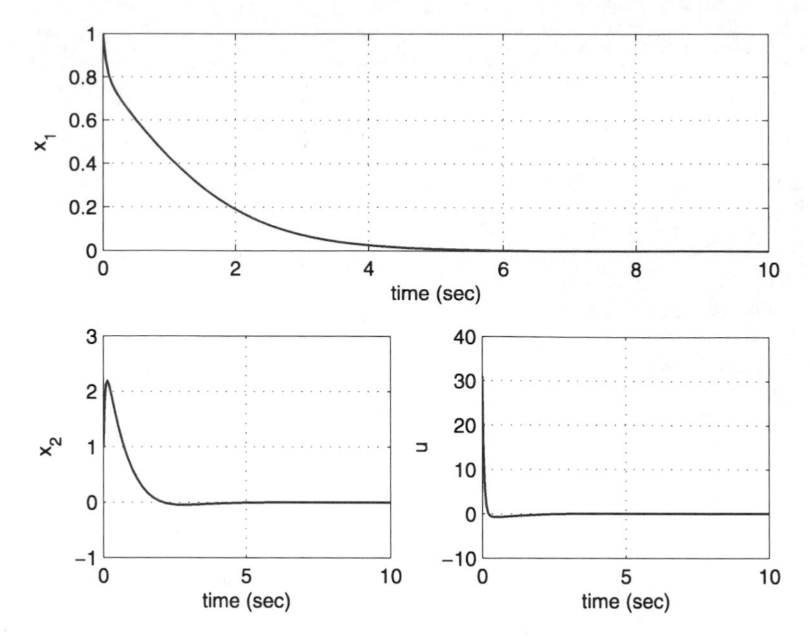

Fig. 2.5 Time responses of x_1, x_2, and u of DSC

Based on Corollary 2.2, the origin in (2.42) is exponentially stable if

$$\gamma < \frac{1}{2\lambda_{\max}(P)\|B_w C_z\|_2} = 0.5026.$$

That is, it is exponentially stable if $2a|x_1| < 0.5026$ from (2.45). Finally, we can define a region of attraction as the domain $\mathscr{D}_1 = \{x \in \Re^2 || x_1| < 0.2513/a \text{ for } a > 0\}$ where the origin is exponentially stable.

Next, we consider the quadratic stability margin to maximize the domain in the sense that γ is maximized in the framework of convex optimization. That is, if the quadratic stability margin is calculated using Algorithm 2.1 (MATLAB Program 2-2), the solutions $\tilde{P}$ and $\alpha = \sqrt{\beta}$ are

$$\alpha = 10, \qquad \tilde{P} = \begin{bmatrix} 109.9980 & -78.5688 & -9.9998 \\ -78.5688 & 144.1719 & 21.4281 \\ -9.9998 & 21.4281 & 9.9998 \end{bmatrix}.$$

Therefore, the origin is exponentially stable if $\gamma < 10$, which is about 20 times larger than the previous result. Then we can define the larger domain such as $\mathscr{D}_2 = \{x \in \Re^2 || x_1| < 5/a \text{ for } a > 0\}$.

Let $a = 5$ and $\{x_1(0),\ x_2(0)\} = \{1,\ 1\}$, thus x is on the boundary of $\mathscr{D}_2$. Figure 2.5 shows the time responses of x and the control input u for the given set of controller gains and initial conditions. It is shown that $x \to 0$, and $\mathscr{D}_2$ is a region of attraction.

2.5.3 Diagonal Norm-Bounded Error Dynamics

Although the perturbation terms $w = J(x)C_z z$ is norm-bounded, $\|w\| \le \gamma \|C_z z\|$, for some cases we can compute a tighter upper bound of w by calculating the componentwise upper bound of w. Suppose there are nonnegative constants γ_i such that

$$\|\bar{J}_i\| \le \gamma_i \quad \text{for all } x \in \mathscr{D}_i \subset \mathscr{D} \tag{2.46}$$

where $\bar{J}_i$ is the ith row of the Jacobian matrix J. Then, the componentwise upper bound of w is

$$\begin{aligned} |w_i| &= \left|\sum_{j=1}^{i} J_{ij}(x) c_{zj} z\right| = \left| [J_{i1}\ J_{i2}\ \cdots\ J_{ii}] \begin{bmatrix} c_{z1} \\ c_{z2} \\ \vdots \\ c_{zi} \end{bmatrix} z \right| \\ &\le \left\| [J_{i1}\ J_{i2}\ \cdots\ J_{ii}] \right\| \left\| \begin{bmatrix} c_{z1} \\ c_{z2} \\ \vdots \\ c_{zi} \end{bmatrix} z \right\| \end{aligned}$$

where the inequality comes from Cauchy–Schwartz inequality. Then, using (2.46),

$$|w_i| \le \gamma_i \|C_{zi} z\| := \|\tilde{C}_{zi} z\| \tag{2.47}$$

where

$$C_{zi} = \left[c_{z1}^T\ c_{z2}^T\ \cdots\ c_{zi}^T\right]^T \in \Re^{n_i \times n_z}, \qquad \tilde{C}_{zi} = \gamma_i C_{zi}.$$

Therefore, since w is bounded componentwise by the function of z, the closed-loop error dynamics is called the diagonal norm-bounded error dynamics and written as

$$\begin{cases} \dot{z} = A_{cl} z + B_w w, \\ w = J(x) C_z z, \\ \|w_i\| \le \|\tilde{C}_{zi} z\|. \end{cases} \tag{2.48}$$

Example 2.5 (Diagonal norm-bound of w) Consider the third-order nonlinear system in Example 2.1 and the control objective is $x(t) \to 0$. Then, the error dynamics is given as

$$\dot{z} = A_{cl} z + B_w w, \qquad w = J C_z z$$

where

$$J(x) := \left[\frac{\partial f}{\partial x}\right] = \begin{bmatrix} 2x_1 & 0 \\ x_2^2 & 2x_1 x_2 \end{bmatrix}.$$

Then, there are nonnegative constants γ_1 and γ_2 such that

$$\begin{aligned} |\bar{J}_1| &= |J_{11}| = |2x_1| \le 2\delta_1 := \gamma_1, \\ |\bar{J}_2| &= \left\|[J_{21}\ J_{22}]\right\|_2 = \sqrt{x_2^4 + 4x_1^2 x_2^2} \le 2x_1^2 + x_2^2 \le 2\delta_1^2 + \delta_2^2 := \gamma_2 \end{aligned}$$

for all $x \in \mathscr{D}_\delta = \{x \in \Re^3 | |x_1| \le \delta_1, |x_2| \le \delta_2\}$. Then,

$$|w_1| = |J_{11} c_{z1} z| \le \gamma_1 |c_{z1} z| = 2\delta_1 |c_{z1} z| := \|\tilde{C}_{z1} z\|_2,$$
$$|w_2| = |J_{21} c_{z1} z + J_{22} c_{z2} z| \le \gamma_2 \left\| \begin{bmatrix} c_{z1} \\ c_{z2} \end{bmatrix} z \right\| := \|\tilde{C}_{z2} z\|_2$$

where $\tilde{C}_{z1} = 2\delta_1 c_{z1} \in \Re^{1\times 5}$, $\tilde{C}_{z2} = (2\delta_1^2 + \delta_2^2)\left[\begin{smallmatrix} c_{z1} \\ c_{z2} \end{smallmatrix}\right] \in \Re^{2\times 5}$, and c_{zi} are defined in Example 2.1.

Theorem 2.3 *Suppose that the diagonal norm-bounded error dynamics* (2.48) *is given for the given set of controller gains, Θ. The nonlinear system* (2.3) *is quadratically stabilizable via DSC for the given Θ on $\mathscr{D}$ if there exist $P > 0$ and $\Sigma = \mathrm{diag}(\sigma_1, \sigma_2, \ldots, \sigma_{n_w}) \ge 0$ such that*

$$\begin{bmatrix} A_{cl}^T P + P A_{cl} + \tilde{C}_z^T \Sigma_B \tilde{C}_z & P B_w \\ B_w^T P & -\Sigma \end{bmatrix} < 0 \tag{2.49}$$

where $\tilde{C}_z = [\tilde{C}_{z1}^T, \ldots, \tilde{C}_{zn_w}^T]^T$, $\tilde{C}_{zi} \in \Re^{n_i \times n_z}$, and $\Sigma_B = \mathrm{diag}(\sigma_1,\ \sigma_2 \mathbf{I}_2 \ldots,\ \sigma_{n_w} \mathbf{I}_{n_w})$ is the diagonal block matrix.

Proof We need to show the existence of a quadratic function that decreases along every nonzero trajectory of (2.23). Let a quadratic Lyapunov function be $V_z(t) = z(t)^T P z(t)$ where $P > 0$. The derivative of the function satisfies

$$\frac{d}{dt} V_z(t) = (A_{cl} z + B_w w)^T P z + z^T P (A_{cl} z + B_w w) < 0 \tag{2.50}$$

for all nonzero z. This is equivalent to

$$\dot{V}(z) = \begin{bmatrix} z \\ w \end{bmatrix}^T \begin{bmatrix} A_{cl}^T P + P A_{cl} & P B_w \\ B_w^T P & 0 \end{bmatrix} \begin{bmatrix} z \\ w \end{bmatrix} < 0$$

for all nonzero (z, w) satisfying (2.53), which is

$$\begin{bmatrix} z \\ w_i \end{bmatrix}^T \begin{bmatrix} -\tilde{C}_{zi}^T \tilde{C}_{zi} & 0 \\ 0 & 1 \end{bmatrix} \begin{bmatrix} z \\ w_i \end{bmatrix} < 0 \quad \text{for } i = 1, \ldots, n_w. \tag{2.51}$$

Using the S-procedure [12], the inequality condition (2.51) holds if there exist nonnegative constants $\sigma_1, \ldots, \sigma_{n_w}$ such that

$$z^T (A_{cl}^T P + P A_{cl}) z + 2 z^T P B_w w - \sum_{i=1}^{n_w} \sigma_i \{ w_i^T w_i - (\tilde{C}_{zi} z)^T (\tilde{C}_{zi} z) \} < 0 \tag{2.52}$$

or

$$\begin{bmatrix} z \\ w \end{bmatrix}^T \begin{bmatrix} A_{cl}^T P + P A_{cl} & P B_w \\ B_w^T P & 0 \end{bmatrix} \begin{bmatrix} z \\ w \end{bmatrix} - \sum_{i=1}^{n_w} \sigma_i \begin{bmatrix} z \\ w_i \end{bmatrix}^T \begin{bmatrix} -\tilde{C}_{zi}^T \tilde{C}_{zi} & 0 \\ 0 & 1 \end{bmatrix} \begin{bmatrix} z \\ w_i \end{bmatrix} < 0.$$

With $\Sigma = \mathrm{diag}(\sigma_1, \sigma_2 \mathbf{I}_2, \ldots, \sigma_{n_w} \mathbf{I}_{n_w})$, this inequality is equivalent to the LMI (2.49). □

Example 2.6 (Global stabilization of a Lipschitz system) Consider the third-order globally Lipschitz nonlinear system:

$$\begin{aligned}\dot{x}_1 &= x_2 + \sin(\omega x_1) := x_2 + f_1,\\ \dot{x}_2 &= x_3 + x_1 \sin(\omega x_2) := x_3 + f_2,\\ \dot{x}_3 &= u\end{aligned} \tag{2.53}$$

where l and ω_i are the known constants. The control objective is $x_1 \to 0$. Then, after applying DSC to the system, the augmented error dynamics is

$$\dot{z} = A_{cl} z + B_w w, \quad w = J C_z z$$

where z, A_{cl}, and B_w are derived in Example 2.1, and w is

$$\begin{aligned} w &= \begin{bmatrix} \dot{f}_1 \\ \dot{f}_2 \end{bmatrix} = \begin{bmatrix} \omega \dot{x}_1 \cos(\omega x_1) \\ \dot{x}_1 \sin(\omega x_2) + \omega \dot{x}_2 \cos(\omega x_2) \end{bmatrix} = J \begin{bmatrix} c_{z1} \\ c_{z2} \end{bmatrix} z \\ &= J C_z z \end{aligned}$$

where

$$J = \begin{bmatrix} \omega \cos(\omega x_1) & 0 \\ \sin(\omega x_2) & \omega \cos(\omega x_2) \end{bmatrix},$$

$$C_z = \begin{bmatrix} -K_1 & 1 & 0 & 1 & 0 \\ 0 & -K_2 & 1 & -\frac{1}{\tau_2} & 1 \end{bmatrix} \in \Re^{2\times 5}.$$

If the Euclidean norm is used, the norm-bound of w is

$$\|w\|_2 \le \|J\|_2 \|C_z z\|_2 \le \gamma \|C_z z\|_2.$$

Since $J(x)$ is not a constant matrix, it is not straightforward to calculate $\|J\|_2$. However, the upper bound of $\|J\|_\infty$ can be calculated using $\|\cdot\|_\infty$ of the vector norm and the induced matrix norm:

$$\|J\|_\infty = \max\{|\omega \cos(\omega x_1)|, |\sin(\omega x_2)| + |\omega \cos(\omega_2 x_2)|\} \le 1 + \omega.$$

Then using norm equivalence, γ can be calculated as

$$\|J\|_2 \le \sqrt{2}\|J\|_\infty \le \sqrt{2}(1+\omega) := \gamma.$$

If the diagonal norm-bound of w for all $x \in \Re^3$ is considered

$$\begin{aligned} |w_1| &= |J_{11} c_{z1} z| \le \omega |c_{z1} z| := |\tilde{C}_{z1} z|, \\ |w_2| &= |J_{21} c_{z1} z + J_{22} c_{z2} z| \le \|[J_{21}\ J_{22}]\|_2 \left\| \begin{bmatrix} c_{z1} \\ c_{z2} \end{bmatrix} z \right\|_2 \\ &\le \sqrt{1+\omega^2} \left\| \begin{bmatrix} c_{z1} \\ c_{z2} \end{bmatrix} z \right\|_2 := \|\tilde{C}_{z2} z\|_2 \end{aligned}$$

where $\tilde{C}_{z1} = \omega c_{z1} \in \Re^{1\times 5}$ and $\tilde{C}_{z2} = \sqrt{1+\omega^2}\left[\begin{smallmatrix} c_{z1} \\ c_{z2} \end{smallmatrix}\right] \in \Re^{2\times 5}$.

Suppose that $\omega = 1$ for simulation. We consider the following sets of controller gains: $\Theta = \{K_1, K_2, K_3, \tau_2, \tau_3\}$ where $\{K_1, K_2, K_3\} = \{4, 6, 6\}$. Then, as derived in Example 2.3, the characteristic equation for the nominal error dynamics is

$$\tau_2 \tau_3 \lambda^4 + (\tau_2 + \tau_3)\lambda^3 + (6\tau_2 + 1)\lambda^2 + 10\lambda + 24 = 0.$$

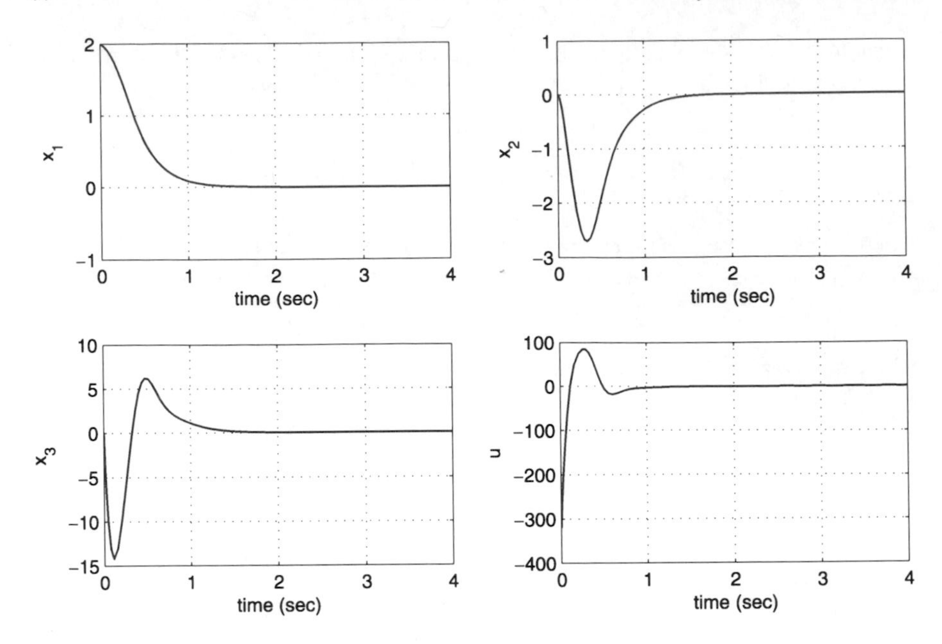

Fig. 2.6 Time responses of x and u for the given initial condition

The inequality condition with respect to τ_2 and τ_3 can be derived for A_{cl} to be Hurwitz using the Routh stability criterion. If $\tau = \tau_2 = \tau_3$, the inequality condition is $\tau \leq 0.2632$. It is interesting to note that the inequality condition is $\tau \leq 0.0263$ for A_{cl} to be Hurwitz when 10 times greater values of K_i are used in Example 2.3. It implies that higher gains K_i may allow smaller τ_i to make A_{cl} Hurwitz. It is intuitive that if higher gains are assigned, larger change of ξ_i is expected and the smaller τ_i makes ξ_i smaller.

When τ is chosen as 0.028, the eigenvalues of A_{cl} are

$$\lambda(A_{cl}) = \{-3.7872, -6, -12.0213 \pm i6.3943, -43.5988\}$$

and A_{cl} is thus Hurwitz. With respect to two different upper bounds of w, either LMI (2.33) or LMI (2.49) can be computed numerically and there exist solutions P for both cases for the given Θ_1. Therefore, it is expected that the origin of the augmented error dynamics is globally exponentially stable. For the initial conditions given as $x_1(0) = 2$ and $x_2(0) = x_3(0) = 0$, the time responses of x and u are plotted in Fig. 2.6. It is shown that x in (2.53) is stabilized as expected.

2.6 Ultimate and Quadratic Boundedness

If we consider a tracking problem such that the control objective is $x_1(t) \to x_{1d}(t)$, the augmented error dynamics from Lemma 2.1 then is

$$\dot{z} = A_{cl}z + B_w w + B_r r, \quad w = JC_z z.$$

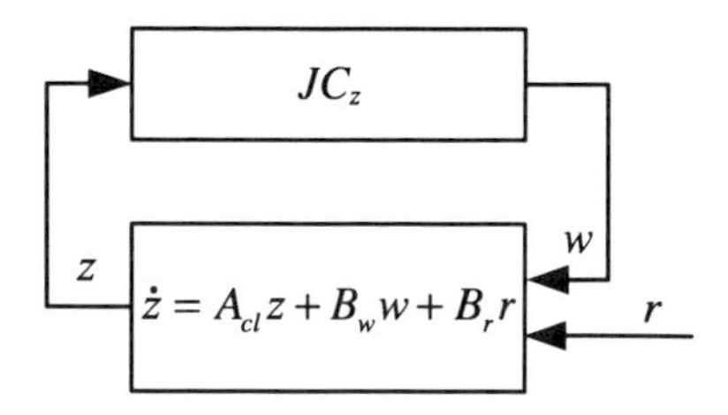

Fig. 2.7 Graphical interpretation for augmented closed-loop error dynamics

Moreover, for some compact and convex set $\mathscr{D}_i \subset \mathscr{D}$,

$$\|J\| \leq \gamma$$

for all $x \in \mathscr{D}_i$, and x_{1d} is the feasible output trajectory as follows:

Definition 2.2 $x_{1d}(t)$ is a feasible output trajectory in $\mathscr{D}$ if $x_{1d}(t)$ is a C^2 function and $[x_{1d}\ \dot{x}_{1d}\ \ddot{x}_{1d}]^T$ is uniformly bounded on the convex and compact set $\mathscr{D}_i \subset \mathscr{D}$.

Suppose the convex and compact set $\mathscr{D}_i := \{x \mid \|x\| \leq c,\ c > 0\}$ is defined in $\mathscr{D}$. Then, r is bounded as

$$\|r\|_2^2 = \left\| \begin{bmatrix} J_1 & 0_{n_w} \\ 0 & 1 \end{bmatrix} \begin{bmatrix} \dot{x}_{1d} \\ \ddot{x}_{1d} \end{bmatrix} \right\|_2^2 = J_1^T J_1 \dot{x}_{1d}^2 + \ddot{x}_{1d}^2 \leq (\gamma^2 + 1)c^2 := r_0^2. \tag{2.54}$$

Therefore, the augmented error dynamics for the tracking problem is

$$\begin{cases} \dot{z} = A_{cl}z + B_w w + B_r r, & w = JC_z z, \\ \|w\| \leq \gamma \|C_z z\|, & \|r\| \leq r_0. \end{cases} \tag{2.55}$$

As shown in Fig. 2.7, r is not the vanishing perturbation in the sense that $r \neq 0$ for $z = 0$ while w is regarded as the vanishing perturbation. If the error dynamics includes the nonvanishing perturbation, we can no longer study stability of the origin in the error dynamics and expect that $z(t) \to 0$ as $t \to \infty$. We may hope that $z(t)$ will be ultimately bounded by a small bound if the nonvanishing perturbation, r, is small in some sense. This concept of ultimate boundedness in [51] is applied to the augmented error dynamics in (2.55) and its ultimate bound will be approximated in the form of a quadratic function, $z^T P z$, using the concept of *quadratic boundedness* in [14, 15].

Definition 2.3 Suppose $V(z)$ is a continuously differentiable, positive definite function and a set $\Omega_c = \{z \in \Re^{n_z} \mid V(z) \leq c\}$ is compact for some $c > 0$. Let $\Delta = \{z \mid \varepsilon \leq V(z) \leq c\}$ for some positive constant $\varepsilon < c$. If the derivative of V along the trajectories of the error dynamics (2.19) satisfies

$$\dot{V}(z) \leq -W(z), \quad \forall z \in \Delta$$

where $W(z)$ is a continuous positive definite function. A set $\Omega_\varepsilon = \{z \mid V(z) \leq \varepsilon\}$ is an uniform ultimate error bound if there exists a positive constant c, and for every $\delta \in (0, c)$ there is a positive constant $T = T(\delta)$ such that

$$z(t_0) \in \Omega_\delta \quad \Longrightarrow \quad z(t) \in \Omega_\varepsilon, \quad \forall t \geq t_0 + T$$

where $\Omega_\delta = \{z \in \Re^{n_z} \mid V(z) \leq \delta\}$.

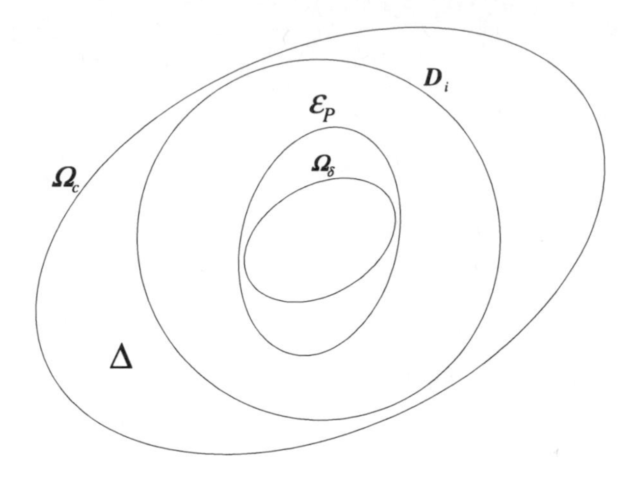

Fig. 2.8 Graphical interpretation of domain

Definition 2.4 Suppose a set of controller gains, Θ, is given, $\|r\| \neq 0$, and x_{id} in r is the feasible output trajectory. The nonlinear system (2.3) is *quadratically trackable* via DSC for a feasible output trajectory if the error dynamics (2.19) is *quadratically bounded with Lyapunov matrix P*, i.e., if there exists $P > 0$ such that

$$z^T Pz > 1 \quad \text{implies } (A_{cl}z + B_w w + B_r r)^T Pz + z^T P(A_{cl}z + B_w w + B_r r) < 0$$

for all nonzero $z \in \mathscr{E}_P = \{z \in \Re^{n_z} | z^T Pz \leq 1\}$. Then, it has the following properties for the set $\mathscr{E}_P$:

(i) The set $\mathscr{E}_P$ is *controlled invariant* via DSC, i.e. if for all $z(0) \in \mathscr{E}_P$ the solution $z(t) \in \mathscr{E}_P$ for all $t > 0$.
(ii) The set $\mathscr{E}_P$ contains the *reachable* set from the origin, i.e., if $z(t)$ is any solution with $z(0) = 0$, then $z(t) \in \mathscr{E}_P$ for all $t > 0$.
(iii) The set $\mathscr{E}_P$ is the uniform ultimate error bound.

All domains defined in Definitions 2.2 to 2.4 are described graphically in Fig. 2.8. In this section, we will compute the smallest quadratic error bound ($\mathscr{E}_P$) in some sense which contains the ultimate error bound (Ω_δ) when the x_{1d} is the feasible trajectory, i.e., $[x_{1d}\ \dot{x}_{1d}\ \dot{x}_{1d}]^T \in \mathscr{D}_i$.

Example 2.7 (Boundedness of the closed-loop error dynamics) Consider the system (2.53) in Example 2.6 with no uncertainty. Suppose the control objective is $x_1 \to x_{1d}(t) = \sin t$. If DSC is applied, we obtain the error dynamics as follows:

$$\begin{aligned} \dot{z} &= A_{cl}z + B_w w + B_r r, \\ w &= JC_z z \end{aligned}$$

where z, A_{cl}, B_w are defined in Example 2.1, and w, J, C_z are derived in Example 2.6. Moreover,

$$r = \begin{bmatrix} J_1 \dot{x}_{1d} \\ \ddot{x}_{1d} \end{bmatrix} = \begin{bmatrix} \omega\cos(\omega x_1)\dot{x}_{1d} \\ \sin(\omega x_2)\dot{x}_{1d} \\ \ddot{x}_{1d} \end{bmatrix}, \qquad B_r = \begin{bmatrix} 0 & 0 & 0 & 1 & 1/\tau_2 \\ 0 & 0 & 0 & 0 & 1 \\ 0 & 0 & 0 & -1 & -1/\tau_2 \end{bmatrix}^T.$$

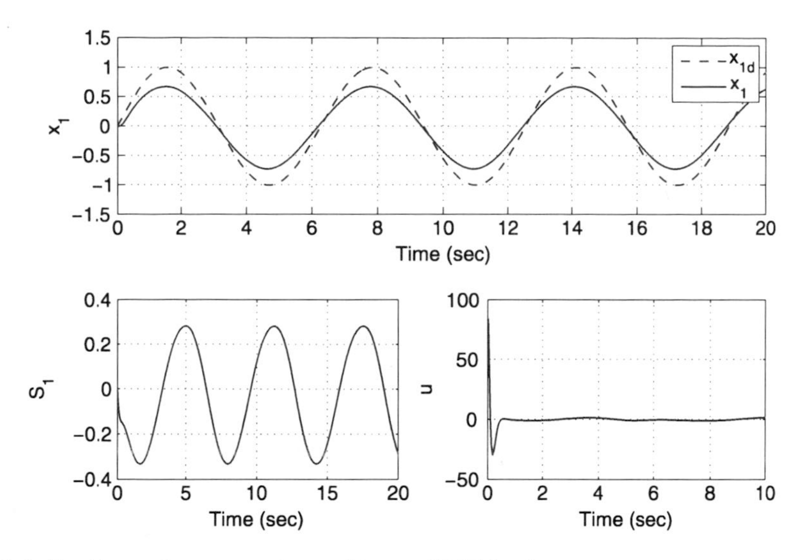

Fig. 2.9 Tracking performance and control input of DSC for Θ_1

Since f is globally Lipschitz and $\dot{x}_{1d}^2 + \ddot{x}_{1d}^2 = 1$,

$$\|r\|_2^2 = \left[\omega^2 \cos(\omega x_1)^2 + \sin(\omega x_2)^2\right]\dot{x}_{1d}^2 + \ddot{x}_{1d}^2 \le \omega^2 + 1 := r_0^2$$

for all $x \in \Re^3$. Then, the diagonal norm-bounded error dynamics is

$$\begin{aligned} &\dot{z} = A_{cl}z + B_w w + B_r r, \quad w = JC_z z, \\ &|w_i| \le \|\tilde{C}_{zi} z\|_2 \quad \text{for } i = 1, 2, \qquad \|r\| \le r_0 \end{aligned} \tag{2.56}$$

where $\tilde{C}_{zi}$ are defined in Example 2.6 and $r_0 = \sqrt{\omega^2 + 1}$.

First consider the same controller gain set $\Theta_1 = \{4, 6, 6, 0.028, 0.028\}$ which is used for the stabilization problem in Example 2.6. Figure 2.9 shows that the maximum error of S_1 reaches up to about 0.3 although it does not diverge. If the surface gains are increased to $\{K_1, K_2, K_3\} = \{40, 60, 60\}$ to reduce the maximum error of S_1, A_{cl} in (2.56) is not Hurwitz, i.e., $\lambda(A_{cl}) = \{0.5623 \pm i42.2411, -36.2766 \pm i19.9836, -60\}$. Therefore, the error vector, z, is diverging due to the eigenvalues with positive real part. Consequently, the control input u also diverges (refer to Fig. 2.1). However, when $\Theta_2 = \{40, 60, 60, 0.026, 0.026\}$, A_{cl} becomes Hurwitz, i.e., $\lambda(A_{cl}) = \{-0.1157 \pm i43.8554, -38.3459 \pm i19.3785, -60\}$. Figure 2.10 shows that the error S_1 converges to about ± 0.05 around the origin with a high frequency oscillation. Therefore, this example motivates the question of how to estimate the ultimate error bound as well as to ensure the quadratic stability outside the error bound for the given controller gains.

If $\tilde{r} := r/r_0$ and $\tilde{B}_r := r_0 B_r$ in (2.55), the error dynamics can rewritten as

$$\begin{aligned} &\dot{z} = A_{cl}z + B_w w + \tilde{B}_r \tilde{r}, \quad w = JC_z z, \\ &\|w\| \le \gamma \|C_z z\|, \qquad \|\tilde{r}\| \le 1. \end{aligned} \tag{2.57}$$

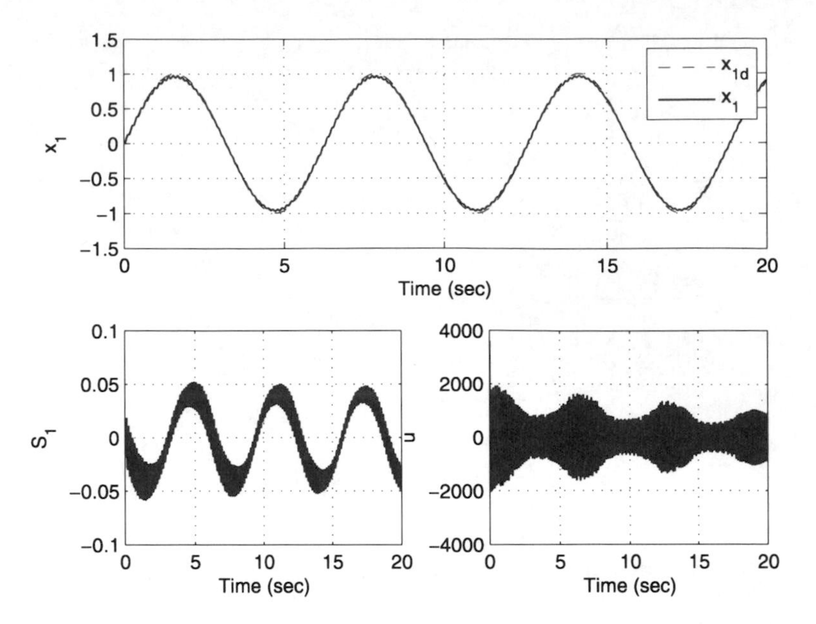

Fig. 2.10 Tracking performance and control input of DSC for Θ_2

Without loss of generality, it can be considered that $\tilde{r}$ is a unit-peak input. Then, the following theorem describes the condition for guaranteeing quadratic tracking as well as the computation of the matrix P for a given set of controller gains.

Theorem 2.4 *For the given set of controller gains, Θ, suppose that the closed-loop error dynamics* (2.57) *is given on the domain $\mathcal{D}_i \subset \mathcal{D}$ and x_{1d} is a feasible output trajectory. The nonlinear system* (2.3) *is quadratically trackable via DSC if there exist $P > 0$, $\sigma \geq 0$, and $\alpha \geq 0$ such that*

$$\begin{bmatrix} A_{cl}^T P + P A_{cl} + \alpha P + \sigma \tilde{C}_z^T \tilde{C}_z & P B_w & P \tilde{B}_r \\ B_w^T P & -\sigma I & 0 \\ \tilde{B}_r^T P & 0 & -\alpha I \end{bmatrix} < 0 \tag{2.58}$$

where $\tilde{C}_z = \gamma C_z$ and $\tilde{B}_r = r_0 B_r$.

Proof Suppose that there exist a function $V_z = z^T P z$, with $P > 0$, such that

$$\frac{d}{dt} V_z(t) = \left(A_{cl} z + B_w w + \tilde{B}_r \tilde{r}\right)^T P z + z^T P \left(A_{cl} z + B_w w + \tilde{B}_r \tilde{r}\right) < 0 \tag{2.59}$$

for every nonzero $(z, w, \tilde{r})$ satisfying $\tilde{r}^T \tilde{r} \leq 1$ and $z^T P z \geq 1$. Using the S-procedure, we see that a sufficient condition for the inequality conditions (2.59) to hold is the existence of $P > 0$, $K \geq 0$, and $\alpha \geq 0$ such that

$$z^T\left(A_{cl}^T P + P A_{cl}\right)z + 2z^T(PB_w w + P\tilde{B}_r\tilde{r}) - \alpha\left(\tilde{r}^T\tilde{r} - z^T P z\right)$$
$$-\sigma\left\{w^T w - (\tilde{C}_z z)^T(\tilde{C}_z z)\right\} < 0. \tag{2.60}$$

The above inequality is equivalent to LMI (2.58). □

As done for the stabilization problem in Sect. 2.5.3, if a componentwise upper bound of w is obtained

$$|w_i| \leq \|\tilde{C}_{zi} z\|_2,$$

the quadratic boundedness can be stated as follows:

Corollary 2.2 *For the given set of controller gains, Θ, suppose that the closed-loop error dynamics* (2.57) *is given on the domain $\mathcal{D}_i \subset \mathcal{D}$, w is diagonally bounded, $|w_i| \leq \|\tilde{C}_z z\|_2$, and x_{1d} is the feasible output trajectory. The nonlinear system* (2.3) *is quadratically trackable via DSC if there exist $P > 0$, $\Sigma = \mathrm{diag}(\sigma_1, \ldots, \sigma_{n_w}) \geq 0$ and $\alpha \geq 0$ such that*

$$\begin{bmatrix} A_{cl}^T P + P A_{cl} + \alpha P + \tilde{C}_z^T \Sigma_B \tilde{C}_z & PB_w & P\tilde{B}_r \\ B_w^T P & -\Sigma & 0 \\ \tilde{B}_r^T P & 0 & -\alpha I \end{bmatrix} < 0 \tag{2.61}$$

where $\tilde{C}_z = [\tilde{C}_{z1}^T \cdots \tilde{C}_{zn_w}^T]^T$, $\tilde{C}_{zi} \in \Re^{n_i \times n_z}$, and $\Sigma_B = \mathrm{diag}(\sigma_1, \sigma_2\mathbf{I}_2, \ldots, \sigma_{n_w}\mathbf{I}_{n_w})$ is the diagonal block matrix.

While the given inequality (2.58) or (2.61) gives an approximation on the ultimate error bound, we would like to find the *smallest* upper bound to accurately estimate the reachable set. Therefore, we need to determine an appropriate measure of size for the ellipsoid, such as the volume or the largest semi-axis. For the purposes of this chapter, the largest semi-axis (or diameter) of the ellipsoid will be used to measure its size. Since w and r in the error dynamics (2.57) enter only the ξ subspace in the sense that the upper block matrices of B_w and B_r are the zero matrix, it is natural that the largest diameter of the ellipsoid may be defined in the ξ subspace and it might be overestimated due to a conservative upper bound of w and r. Thus if the largest diameter is minimized, a smaller quadratic error bound can be computed. Although the inequality (2.58) or (2.61) is not an LMI as stated, it can be posed as a convex optimization problem for the fixed α, so we have to fix the gain α and obtain the smallest ellipsoid by maximizing the smallest eigenvalue of P. Therefore, the LMI relaxation problem is written as

Algorithm 2.2 For a fixed $\alpha \in [a,\ b]$,

$$\begin{aligned} &\text{maximize} && \lambda_{\min}(P) \\ &\text{subject to} && P > 0, \quad \sigma \geq 0, \quad \text{LMI (2.58)} \\ &&& \text{or} \\ &&& P > 0, \quad \Sigma \geq 0, \quad \text{LMI (2.61).} \end{aligned} \tag{2.62}$$

As shown in MATLAB Program 2-3, a specific α which gives the smallest ellipsoid can be determined by iterating over a range of values. It should be noted that the objective function can be changed depending on what we want to minimize, e.g., a volume or the largest semi-axis of the ellipsoidal set [12]. It is also remarked that other types of input such as unit energy, componentwise unit energy and componentwise peak inputs can be also considered in the proposed framework for quadratic tracking or quadratic boundedness [12].

MATLAB Program 2-3

```
%***** Define the dimension of matrices *****
n = size(Acl, 1);
m = size(Bw, 2);
nr = size(Br, 2);

%***** Define the interval of alpha *****
alpha = logspace(-2,2,40);

for k = 1:length(alpha),
%***** cvx version *****
  cvx_begin sdp
    variable P(n,n) symmetric;
    variable beta;
    maximize( lambda_min(P) );
    beta >= 0;
    P == semidefinite(n);
    -[Acl'*P + P*Acl + alpha(K)*P + beta*Cz'*Cz P*Bw P*Br;
    Bw'*P -beta*eye(m) zeros(m,nr);
    Br'*P zeros(nr,m) -alpha(k)*eye(nr)] == semidefinite(n+m+nr);
  cvx_end
end
```

Example 2.8 (Quadratic ultimate bound for a second-order nonlinear system) Consider the second-order nonlinear system:

$$\begin{aligned} \dot{x}_1 &= x_2 - x_1^2, \\ \dot{x}_2 &= u - x_2. \end{aligned} \tag{2.63}$$

The control objective is $x_1 \to x_{1d} = \sin t$.

Let $\mathscr{D}_i = \{x \in \Re^2 \,|\, |x_i| \le c \text{ for } c \ge 1\}$. Then, the x_{1d} is a feasible trajectory in the sense that $[x_{1d}\ \dot{x}_{1d}\ \ddot{x}_{1d}]^T = [\sin t\ \cos t\ -\sin t]^T$ is uniformly bounded on $\mathscr{D}_0$. If DSC is applied to the system, the augmented error dynamics is

$$\begin{aligned} \dot{z} &= A_{cl} z + B_w w + B_r r, \\ w &= -2x_1 C_z z := J_{11} C_z z, \\ |w| &= 2|x_1||C_z z| \le 2c|C_z z| := \gamma |C_z z| \quad \text{for } \forall x \in \mathscr{D}_0 \end{aligned} \tag{2.64}$$

where z, A_{cl}, B_w, and C_z are given in (2.42) of Example 2.4,

$$r = [J_{11}\dot{x}_{1d}\ \ddot{x}_{1d}]^T = [J_{11}\cos t\ -\sin t]^T \in \Re^2,$$

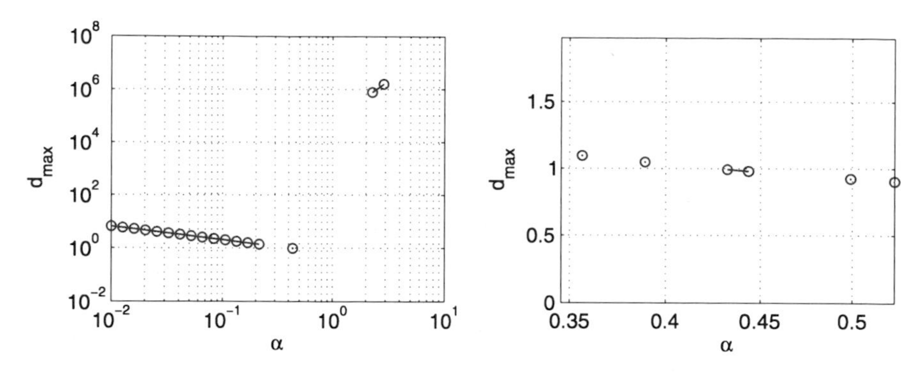

Fig. 2.11 Maximum radius of ellipsoid along line search of α

$$B_r = \begin{bmatrix} 0_2 & 0_2 \\ 1 & -1 \end{bmatrix} \in \Re^{3\times 2}.$$

Furthermore,

$$\|r\|_2^2 = J_{11}^2 \cos^2 t + \sin^2 t \le 4c^2 + 1 := r_0^2$$

for all $x \in \mathscr{D}_i$. Then, the augmented error dynamics (2.64) can be rewritten as

$$\begin{aligned} \dot{z} &= A_{cl}z + B_w w + \tilde{B}_r \tilde{r}, \\ |w| &\le \gamma |C_z z| = |\tilde{C}_z z|, \qquad \|\tilde{r}\| < 1 \end{aligned} \tag{2.65}$$

where $\tilde{B}_r = r_0 B_r = \sqrt{4c^2+1}B_r$, and $\tilde{r} = r/r_0 = r/\sqrt{4c^2+1}$.

Suppose that $c = 1$ for computation of algorithm (2.62) and a set of controller gains is $\Theta_1 = \{K_1, K_2, \tau_2\} = \{1, 2, 0.02\}$ which makes $\lambda(A_{cl}) = \{-1.0208, -2, -48.9792\}$ Hurwitz. To compute the ultimate bound of the closed-loop error dynamics, the LMI (2.62) is solved iteratively for fixed α. That is, after the 40 logarithmically equally spaced points between 10^{-2} and 10^2 are generated for α's, the minimum of the maximum diameter, which is $d_{\max} = 2/\sqrt{\lambda_{\min}(P)}$, is obtained when $\alpha = 0.4375$ (see in the left plot of Fig. 2.11). Then the 20 linearly equally spaced points between 0.3455 and 0.5541 are generated and the iterative computation of LMI (2.62) is performed for each α. Finally, for $\alpha = 0.5212$, the corresponding solution P is

$$P = \begin{bmatrix} 0.1655 & -0.1578 & -0.0032 \\ -0.1578 & 1.1185 & 0.0030 \\ -0.0032 & 0.0030 & 0.0049 \end{bmatrix} \times 10^3,$$

and the corresponding maximum diameter of the ellipsoid, $d_{\max}$, is 0.4528 which is the semi-axis in the ξ_2 axis.

To validate the computed quadratic ultimate error bound, Fig. 2.12 shows the time responses of x and u for the given Θ_1 after simulation. The upper plot in Fig. 2.13 shows that the error trajectory stays in the quadratic error bound which is calculated above in the sense that $z(t)^T P z(t) \le 1$ after 2.1846 seconds and the

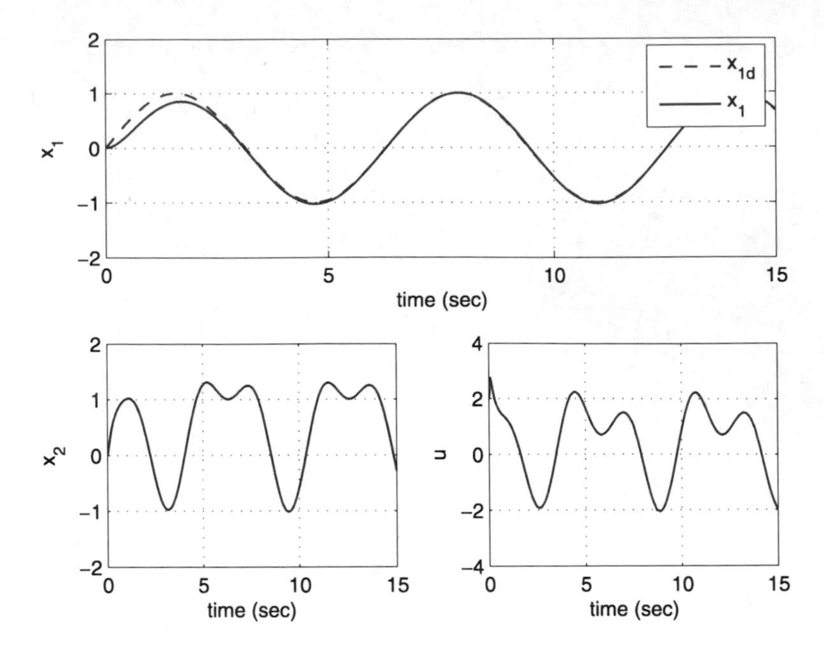

Fig. 2.12 Time responses of x_1, x_2, and u for Θ_1

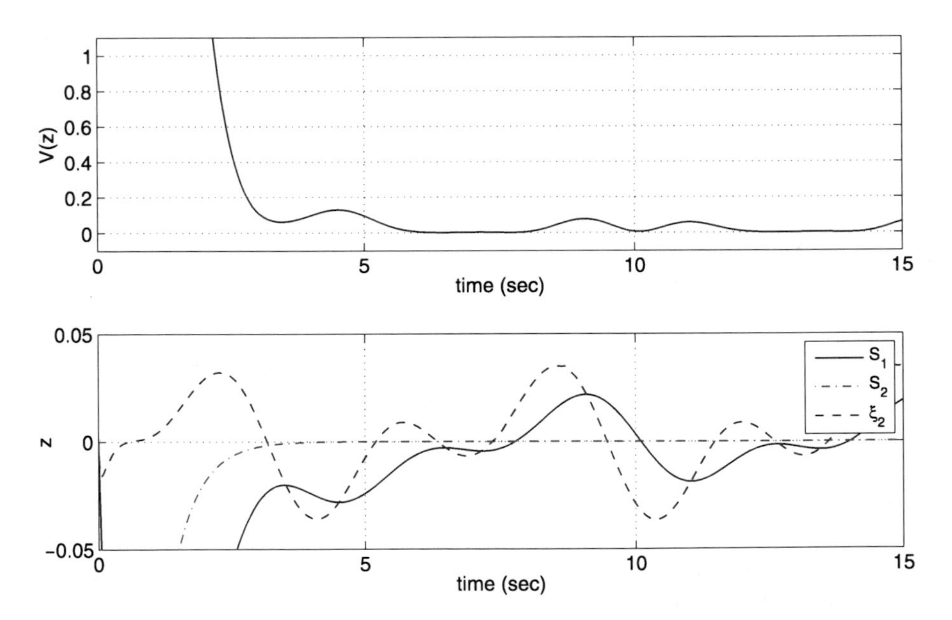

Fig. 2.13 Quadratic function level $V(z) = z^T P z$ and time responses of z

bottom plot shows that all errors are less than $d_{\max}/2$ after a certain time. It is interesting to note that the semi-axes of the quadratic error bound are $\{d_{S_1}, d_{S_2}, d_{\xi_2}\} =$

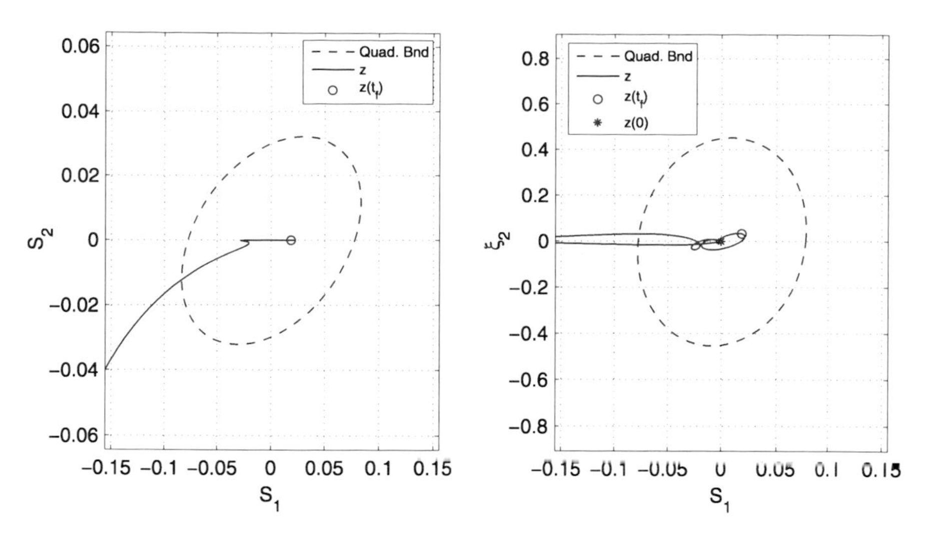

Fig. 2.14 Quadratic bound and z in S_1–S_2 and S_1–ξ_2 planes

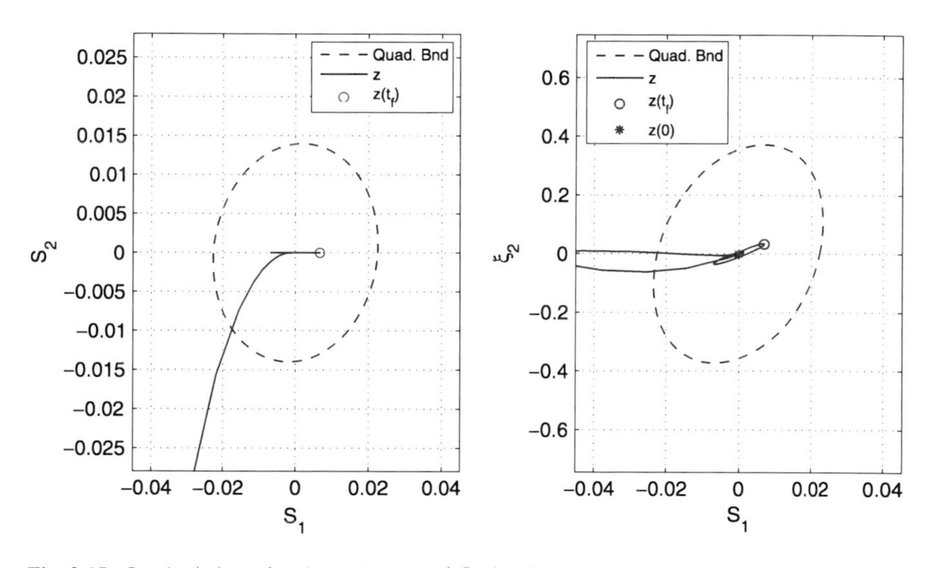

Fig. 2.15 Quadratic bound and z in S_1–S_2 and S_1–ξ_2 planes

$\{0.1555, 0.0643, 0.9056\}$ and d_{ξ_2} is the most overestimated. As expected, both w and r enter through the ξ subspace in the sense that the upper block matrices of B_w and B_r are a zero matrix. So d_{ξ_2} may be estimated in a conservative way, thus this is the reason to minimize the maximum diameter of the quadratic error bound in the framework of convex optimization. If the quadratic error bound is projected onto S_1–S_2 or S_1–ξ_2 plane, it is shown in Fig. 2.14 that z stays in the quadratic bound once it reaches the bound.

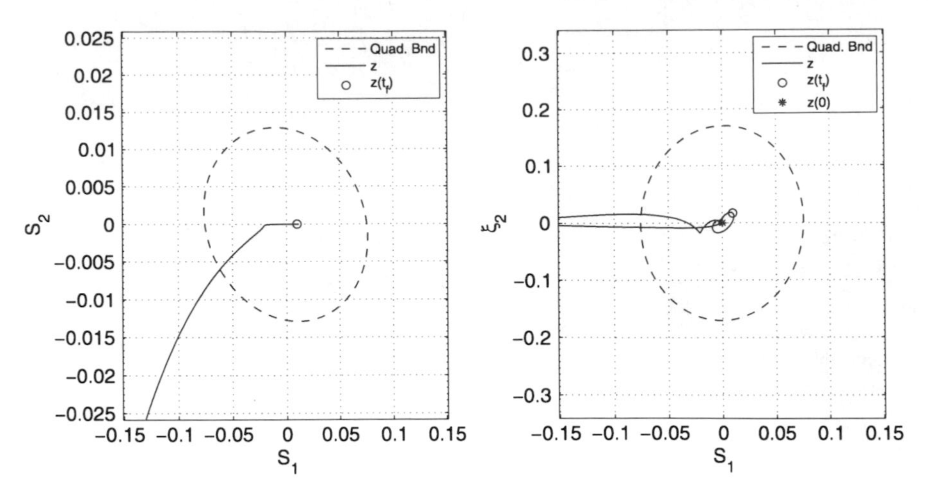

Fig. 2.16 Quadratic bound and z in S_1–S_2 and S_1–ξ_2 planes

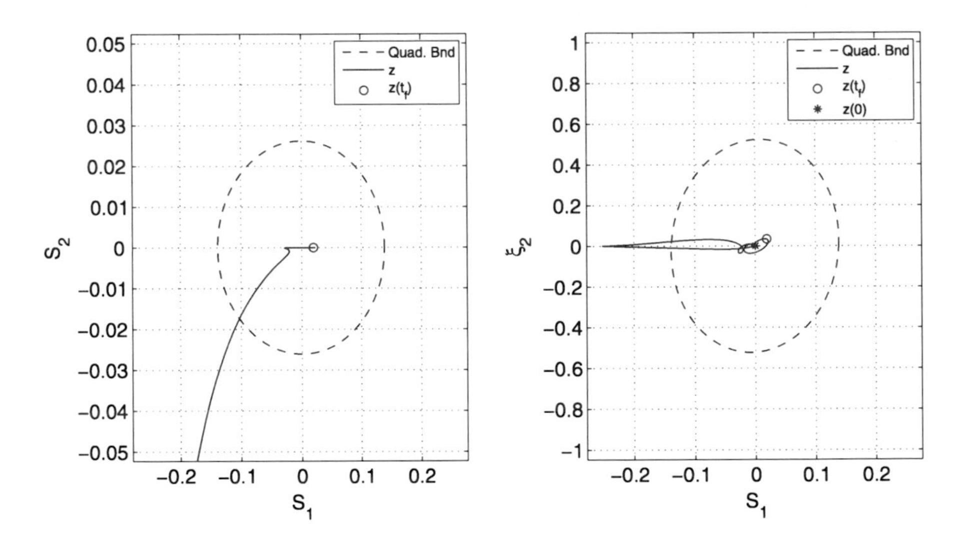

Fig. 2.17 Quadratic bound and z in S_1–S_2 and S_1–ξ_2 planes

Let us consider that the gains K_i are larger. Suppose the second set of controller gains are $\Theta_2 = \{5, 10, 0.02\}$ for which $\lambda(A_{cl}) = \{-5.6351, -10, -44.3649\}$ is Hurwitz. Similarly the LMI (2.62) can be solved iteratively for a fixed α. Then, the minimum of a maximum diameter of a quadratic error bound is obtained as 0.7442 for $\alpha = 0.8358$ and the corresponding semi-axis diameters are $\{d_{S_1}, d_{S_2}, d_{\xi_2}\} = \{0.0451, 0.028, 0.7441\}$ as shown in Fig. 2.15. As expected, a smaller ultimate bound is estimated for Θ_2 and especially the semi-axes in S_1 and S_2 axis are much reduced such that $\frac{d_{S_1}(\Theta_1)}{d_{S_1}(\Theta_2)} = \frac{0.1555}{0.0451} = 3.4479$, $\frac{d_{S_2}(\Theta_1)}{d_{S_2}(\Theta_2)} = \frac{0.0643}{0.028} = 2.2963$. If the time constant τ_2 becomes 0.01, i.e. $\Theta_3 = \{1, 2, 0.01\}$, the semi-axis diameters are calculated as $\{d_{S_1}, d_{S_2}, d_{\xi_2}\} = \{0.151, 0.0258, 0.3408\}$ for $\alpha = 1.8326$ as shown in Fig. 2.16. The semi-axes in ξ_2 and S_2 axis are much reduced such that $\frac{d_{\xi_2}(\Theta_1)}{d_{\xi_2}(\Theta_3)} = \frac{0.9056}{0.3408} = 2.6573$, $\frac{d_{S_2}(\Theta_1)}{d_{S_2}(\Theta_3)} = \frac{0.0643}{0.0258} = 2.4922$. Therefore, the results imply that higher surface gains (K_i) and smaller time constants (τ_i) improve the tracking performance. However, as seen in Example 2.3, higher surface gains may result in instability of the closed-loop system. Consequently, the tradeoff between stability and tracking performance should be considered to determine the control gains.

It is also important to note that the upper bounds of w and r affect the performance of the error bound estimation. In this example, $c = 1$ is used under the assumption that x_1 tracks x_{1d} with very small error after a certain time. If $c = 2$ is assigned in a more conservative way, a larger error bound will be estimated as shown in Fig. 2.17. If semi-axis diameters in the S_1 axis are compared, it is shown that $\frac{d_{S_1}(c=1)}{d_{S_1}(c=2)} = \frac{0.1555}{0.2771} = 0.5612$. Therefore, it is necessary to have tighter upper bounds of w and r for better estimation of an error bound.

Chapter 3
Robustness to Uncertain Nonlinear Systems

This chapter extends the results of Chap. 2 to uncertain nonlinear systems with modeling uncertainty. That is, after DSC is applied to the uncertain system, the augmented closed-loop error dynamics is derived and decoupled into linear and nonlinear terms. Then, the nonlinear term is classified into either a vanishing or a nonvanishing perturbation. If some inequality constraints are imposed to the perturbation terms, they are formulated in the form of LMI to discuss stability and performance. Finally the stability and/or performance problem is solved numerically in the framework of convex optimization.

Two classes of uncertain nonlinear systems are considered for DSC design in this chapter, i.e., globally and locally Lipschitz nonlinear systems. Quadratic stability and tracking is discussed for the globally Lipschitz system in Sect. 3.1. In Sect. 3.2, using the concept of nonlinear damping, a robust DSC system is designed for the locally Lipschitz system based on an LMI approach and a quadratic ultimate error bound is computed as a result of the convex optimization. In Sect. 3.3, a concept of stability in the input-output sense called $\mathscr{L}$ stability is introduced and the calculation of the $\mathscr{L}_2$ gain is discussed for the error dynamics.

3.1 Uncertain Lipschitz Systems

3.1.1 Problem Statement

Consider a class of nonlinear systems with mismatched uncertainties:

$$\begin{aligned} \dot{x}_i &= x_{i+1} + f_i(x_1, \ldots, x_i) + \Delta f_i(x_1, \ldots, x_i) \quad \text{for } i = 1, \ldots, n-1, \\ \dot{x}_n &= u + f_n(x) \end{aligned} \tag{3.1}$$

where the state $x \in \Re^n$, the control input $u \in \Re$, and $f(x)$ and $[\partial f/\partial x](x)$ are continuous for all $x \in \Re^n$, and there exist constants γ_{ij} such that $|J_{ij}| \le \gamma_{ij}$ for all $x \in \Re^n$. Then, f is globally Lipschitz in x. Furthermore, the uncertainty Δf is also

B. Song, J.K. Hedrick, *Dynamic Surface Control of Uncertain Nonlinear Systems*,
Communications and Control Engineering,
DOI 10.1007/978-0-85729-632-0_3,

globally Lipschitz on $\Re^n$, f and Δf is thus diagonally norm-bounded, i.e., there exist nonnegative constants m_{ij} and n_{ij} such that

$$|f_i(x_1,\dots,x_i)| \le \sum_{j=1}^{i} m_{ij}|x_j|, \qquad |\Delta f_i(x_1,\dots,x_i)| \le \sum_{j=1}^{i} n_{ij}|x_j|.$$

3.1.2 Quadratic Stability and Tracking

If the DSC laws outlined in Sect. 2.3 are applied to the uncertain nonlinear system (3.1), as done in Sect. 2.4, similarly the augmented error dynamics can be given as

$$\begin{bmatrix} \mathbf{I}_n & \mathbf{0}_{nn_w} \\ -K & T_\xi \end{bmatrix} \begin{bmatrix} \dot{S} \\ \dot{\xi} \end{bmatrix} = \begin{bmatrix} A_{11} & A_{12} \\ \mathbf{0}_{n_w n} & A_{22} \end{bmatrix} \begin{bmatrix} S \\ \xi \end{bmatrix} + \begin{bmatrix} \mathbf{0}_{nn_w} \\ \mathbf{I}_{n_w} \end{bmatrix} \dot{f} + \begin{bmatrix} \mathbf{I}_{nn_u} & 0_n \\ \mathbf{0}_{n_w n_u} & -b_r \end{bmatrix} \begin{bmatrix} \Delta f \\ \ddot{x}_{1d} \end{bmatrix} \tag{3.2}$$

where $\dot{f} = [\dot{f}_1 \ \cdots \ \dot{f}_{n-1}]^T \in \Re^{n-1} := \Re^{n_w}$, $\Delta f = [\Delta f_1 \ \cdots \ \Delta f_{n-1}]^T \in \Re^{n-1} := \Re^{n_u}$, and all other definitions of vectors and matrices are given in (2.17). It is noted that the uncertain term Δf is only added in the last term and the corresponding matrix is expanded if (3.2) is compared with (2.17).

If an inverse matrix of the first block matrix in (3.2):

$$T^{-1} = \begin{bmatrix} \mathbf{I}_n & \mathbf{0}_{nn_w} \\ T_\xi^{-1} K & T_\xi^{-1} \end{bmatrix}$$

is multiplied on both sides, the augmented closed-loop error dynamics with DSC is rewritten as

$$\dot{z} = A_{cl} z + B_w \dot{f} + B_{ue} r_e \tag{3.3}$$

where $z = [S \ \xi]^T \in \Re^{n_z}$, $r_e = [\Delta f \ \ddot{x}_{1d}]^T \in \Re^{n_u+1}$, A_{cl} and B_w are defined in (2.18), and

$$B_{ue} = \begin{bmatrix} \mathbf{I}_{nn_u} & 0_n \\ T_\xi^{-1} K & -T_\xi^{-1} b_r \end{bmatrix} = [B_u \ B_e], \quad B_u = \begin{bmatrix} \mathbf{I}_{n(n-1)} \\ T_\xi^{-1} K \end{bmatrix} \in \Re^{n_z \times n_u}.$$

It is remarked again that the only difference between (3.3) and (2.18) in Sect. 2.4 is that the last term is expanded to $r_e \in \Re^{(n_u+1}$ due to the inclusion of model uncertainty (Δf).

As done in Lemma 2.1, we will decompose $\dot{f}$ and Δf into vanishing and nonvanishing perturbation terms with inequality constraints. Then, the augmented closed-loop error dynamics can be summarized as follows:

Lemma 3.1 *For the given uncertain nonlinear system in* (3.1), *the augmented closed-loop error dynamics with DSC is*

$$\dot{z} = A_{cl} z + B_{wu} w_u + B_{ru} r_u \tag{3.4}$$

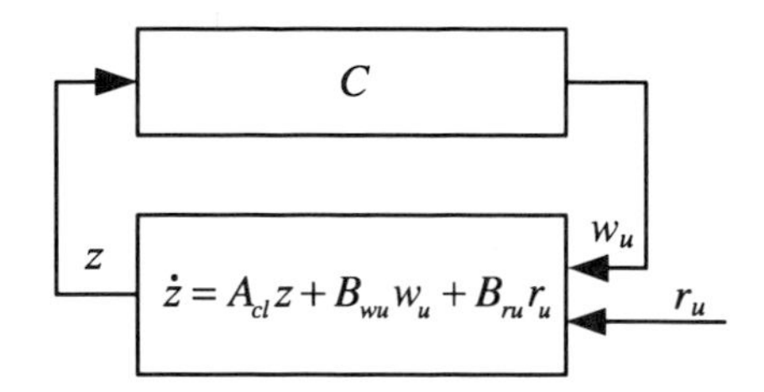

Fig. 3.1 Graphical interpretation for augmented closed-loop error dynamics

where

$$w_u = \begin{bmatrix} p \\ Jp \\ w \end{bmatrix} \in \Re^{n_u+n_w+n_w} := \Re^{n_{wu}}, \qquad r_u = \begin{bmatrix} q \\ Jq \\ r \end{bmatrix} \in \Re^{n_u+n_w+n_r} := \Re^{n_{ru}},$$

$$\Delta f = p + q \in \Re^{n_u},$$

$$B_{wu} = [B_u \ B_w \ B_w] \in \Re^{n_z \times n_{wu}}, \qquad B_{ru} = [B_u \ B_w \ B_r] \in \Re^{n_z \times n_{ru}},$$

B_u is defined in (3.3), *and all other vectors and matrices are defined in Lemma* 2.1. *Furthermore, there are C_{zi} such that*

$$|w_{ui}| \leq \|C_{zi}z\| \tag{3.5}$$

for $i = 1, \ldots, n_{wu}$.

The proof is given in Appendix A.2 and the result of Lemma 3.1 is summarized graphically in Fig. 3.1. It is remarked that the perturbation terms w_u and r_u are expanded linearly due to inclusion of model uncertainty (Δf) when they are compared with w and r in Lemma 2.1. Therefore, the augmented error dynamics can be regarded as diagonal norm-bounded LDI (DNLDI) as discussed in Sect. 2.5.3.

If x_{1d} is a feasible output trajectory on a domain $\mathscr{D}_i$ as stated in Definition 2.2, e.g.,

$$\mathscr{D}_i = \{x \,|\, \|x\| \leq c, c > 0\},$$

the norm-bound of the nonvanishing perturbation r_u is

$$\|r_u\|_2^2 = q^T q + q^T J^T J q + r^T r \leq (1+\gamma^2) \sum_{i=1}^{n-1} \|D_{ui} r_d\|_2^2 + \gamma^2 \dot{x}_{1d}^2 + \ddot{x}_{1d}^2$$

where $\gamma = \max(\gamma_{ij})$ for all i, j. Since there exist α_i such that $\|D_{ui}\| \leq \alpha_i$,

$$\begin{aligned} \|r_u\|_2^2 &\leq (1+\gamma^2) \sum_{i=1}^{n-1} \alpha_i^2 (x_{1d}^2 + \dot{x}_{1d}^2) + (1+\gamma^2)(\dot{x}_{1d}^2 + \ddot{x}_{1d}^2) \\ &\leq (1+\gamma^2) c^2 \left(\sum_{i=1}^{n-1} \alpha_i^2 + 1 \right) := r_{u0}^2. \end{aligned} \tag{3.6}$$

As discussed in Sect. 2.6, the quadratic boundedness can be analyzed for the error dynamics (3.4) with (3.5) and (3.6).

Example 3.1 (Error dynamics of second order uncertain nonlinear system) Consider the second order uncertain nonlinear system:

$$\begin{aligned}\dot{x}_1 &= x_2 + l\sin(\omega x_1) + \Delta f_1(x_1) := x_2 + f_1 + \Delta f_1,\\ \dot{x}_2 &= u + l\sin(\omega x_2) := u + f_2\end{aligned} \tag{3.7}$$

where l and ω are known as a constant and Δf_1 is an unknown Lipschitz function but bounded by $|\Delta f_1| \le |x_1|$. The control objective is $x_1(t) \to x_{1d}$ where x_{1d} is feasible on a domain $\mathscr{D}_i$

$$\mathscr{D}_i = \{x \in R^3 \,|\, \|[x_{1d}\ \dot{x}_{1d}\ \ddot{x}_{1d}]\| \le R,\ R > 0\}.$$

After defining $S_1 = x_1 - x_{1d}$ and differentiating it, $\bar{x}_2$ and x_{2d} are, respectively,

$$\dot{S}_1 = \dot{x}_1 - \dot{x}_{1d} = x_2 + f_1 + \Delta f_1 - \dot{x}_{1d},$$

$$\bar{x}_2 = -f_1 - K_1 S_1 + \dot{x}_{1d}, \qquad \tau_2 \dot{x}_{2d} + x_{2d} = \bar{x}_2.$$

In sequence, define $S_2 = x_2 - x_{2d}$ and differentiate it again. Then, the control input u is

$$\dot{S}_2 = \dot{x}_2 - \dot{x}_{2d} = u + f_2 - \dot{x}_{2d},$$

$$u = -f_2 - K_2 S_2 + \dot{x}_{2d}.$$

Using (3.3), the augmented error dynamics is

$$\dot{z} = A_{cl} z + B_w \dot{f}_1 + B_{ue} r_e \tag{3.8}$$

where $z = [S_1\ S_2\ \xi_2]^T \in \Re^3$, A_{cl} and B_w are given in (2.44) of Example 2.4,

$$B_{ue} = [B_u\ B_e] = \left[\begin{array}{c|c} 1 & 0\\ 0 & 0\\ K_1 & -1\end{array}\right], \quad \text{and} \quad r_e = \begin{bmatrix}\Delta f_1\\ \ddot{x}_{1d}\end{bmatrix}.$$

Furthermore, $\dot{f}_1$ and Δf_1 can be written as a function of z and x_{1d}.

$$\begin{aligned}\dot{f}_1 &= \frac{\partial f_1}{\partial x_1}\dot{x}_1 = J_{11}(x_2 + f_1 + \Delta f_1) = J_{11} c_{z1} z + J_{11}\dot{x}_{1d} + J_{11}\Delta f_1\\ &= w + J_{11}\dot{x}_{1d} + J_{11}\Delta f_1\end{aligned}$$

where $c_{z1} = [-K_1\ 1\ 1] \in \Re^{1\times 3}$ and

$$|\Delta f_1| \le |x_1| = |S_1 + x_{1d}| \le |S_1| + |x_{1d}|.$$

Then, after using the inequality condition above and defining the functions p_1 and q_1 such that

$$\Delta f_1 = p_1 + q_1, \quad |p_1| \le |S_1|,\ |q_1| \le |x_{1d}|,$$

the error dynamics in (3.8) is rewritten as

$$\begin{aligned}
\dot{z} &= A_{cl}z + B_w(J_{11}c_{z1}z + J_{11}\dot{x}_{1d} + J_{11}\Delta f_1) + B_u\Delta f_1 + B_e\ddot{x}_{1d} \\
&= A_{cl}z + (B_w w + B_w J_{11}p_1 + B_u p_1) + (B_w J_{11}\dot{x}_{1d} + B_w J_{11}q_1 + B_u q_1 + B_e\ddot{x}_{1d}) \\
&= A_{cl}z + [B_u\ B_w\ B_w]\begin{bmatrix} p_1 \\ J_{11}p_1 \\ w \end{bmatrix} + [B_u\ B_w\ B_r]\begin{bmatrix} q_1 \\ J_{11}q_1 \\ r \end{bmatrix} \\
&:= A_{cl}z + B_{wu}w_u + B_{ru}r_u
\end{aligned} \tag{3.9}$$

where $B_r = [B_w\ B_e]$ and r are defined in Lemma 2.1.

Next, we need to derive the inequality conditions for the vanishing perturbation w_u and the nonvanishing term r_u. The component of w_u is bounded by a linear function of z such that

$$\begin{cases} |w_{u1}| = |p_1| \le |S_1| = |C_{z1}z|, \\ |w_{u2}| = |J_{11}p_1| \le \gamma|S_1| = |C_{z2}z|, \\ |w_{u3}| = |w| \le \gamma|c_{z1}z| = |C_{z3}z| \end{cases} \tag{3.10}$$

where $C_{z1} = [1\ 0\ 0] \in \Re^{1\times 3}$, $C_{z2} = \gamma \cdot C_{z1}$, $C_{z3} = \gamma c_{z1}$, and γ is chosen to satisfy

$$|J_{11}| = \left|l\omega\cos(\omega x_1)\right| \le l\omega = \gamma.$$

Therefore, the augmented error dynamics in (3.9) with the inequality conditions (3.10) can be written in the form of the result of Lemma 3.1. In addition, r_u is bounded by a constant r_{u0} such that

$$r_u^T r_u \le (1+\gamma^2)q_1^T q_1 + r^T r \le (1+\gamma^2)\left(x_{1d}^2 + \dot{x}_{1d}^2 + \ddot{x}_{1d}^2\right) = (1+\gamma^2)R^2 := r_{u0}^2.$$

Due to the similarity of the closed-loop error dynamics in the form of DNLDI, we can use Corollary 2.2 to discuss the quadratic tracking to the new closed-loop error dynamics of Lipschitz nonlinear system with uncertainty.

Theorem 3.1 *Suppose that the closed-loop error dynamics* (3.4) *with the inequality constraint* (3.5) *is given on the domain* $\Re^n$ *for the given set of controller gains,* Θ *and* x_{1d} *is a feasible output trajectory on* $\mathscr{D} \subset \Re^n$, *thus* r_u *is norm-bounded, i.e.,*

$$\|r_u\| \le r_{u0}.$$

The nonlinear system (3.1) *with mismatched Lipschitz uncertainty is quadratically trackable via DSC if there exist* $P > 0$, $\Sigma = \mathrm{diag}(\sigma_1, \ldots, \sigma_{n_u}, \sigma_{n_u+1}, \ldots, \sigma_{n_u+n_w}) \ge 0$ *and* $\alpha \ge 0$ *such that*

$$\begin{bmatrix} A_{cl}^T P + PA_{cl} + \alpha P + \tilde{C}^T\Sigma_B\tilde{C} & PB_{wu} & P\tilde{B}_{ur} \\ B_{wu}^T P & -\Sigma & 0 \\ \tilde{B}_{ur}^T P & 0 & -\alpha I \end{bmatrix} < 0 \tag{3.11}$$

where $\tilde{B}_{ur} = r_{u0}B_{ru}$,

$$\tilde{C} = \mathrm{diag}\left(\tilde{C}_{u1}, \ldots, \tilde{C}_{un_u}, \tilde{C}_{z1}, \ldots, \tilde{C}_{zn_w}\right) \quad \textit{and}$$
$$\Sigma_B = \mathrm{diag}(\sigma_1\mathbf{I}, \ldots, \sigma_{n_u}\mathbf{I}, \sigma_{n_u+1}\mathbf{I}, \ldots, \sigma_{n_u+n_w}\mathbf{I})$$

are block diagonal matrices.

Remark 3.1 For clarity of understanding, consider the stabilization problem of the system with mismatched uncertainty with the control objective $x_1 \to 0$. Then the error dynamics is simplified to

$$\begin{aligned} \dot{z} &= A_{cl}z + B_{wu}w_u, \\ |w_{ui}| &\le \|C_{zi}z\|. \end{aligned} \tag{3.12}$$

The quadratic stability can be analyzed using Theorem 2.3 for (3.12). Furthermore, if there is no uncertainty, $\Delta f = 0$, the error dynamics (3.12) is then equal to (2.48).

Remark 3.2 If both f and Δf are not globally Lipschitz but locally Lipschitz, there may not exist $\tilde{C_{ui}}$ and $\tilde{D}_{ui}$ such that $\|\Delta f_i\| \le \|\tilde{C}_{ui}z\| + \|\tilde{D}_{ui}r_d\|$. This motivates us to use nonlinear damping to bound Δf_i in next section.

Example 3.2 (Quadratic tracking of the third order uncertain system) Consider the third order nonlinear system subject to mismatched Lipschitz uncertainty:

$$\begin{aligned} \dot{x}_1 &= x_2 + \Delta f_1, \\ \dot{x}_2 &= x_3 + \Delta f_2, \\ \dot{x}_3 &= u \end{aligned} \tag{3.13}$$

where Δf_i are an unknown function but globally Lipschitz, i.e., $|\Delta f_i| \le |x_i|$ for all $x \in \Re^3$. The control objective is $x_1 \to x_{1d} := \sin t$.

If DSC is applied to the system, the design procedure is

$$\begin{aligned} S_1 &:= x_1 - x_{1d}, \\ \dot{S}_1 &= \dot{x}_1 - \dot{x}_{1d} = x_2 + \Delta f_1 - \dot{x}_{1d}, \\ \bar{x}_2 &:= \dot{x}_{1d} - K_1 S_1, \\ \tau_2 \dot{x}_{2d} + x_{2d} &= \bar{x}_2, \quad x_{2d}(0) := \bar{x}_2(0). \end{aligned}$$

After defining $S_2 := x_2 - x_{2d}$, similarly

$$\begin{aligned} \dot{S}_2 &= \dot{x}_2 - \dot{x}_{2d} = x_3 + \Delta f_2 - \dot{x}_{2d}, \\ \bar{x}_3 &:= \dot{x}_{2d} - K_2 S_2, \\ \tau_3 \dot{x}_{3d} + x_{3d} &= \bar{x}_3, \quad x_{3d}(0) := \bar{x}_3(0). \end{aligned}$$

Finally, the control input u is

$$\begin{aligned} S_3 &:= x_3 - x_{3d}, \\ \dot{S}_3 &= u - \dot{x}_{3d}, \\ u &:= \dot{x}_{3d} - K_2 S_3. \end{aligned}$$

Since $f = 0$ and thus $B_w = 0$, the augmented error dynamics is

$$\dot{z} = A_{cl}z + B_{ru}r_u = A_{cl}z + B_u \Delta f + B_r \ddot{x}_{1d} \tag{3.14}$$

where z and A_{cl} are equal to the closed-loop error dynamics for the nominal system, which is derived in Example 2.3, and

$$r_u = \begin{bmatrix} \Delta f_1 \\ \Delta f_2 \\ \ddot{x}_{1d} \end{bmatrix}^T := \begin{bmatrix} \Delta f \\ \ddot{x}_{1d} \end{bmatrix} \in \Re^3,$$

$$B_{ru} = \begin{bmatrix} \mathbf{I}_{3\times 2} & 0_3 \\ T_\xi^{-1} K \mathbf{I}_{3\times 2} & -T_\xi^{-1} b_r \end{bmatrix} = [B_u \ B_r] \in \Re^{5\times 3},$$

where

$$T_\xi^{-1} = \begin{bmatrix} 1 & 0 \\ 1/\tau_2 & 1 \end{bmatrix}, \qquad K = \begin{bmatrix} K_1 & 0 & 0 \\ 0 & K_2 & 0 \end{bmatrix}, \qquad b_r = [1 \ 0]^T.$$

Moreover, since the uncertainties are globally Lipschitz,

$$|\Delta f_1| \le |x_1| = |S_1 + x_{1d}| \le |C_{u1} z| + |D_{u1} r_d|,$$
$$|\Delta f_2| \le |x_2| = |S_2 + \xi_2 + \bar{x}_2| = |S_2 + \xi_2 - K_1 S_1 + \dot{x}_{1d}| \le |C_{u2} z| + |D_{u2} r_d|.$$

Therefore, there exist C_{ui} and D_{ui} such that

$$|\Delta f_i| \le |C_{ui} z| + |D_{ui} r_d|$$

where

$$r_d = \begin{bmatrix} x_{1d} \\ \dot{x}_{1d} \end{bmatrix}, \qquad C_u = \begin{bmatrix} C_{u1} \\ C_{u2} \end{bmatrix} = \begin{bmatrix} 1 & 0 & 0 & 0 & 0 \\ -K_1 & 1 & 0 & 1 & 0 \end{bmatrix},$$

$$D_u = \begin{bmatrix} D_{u1} \\ D_{u2} \end{bmatrix} = \begin{bmatrix} 1 & 0 \\ 0 & 1 \end{bmatrix}.$$

If $\Delta f_i := p_i + q_i$ such that $|p_i| \le |C_{ui} z|$ and $|q_i| \le |D_{ui} r_d|$. The error dynamics (3.14) can be rewritten as

$$\dot{z} = A_{cl} z + B_{wu} w_u + B_{ru} r_u,$$
$$|w_{ui}| \le |C_{ui} z|$$

where $w_u = p \in \Re^2$, $B_{wu} = [B_u \ B_w] = B_u \in \Re^{5\times 2}$, $B_{ru} = [B_u \ B_r] \in \Re^{5\times 3}$, and

$$\|r_u\|_2^2 = \left\| \begin{bmatrix} q \\ \ddot{x}_{1d} \end{bmatrix} \right\|_2^2 = q^T q + \ddot{x}_{1d}^2 \le x_{1d}^2 + \dot{x}_{1d}^2 + \ddot{x}_{1d}^2 \le 2 := r_0^2.$$

Finally,

$$\begin{aligned} \dot{z} &= A_{cl} z + B_{wu} w_u + \tilde{B}_{ru} \tilde{r}_u, \\ |w_{ui}| &\le |C_{ui} z|, \\ \|\tilde{r}_u\| &\le 1 \end{aligned} \tag{3.15}$$

where $\tilde{B}_{ru} = \sqrt{2} B_{ru}$ and $\tilde{r}_u = \frac{1}{\sqrt{2}} r_u$.

Suppose that the uncertainties are given as $\Delta f_i = -x_i \sin(\omega x_i)$ for $i = 1, 2$ where $\omega = 10$ for simulations and the first set of controller gains is $\Theta_1 = \{K_1, K_2, K_3, \tau_2, \tau_3\} = \{10, 20, 20, 0.01, 0.01\}$, thus A_{cl} is Hurwitz. Next, we can

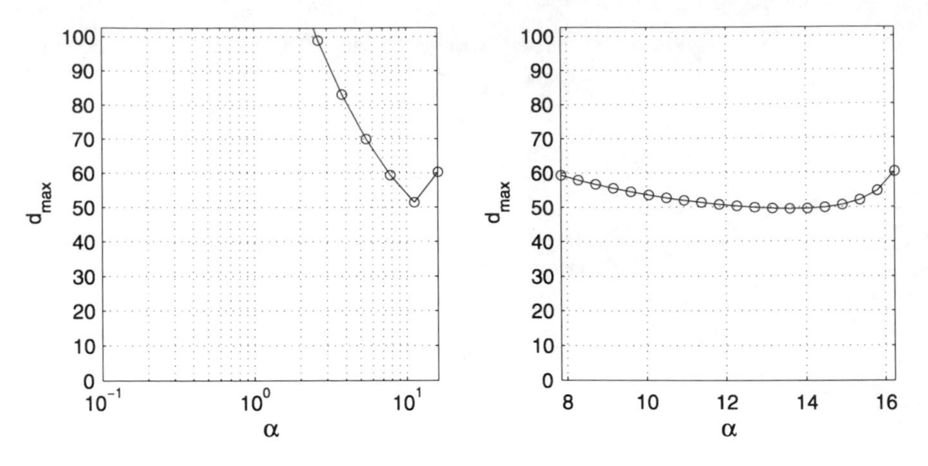

Fig. 3.2 Maximum radius of ellipsoid along line search of α

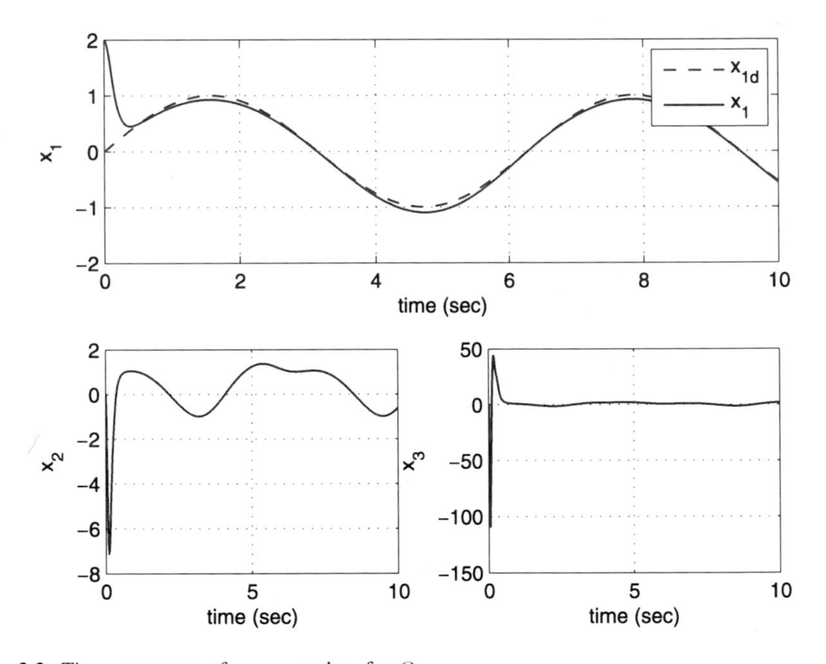

Fig. 3.3 Time responses of x_1, x_2, and x_3 for Θ_1

calculate the quadratic error bound of z using LMI (3.25) and the algorithm (3.27). Figure 3.2 shows the minimum of the largest diameter of the ellipsoidal bound $\mathscr{E}_P = \{z|z^T Pz \leq 1\}$ along the line search of α, i.e. to maximize $\lambda_{\min}(P)$ in the framework of convex optimization using CVX [32] (refer to MATLAB Program 2-3). When $\alpha = 13.5882$, the largest diameter of the ellipsoidal bound is calculated as 49.3904 and it corresponds to the diameter of ξ_3 as shown in the right plot of Fig. 3.5.

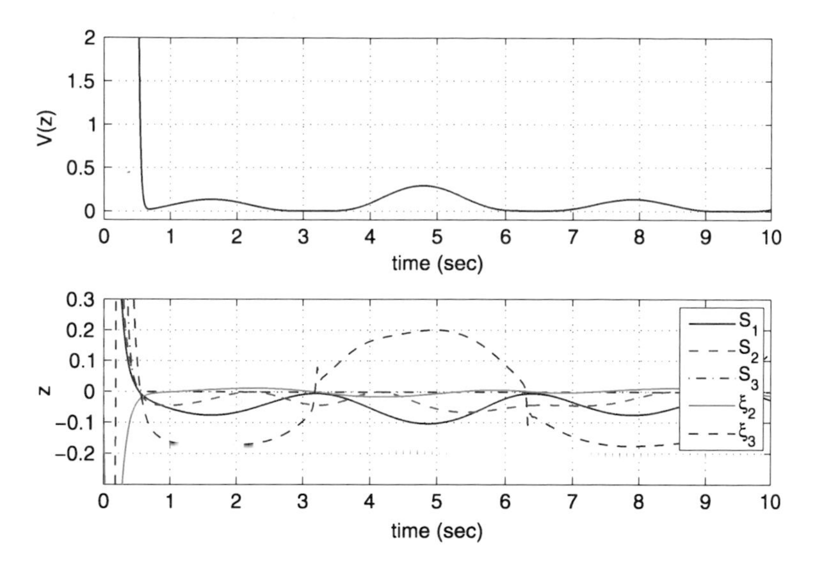

Fig. 3.4 Quadratic function level $V(z) = z^T Pz$ (*top*) and time responses of z (*bottom*)

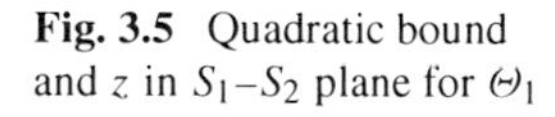

Fig. 3.5 Quadratic bound and z in S_1–S_2 plane for Θ_1

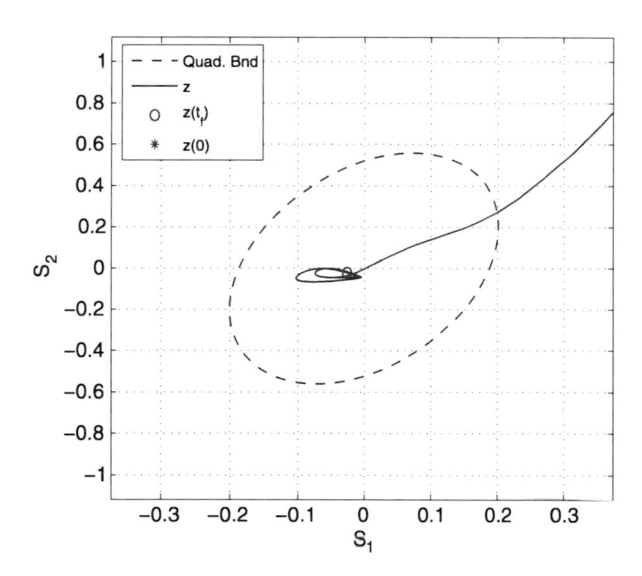

Figure 3.4 shows that the error z stays within the ellipsoidal bound in the sense that $z^T Pz < 1$ after a certain time, while the time responses of x and u are shown in Fig. 3.3. When K_i is doubled, i.e., $\Theta_2 = \{20, 40, 40, 0.01, 0.01\}$, the largest diameter of the ellipsoidal bound is calculated as 91.6678 for $\alpha = 27.2022$. Figure 3.6 shows that S_1 is reduced about half for the given Θ_2 and the diameter of the estimated tracking error bound in S_1 axis also becomes smaller than half of the diameter calculated for Θ_1.

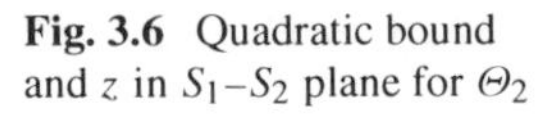

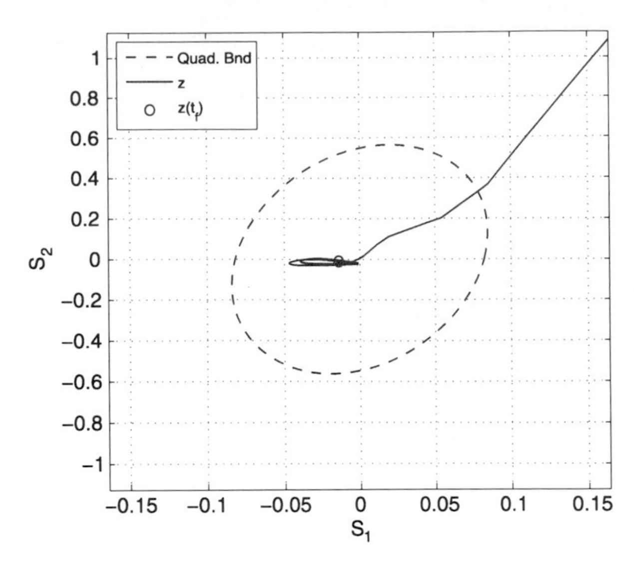

Fig. 3.6 Quadratic bound and z in S_1–S_2 plane for Θ_2

3.2 DSC with Nonlinear Damping

3.2.1 Problem Statement

Consider the class of nonlinear systems with mismatched uncertainties:

$$\begin{aligned} \dot{x}_i &= x_{i+1} + f_i(x) + \Delta f_i(x) \quad \text{for } i = 1, \ldots, n-1, \\ \dot{x}_n &= u + f_n(x) \end{aligned} \tag{3.16}$$

where the state $x \in \Re^n$, the control input $u \in \Re$, and $f(x)$, f_n, and $[\partial f/\partial x](x)$ are continuous for all x in a domain $\mathscr{D} \subset \Re^n$ and there exists $\gamma \geq 0$ such that

$$\|J(x)\| = \left\| \frac{\partial f(x)}{\partial x} \right\| \leq \gamma$$

where $J(x)$ is a Jacobian matrix of f and $J_{ij} = \partial f_i/\partial x_j$. Then, the f is locally Lipschitz on $\mathscr{D}$. Moreover, Δf is an unknown but locally Lipschitz nonlinearity to guarantee the existence and uniqueness of the solution of (3.16) [51]. Moreover, each component function of Δf is bounded by a known $\mathscr{C}^1$ function, $\rho_i(x)$, such that

$$|\Delta f_i(x)| \leq \rho_i(x) \quad \text{and} \quad \rho_i(0, \ldots, 0) = 0$$

for all x on $\mathscr{D}$ and there exists a nonnegative constant δ such that

$$\|\Phi(x)\| = \left\| \frac{\partial \rho(x)}{\partial x} \right\| \leq \delta$$

where $\Phi_{ij} = \partial \rho_i/\partial x_j$.

3.2.2 Stabilization and Quadratic Boundedness

The standard design procedure of DSC with nonlinear damping is described in [94]. An outline of this procedure is as follows: Define the first error surface as $S_1 := x_1 - x_{1d}(t)$. Then

$$\begin{aligned} \dot{S}_1 &= x_2 + f_1(x) - \dot{x}_{1d}(t) + \Delta f_1(x), \\ \bar{x}_2 &:= -f_1(x) + \dot{x}_{1d}(t) - K_1 S_1 - \frac{S_1 \rho_1^2(x)}{2\varepsilon} \end{aligned} \tag{3.17}$$

where $\frac{S_1\rho_1^2}{2\varepsilon}$ is the nonlinear damping term, which makes x_1 remain in an arbitrarily small boundary layer around x_{1d} after some time in the presence of the locally Lipschitz uncertainty, Δf_1 [55]. That is, suppose $V_1 = S_1^2/2$. The derivative of V_1 is

$$\dot{V}_1 = S_1 \dot{S}_1 = S_1(x_2 + f_1 - \dot{x}_{1d} + \Delta f_1).$$

If x_2 tracks $\bar{x}_2$ asymptotically ($x_2 \to \bar{x}_2$), the surface error S_1 will converge to some boundary layer around zero, i.e., using (3.17),

$$\begin{aligned} S_1 \dot{S}_1 &= S_1 \big\{ (x_2 - \bar{x}_2) + \bar{x}_2 + f_1 - \dot{x}_{1d} + \Delta f_1 \big\} \approx S_1 \left(-K_1 S_1 - \frac{S_1 \rho_1^2(x)}{2\varepsilon} + \Delta f_1 \right) \\ &\le -K_1 S_1^2 + \varepsilon/2 \end{aligned}$$

where the last inequality follows from Young's inequality,

$$\frac{S_1^2 \rho_1^2}{2\varepsilon} + \frac{\varepsilon}{2} \ge |S_1|\rho_1 \ge |S_1||\Delta f_1| \ge |S_1|\Delta f_1 \tag{3.18}$$

and ε is arbitrary. Therefore, S_1 is ultimately uniformly bounded if $x_2 = \bar{x}_2$.

The next step is to force $x_2 \to \bar{x}_2$, so define $S_2 := x_2 - x_{2d}$ where x_{2d} equals $\bar{x}_2$ passed through a first order low-pass filter, i.e.,

$$\tau_2 \dot{x}_{2d} + x_{2d} = \bar{x}_2, \quad x_{2d}(0) := \bar{x}_2(0).$$

Similarly, if we choose $\bar{x}_3$ as

$$\bar{x}_3 = -f_2(x) + \dot{x}_{2d} - K_2 S_2 - \frac{S_2 \rho_2^2(x)}{2\varepsilon}$$

and force $x_3 \to \bar{x}_3$. Continuing this process for each consecutive state, define the $(i-1)$th error surface as $S_{i-1} = x_{i-1} - x_{(i-1)d}$ and $\bar{x}_i$ is

$$\bar{x}_i = -f_{i-1}(x) + \dot{x}_{(i-1)d} - K_{i-1} S_{i-1} - \frac{S_{i-1} \rho_{i-1}^2(x)}{2\varepsilon}. \tag{3.19}$$

Then, x_{id} is obtained by filtering $\bar{x}_i$, i.e.,

$$\tau_i \dot{x}_{id} + x_{id} = \bar{x}_i, \quad x_{id}(0) := \bar{x}_i(0). \tag{3.20}$$

After continuing this procedure for $2 \le i \le n-1$, define $S_n := x_n - x_{nd}$. Finally, the control input is chosen as

$$u = -f_n(x) + \dot{x}_{nd} - K_n S_n = -f_n(x) + \frac{\bar{x}_n - x_{nd}}{\tau_n} - K_n S_n. \tag{3.21}$$

The structural design procedure of DSC with nonlinear damping can be summarized using the graphical representation as shown in Fig. 3.7.

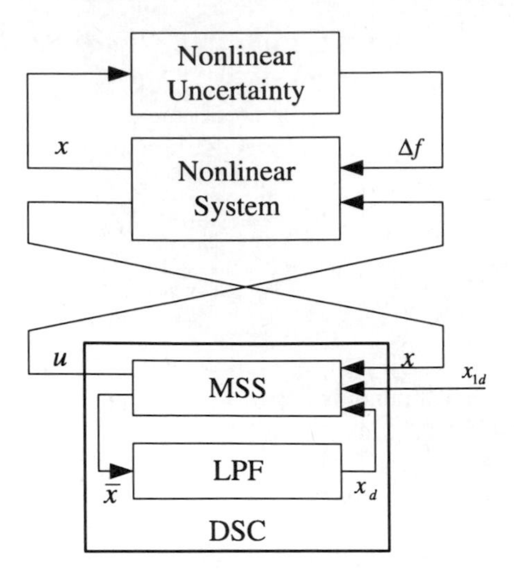

Fig. 3.7 Graphical schematic for DSC system subject to uncertainty

Lemma 3.2 *For the given autonomous nonlinear system* (3.16) *with mismatched uncertainties, the augmented error dynamics with DSC is written as*

$$\dot{z} = A_{cl} z + B_w w + B_{rd} r_d + B_{nd} n_d \tag{3.22}$$

where z, A_{cl}, *and* B_w *are defined in* (2.18),

$$r_d = \begin{bmatrix} r \\ D_2 \Phi d \\ J d \end{bmatrix} \in \Re^{n_{rd}}, \qquad n_d = \begin{bmatrix} d \\ D_1 d \end{bmatrix} \in \Re^{n_{nd}},$$

$$B_{rd} = [B_r \ B_w \ B_w] \in \Re^{n_z \times n_{rd}}, \quad \text{and} \quad B_{nd} = [B_u \ B_w] \in \Re^{n_z \times n_{nd}}$$

where r, B_r, *and* J *are defined in Lemma* 2.1, B_u *is defined in* (3.3),

$$d = [\Delta f_1 - h_1 \ \cdots \ \Delta f_{n-1} - h_{n-1}]^T \in \Re^{n-1} := \Re^{n_u},$$

$$h = [h_1 \ \cdots \ h_{n-1}]^T \in \Re^{n_u}, \quad h_i = \frac{S_i \rho_i^2}{2\varepsilon},$$

$$D_1 = \operatorname{diag}\left(\left[\frac{\rho_1^2}{2\varepsilon} \ \cdots \ \frac{\rho_{n-1}^2}{2\varepsilon}\right]\right) \in \Re^{n_u \times n_u}, \quad \text{and}$$

$$D_2 = \operatorname{diag}\left(\left[\frac{S_1 \rho_1}{\varepsilon} \ \cdots \ \frac{S_{n-1} \rho_{n-1}}{\varepsilon}\right]\right) \in \Re^{n_u \times n_u}.$$

Furthermore, if all assumptions in (3.16) *are satisfied for all* $x \in \mathscr{D}$, x_{1d} *is the feasible output trajectory in* $\mathscr{D}$, *and* z *is bounded on the compact and convex set* Ω_c, *there exist* $\tilde{C}_{wi}$ *and* $n_0 > 0$ *for a given* ε *such that*

$$|w_i| \leq \left\| \tilde{C}_{wi}(\varepsilon) z \right\| \quad \text{and} \quad z^T E n_d \leq n_0(\varepsilon) \tag{3.23}$$

where $E = [\mathbf{I}_{n_z \times n_u} \ \mathbf{I}_{n_z \times n_u}] \in \Re^{n_z \times n_{nd}}$.

It is noted that the feasible output trajectory and the set Ω_c are stated in Definitions 2.2 and 2.3, respectively.

Remark 3.3 As discussed in Sects. 2.6 and 3.1, if r_d in (3.22) is norm-bounded, i.e., there exists $r_{d0} > 0$ such that $\|r_d\| \leq r_{d0}$, the error dynamics (3.22) with both (3.23) and the above inequality can be summarized as

$$\begin{cases} \dot{z} = A_{cl}z + B_w w + \tilde{B}_r \tilde{r}_d + B_{nd} n_d, \\ |w_i| \leq \|\tilde{C}_{wi} z\|, \qquad \|\tilde{r}_d\| \leq 1, \qquad z^T \tilde{E} n_d \leq 1 \end{cases} \tag{3.24}$$

where $\tilde{C}_{wi} = \mu_i C_{wi}$, $\tilde{r}_d = r_d / r_{d0}$, $\tilde{B}_r = r_{d0} B_r$, and $\tilde{E} = E/n_0$. For instance, if $\Omega = \{z \in \Re^{n_z} \,|\, \|z\| \leq c, c > 0\}$ is considered, the norm-bound of r_d can be calculated as

$$\begin{aligned} r_d^T r_d &= r^T r + \|D_2 \Phi d\|^2 + \|J d\|^2 \\ &\leq r_0^2 + r_1^2 + r_2^2 := r_{d0}^2 \end{aligned}$$

where r_0 is defined in (2.54),

$$\begin{aligned} \|D_2 \Phi d\| &= \frac{1}{\varepsilon} \left\| \begin{bmatrix} \rho_1 & \cdots & 0 \\ \vdots & \ddots & \vdots \\ 0 & \cdots & \rho_{n-1} \end{bmatrix} \Phi \begin{bmatrix} S_1 d_1 \\ \vdots \\ S_{n-1} d_{n-1} \end{bmatrix} \right\| \leq \frac{\rho_{\max}}{\varepsilon} \|\Phi\| \frac{\varepsilon \sqrt{n_u}}{2} \\ &\leq \rho_{\max} \delta \sqrt{n_u}/2 := r_1, \\ \|J d\| &\leq \|J\| \|d\| \leq \gamma \|d\| \leq \gamma d_{\max} \sqrt{n_u} := r_2 \end{aligned}$$

where $|d_i| \leq |\Delta f_i| + |S_1 \rho_i^2|/2\varepsilon \leq |\rho_i|(1 + |S_1 \rho_i|/2\varepsilon) \leq \rho_{\max}(1 + c\rho_{\max}/2\varepsilon) := d_{\max}$. Therefore, the error dynamics in (3.24) is in the form of DNLDI subject to inequality constraints of the nonvanishing perturbation terms r_d and n_d.

As discussed in Sects. 2.6 and 3.1, similarly the quadratic boundedness of (3.24) can be addressed as follows:

Theorem 3.2 *For the given set of controller gains, Θ, suppose that the closed-loop error dynamics* (3.24) *is given on the set Ω_c and x_{1d} is a feasible output trajectory in the domain $\mathscr{D}$. The nonlinear system* (3.16) *is quadratically trackable via DSC in $\mathscr{D}$ if there exist $P > 0$, $\Sigma = \mathrm{diag}(\sigma_1, \ldots, \sigma_{n_u}) \geq 0$ and $\alpha \geq 0$ such that*

$$\begin{bmatrix} A_{cl}^T P + P A_{cl} + 2\alpha P + \tilde{C}_w^T \Sigma_B \tilde{C}_w & P B_w & P \tilde{B}_{rd} & P B_{nd} - \frac{\alpha}{2} \tilde{E} \\ B_w^T P & -\Sigma & 0 & 0 \\ \tilde{B}_{rd}^T P & 0 & -\alpha I & 0 \\ B_{nd}^T P - \frac{\alpha}{2} \tilde{E}^T & 0 & 0 & -\alpha I \end{bmatrix} < 0 \tag{3.25}$$

where $\tilde{C}_w = \mathrm{diag}(\tilde{C}_{w1}, \ldots, \tilde{C}_{wn_u})$ and $\Sigma_B = \mathrm{diag}(\sigma_1 \mathbf{I}_{n_u}, \sigma_2 \mathbf{I}_{n_u}, \ldots, \sigma_{n_u} \mathbf{I}_{n_u})$ is a block diagonal matrix.

Proof Suppose that there exist a function $V_z = z^T P z$ with $P > 0$ such that

$$\begin{aligned}\frac{d}{dt}V_z(t) &= \left(A_{cl}z + B_w w + \tilde{B}_r \tilde{r}_d + B_{nd}n_d\right)^T P z \\ &\quad + z^T P\left(A_{cl}z + B_w w + \tilde{B}_r \tilde{r}_d + B_{nd}n_d\right) < 0 \end{aligned} \tag{3.26}$$

for all $(z, w, \tilde{r}_d, n_d)$ satisfying (3.24) and $z^T P z \geq 1$. Using the S-procedure, we see that a sufficient condition for the inequality conditions (3.26) to hold is the existence of $P > 0$, $\sigma_i \geq 0$, and $\alpha \geq 0$ such that

$$\begin{aligned} &z^T\left(A_{cl}^T P + P A_{cl}\right)z + 2z^T\left(P B_w w + P \tilde{B}_{rd}\tilde{r}_d + P B_{nd} n_d\right) - \alpha\left\{z^T \tilde{E} n_d - z^T P z\right\} \\ &\quad - \alpha\left\{\tilde{r}_d^T \tilde{r}_d - z^T P z\right\} - \sum_{i=1}^{n_u} \sigma_i\left\{ w_i^T w_i - \left(\tilde{C}_{wi}z\right)^T\left(\tilde{C}_{wi}z\right)\right\} - \alpha n_d^T n_d < 0. \end{aligned}$$

The above inequality is equivalent to LMI (3.25). □

As discussed in Sect. 2.6, the LMI (3.25) is a quasiconvex optimization problem if the objective is given to maximize $\lambda_{\min}(P)$. However, for the fixed α, the problem becomes a convex optimization problem, thus we try to find P to maximize $\lambda_{\min}(P)$ and satisfy the LMI constraints along the line of α as follows: For a fixed $\alpha \in [a, b]$,

$$\begin{aligned} &\text{maximize} \quad \lambda_{\min}(P) \\ &\text{subject to} \quad P > 0, \qquad \Sigma \geq 0, \quad \text{LMI (3.25)}. \end{aligned} \tag{3.27}$$

Example 3.3 (Quadratic tracking of a second order uncertain system) Consider the second order nonlinear system subject to mismatched uncertainty:

$$\begin{aligned} \dot{x}_1 &= x_2 - x_1^2 + \Delta f_1 := x_2 + f_1 + \Delta f_1, \\ \dot{x}_2 &= u - x_2 \end{aligned} \tag{3.28}$$

where Δf_1 is an unknown function but bounded by $|\Delta f_1(x)| \leq \rho_1(x) := x_1^2$ for all $x \in \mathscr{D} = \{x \in \Re^2 | |x_1| \leq 1\}$. The control objective is $x_1 \to x_{1d}$ where $x_{1d} = \sin t$.

If DSC with nonlinear damping is applied to the system, the design procedure is

$$\begin{aligned} S_1 &= x_1 - x_{1d}, \\ \dot{S}_1 &= \dot{x}_1 - \dot{x}_{1d} = x_2 - x_1^2 + \Delta f_1 - \dot{x}_{1d}, \\ \bar{x}_2 &:= x_1^2 + \dot{x}_{1d} - K_1 S_1 - \frac{S_1 \rho_1^2}{2\varepsilon}, \\ \tau_2 \dot{x}_{2d} + x_{2d} &= \bar{x}_2, \quad x_{2d}(0) = \bar{x}_2(0), \\ S_2 &= x_2 - x_{2d}, \\ \dot{S}_2 &= u - x_2 - \dot{x}_{2d}, \\ u &:= x_2 + \dot{x}_{2d} - K_2 S_2. \end{aligned}$$

Then, the augmented error dynamics is

$$\dot{z} = A_{cl}z + B_w g + B_e \ddot{x}_{1d} + B_u d \tag{3.29}$$

where $z = [S_1\ S_2\ \xi_2]^T \in \Re^3$, $g = \dot{f}_1 + \dot{d}_1$, $d = \Delta f_1 - \frac{S_1\rho_1^2}{2\varepsilon}$, and the matrices are

$$A_{cl} = \left[\begin{array}{cc|c} -K_1 & 1 & 1 \\ 0 & -K_2 & 0 \\ \hline -K_1^2 & K_1 & K_1 - \frac{1}{\tau_2} \end{array}\right], \qquad B_w = \begin{bmatrix} 0 \\ 0 \\ 1 \end{bmatrix},$$

$$B_e = \begin{bmatrix} 0 \\ 0 \\ -1 \end{bmatrix}, \qquad B_u = \begin{bmatrix} 1 \\ 0 \\ K_1 \end{bmatrix} \in \Re^{3\times 2}.$$

Furthermore,

$$\begin{aligned}
\dot{g} &= -2x_1\dot{x}_1 + \frac{\rho_1^2}{2\varepsilon}\dot{S}_1 + \frac{S_1\rho_1}{\varepsilon}\frac{\partial \rho_1}{\partial x_1}\dot{x}_1 \\
&= \frac{\rho_1^2}{2\varepsilon}(c_{z1}z + d) + \left(-2x_1 + \frac{2S_1\rho_1}{\varepsilon}x_1\right)(c_{z1}z + \dot{x}_{1d} + d) \\
&= \left(\frac{\rho_1^2}{2\varepsilon} - 2x_1 + \frac{2S_1\rho_1}{\varepsilon}x_1\right)c_{z1}z + \frac{2x_1\rho_1\dot{x}_{1d}}{\varepsilon}S_1 - 2x_1\dot{x}_{1d} \\
&\quad + \left(\frac{\rho_1^2}{2\varepsilon} - 2x_1 + \frac{2\rho_1 x_1 S_1}{\varepsilon}\right)d \\
&= w + J_{11}(\dot{x}_{1d} + d) + \frac{\rho_1\Phi_{11}}{\varepsilon}S_1 d + \frac{\rho_1^2}{2\varepsilon}d
\end{aligned}$$

where $c_{z1} = [-K_1\ 1\ 1] \in \Re^{1\times 3}$, $J_{11} = \partial f_1/\partial x_1 = -2x_1$, $\Phi_{11} = \partial \rho_1/\partial x_1$, and

$$w = \begin{bmatrix} \frac{\rho_1^2}{2\varepsilon} & -2x_1 & \frac{2S_1\rho_1 x_1}{\varepsilon} & \frac{2x_1\rho_1\dot{x}_{1d}}{\varepsilon} \end{bmatrix} \begin{bmatrix} c_{z1}z \\ c_{z1}z \\ c_{z1}z \\ e_1^T z \end{bmatrix} = a^T C_w z.$$

Therefore, (3.29) can be written as

$$\begin{aligned}
\dot{z} &= A_{cl}z + B_w w + [B_w\ B_e\ B_w\ B_w]\begin{bmatrix} J_{11}\dot{x}_{1d} \\ \ddot{x}_{1d} \\ \frac{\rho_1\Phi_{11}}{\varepsilon}S_1 d \\ J_{11}d \end{bmatrix} + [B_u\ B_w]\begin{bmatrix} d \\ \frac{\rho_1^2}{2\varepsilon}d \end{bmatrix} \\
&= A_{cl}z + B_w w + B_{rd}r_d + B_{nd}n_d.
\end{aligned} \tag{3.30}$$

Next, since $|\dot{x}_{1d}| \le 1$, $|\ddot{x}_{1d}| \le 1$, and $x \in \mathscr{D}$, $|J_{11}| \le 2$, $|\Phi_{11}| \le 2$, and $|\rho_1| \le 1$. Furthermore, define $\Omega = \{z \in \Re^3 | \|z\| \le 1\}$ and thus $|S_1| \le 1$ for all $z \in \Omega$. Using these inequality constraints, there exist μ, r_{d0}, and n_{d0} such that

$$\begin{cases} \|a\| \le \{1/(4\varepsilon^2)+4+8/\varepsilon^2\}^{1/2} := \mu, \\ \|r_d\| \le (4+1+1+4d^2)^{1/2} \le \{6+4(1+1/2\varepsilon)^2\}^{1/2} := r_{d0}, \\ z^T E n = S_1[1\ 1]n = (1+\frac{\rho_1^2}{2\varepsilon})S_1 d \le \frac{\varepsilon}{2} + \frac{\rho_1^2}{4} \le \varepsilon/2 + 1/4 := n_{d0} \end{cases} \tag{3.31}$$

where the first inequality of the vector a is defined in componentwise and $S_1 d \le \varepsilon/2$ in (3.18) is used for the second and third inequalities. Therefore, the error dynamics (3.30) with (3.31) is summarized as

$$\begin{cases} \dot{z} = A_{cl} z + B_w w + \tilde{B}_{rd}\tilde{r}_d + B_{nd} n_d, \\ |w| \le \|\tilde{C}_w z\|, \qquad \tilde{r}_d^T \tilde{r}_d \le 1, \qquad z^T \tilde{E} n_d \le 1 \end{cases}$$

where $\tilde{r}_d = r_d/r_{d0}$, $\tilde{B}_{rd} = r_{d0} B_{rd}$, $\tilde{C}_w = \mu C_w$, and $\tilde{E} = E/n_{d0}$. This is written in the form of (3.22) in Lemma 3.2.

Using the result of Theorem 3.2, the quadratic ultimate bound of the error dynamics can be computed as follows:

Algorithm 3.1 For a fixed $\alpha \in [10^{-1}, 10]$,

$$\begin{aligned} &\text{maximize} \quad \lambda_{\min}(P) \\ &\text{subject to} \quad P > 0, \qquad \sigma_1 \ge 0, \\ &\begin{bmatrix} A_{cl}^T P + P A_{cl} + 2\alpha P + \sigma_1 \tilde{C}_w^T \tilde{C}_w & P B_w & P\tilde{B}_{rd} & P B_n - \frac{\alpha}{2}\tilde{E} \\ B_w^T P & -\sigma_1 & 0 & 0 \\ \tilde{B}_{rd}^T P & 0 & -\alpha I_4 & 0 \\ B_n^T P - \frac{\alpha}{2}\tilde{E}^T & 0 & 0 & -\alpha I_2 \end{bmatrix} < 0. \end{aligned} \tag{3.32}$$

Furthermore, COP (3.32) can be solved using MATLAB® and CVX [32] as shown in MATLAB Program 3-1.

Suppose that the uncertainty is given as $\Delta f_1 = -x_1^2 \sin \omega t$ where $\omega = 10$ for simulations and $\varepsilon = 0.2$. If the set of controller gains is $\Theta = \{K_1, K_2, \tau_2\} = \{5, 10, 0.01\}$,

$$\lambda\{A_{cl}\} = \{-5.2786, -10, -94.7214\}$$

for Θ and thus A_{cl} is Hurwitz. Then, we can calculate the quadratic error bound of z using the above algorithm. Figure 3.8 shows a line search of α to find the minimum of the largest diameter of the ellipsoidal bound $\mathcal{E}_P = \{z | z^T P z \le 1\}$. When $\alpha = 2.0996$, the largest diameter of the ellipsoidal bound is calculated as 4.5756 and it corresponds to the diameter of ξ_2 as shown in the right plot of Fig. 3.10. Figure 3.10 shows that the error z stays within the ellipsoidal bound in the sense that $z^T P z < 1$ after a certain time, while the time responses of x and u are shown in Fig. 3.9. It is interesting to remark that a smaller ultimate error bound may be

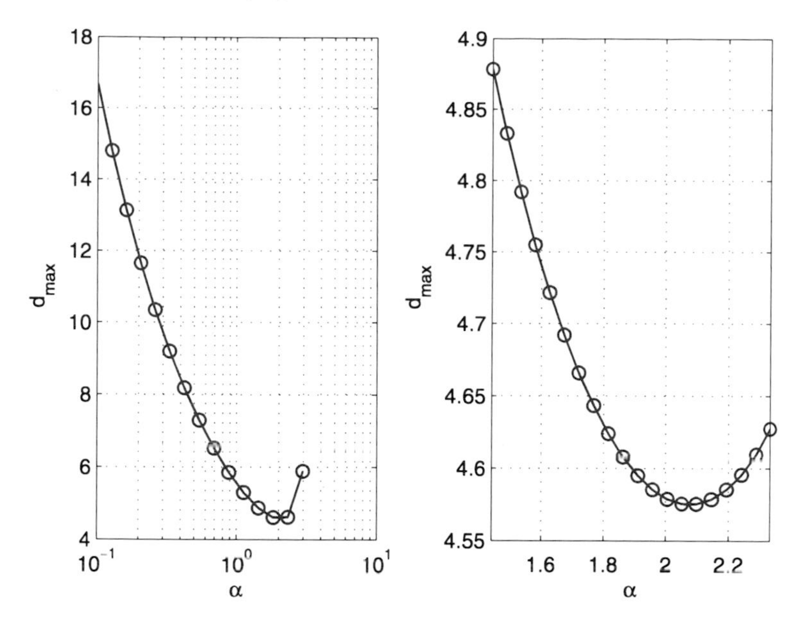

Fig. 3.8 Maximum radius of ellipsoid along line search of α

calculated if we provide a more tight upper bound of w and r_d. For example, once the quadratic error bound $\mathscr{E}_P = \{z|z^T Pz \leq 1\}$ is computed, this error bound is used to calculate μ_1 and γ_1 in (3.31).

MATLAB Program 3-1

```
%***** Define the dimension of matrices *****
n = size(Acl, 1);
m = size(Bw, 2);
nd = size(Bd, 2);
nr = size(Br, 2);

%***** cvx version *****
cvx_begin
   variable P(n,n) symmetric;
   variable beta;
   maximize(lambda_min(P));
   beta >= 0;
   P == semidefinite(n);
   -[Acl'*P + P*Acl + 2*alpha(k)* P + beta*Cw'*Cw P*Bw P*Br P*Bd-
   0.5*alpha(k)*D;
   Bw'*P -beta*eye(m) zeros(m,nr) zeros(m,nd);
   Br'*P zeros(nr,m) -alpha(k)*eye(nr) zeros(nr,nd);
   Bd'*P-0.5*alpha(k)*D' zeros(nd,m) zeros(nd,nr) -alpha(k)*eye(nd)]
   == semidefinite(n+m+nr+nd);
cvx_end
```

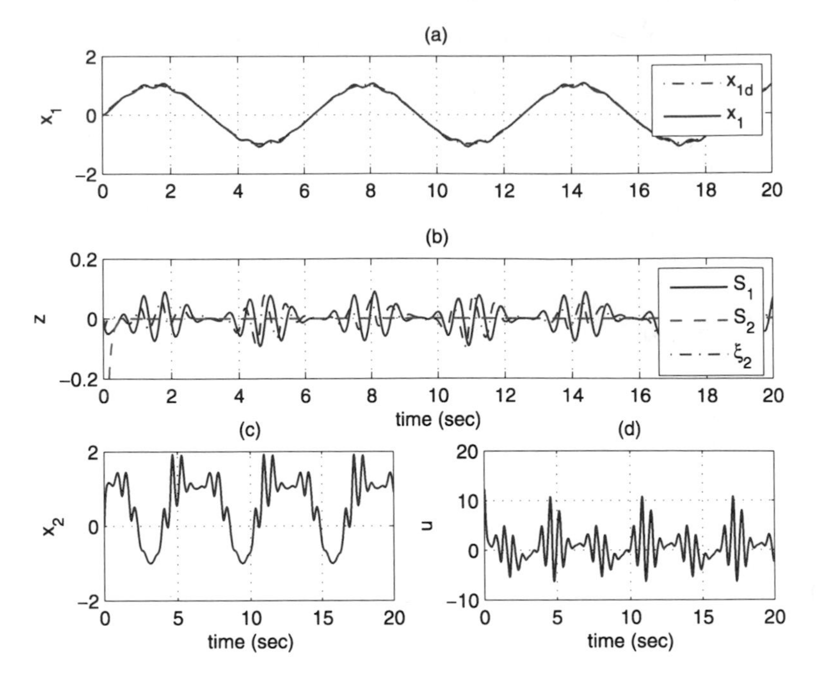

Fig. 3.9 Time responses of x, z, and u for Θ

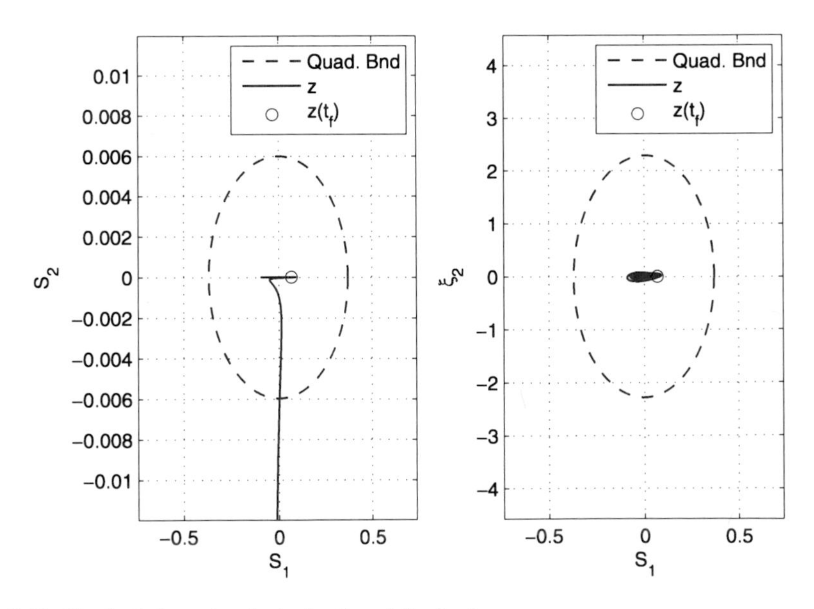

Fig. 3.10 Quadratic bound and z in S_1–S_2 and S_1–ξ_2 planes

3.3 Input–Output Stability

For the given nonlinear uncertain system (3.16), the augmented error dynamics is the following, based on Lemma 3.2:

$$\begin{aligned}\dot{z} &= A_{cl}z + B_w w + B_{rd}r_d + B_{nd}n_d = A_{cl}z + B_w\tilde{w} + [B_{nd}\ B_{rd}]\begin{bmatrix} n_d \\ r_d \end{bmatrix} \\ &= A_{cl}z + B_w w + B_l r_l, \\ y &= C_y z\end{aligned} \tag{3.33}$$

where $r_l \in \Re^{n_{nd}+n_{rd}} := \Re^{n_{rl}}$, $B_l \in \Re^{n_z \times n_{rl}}$, and y is the output of the error dynamics. For example, if the control objective is to make S_{1i} go to zero, the output is defined as $y := S_1$, thus $C_y := [1\ 0\ \cdots\ 0] \in \Re^{1\times n_z}$.

The quadratic stability of DSC was proposed to guarantee the stability of DSC systems and the upper bound containing reachable sets of errors was estimated to predict the performance of DSC systems in Sect. 3.2.2. The quadratic error bound is computed by minimizing the largest diameter of ellipsoids containing the reachable sets. However, it is required to calculate the upper bound of r_d while n_d has a known bound due to nonlinear damping. There is an alternative approach to discuss a concept of stability in the input-output sense. More specifically, the induced $\mathcal{L}_2$ gain, signified as $\|H_{r_l\to y}\|_\infty$, for the augmented error dynamics is considered as follows:

Definition 3.1 When the induced $\mathcal{L}_2$ gain of the augmented error dynamics described in (3.33) is defined as

$$\|H_{r_l\to y}\|_\infty = \sup_{\|r_l\|_2 \neq 0} \frac{\|y\|_2}{\|r_l\|_2}$$

where the $\mathcal{L}_2$ norm of x is $\|x\|_2^2 = \int_0^\infty x^T x\, dt$, and the supremum is taken over all nonzero trajectories of the augmented error dynamics, starting from $z(0) = 0$.

Suppose there exists a quadratic function $V(z) = z^T P z$, $P > 0$, and $\kappa \geq 0$ such that for all t,

$$\frac{d}{dt}V(z) + y^T y - \kappa^2 r_l^T r_l \leq 0 \tag{3.34}$$

for all z, w, and r_l satisfying (3.33). Then the $\mathcal{L}_2$ gain of the augmented error dynamics is less than κ [12].

Theorem 3.3 *Suppose that the closed-loop error dynamics in* (3.33) *is given for a set of controller gains, Θ, in a domain $\mathcal{D}$ and $r_l \in \mathcal{L}_2^{n_{rl}}[0,\infty)$. The augmented error dynamics in* (3.33) *has $\|H_{r_l\to y}\|_\infty \leq \kappa$ if there exist $P > 0$, $\Sigma =$*

$\mathrm{diag}(\sigma_1, \ldots, \sigma_{n_u}) \geq 0$, *and* $\kappa \geq 0$ *such that*

$$\begin{bmatrix} A_{cl}^T P + P A_{cl} + C_y^T C_y + \tilde{C}_w^T \Sigma_B \tilde{C}_w & P B_w & P B_l \\ B_w^T P & -\Sigma & 0 \\ B_l^T P & 0 & -\kappa^2 I \end{bmatrix} < 0 \tag{3.35}$$

where $\tilde{C}_w = \mathrm{diag}(\tilde{C}_{w1}, \ldots, \tilde{C}_{wn_u})$ *and* $\Sigma_B = \mathrm{diag}(\sigma_1 \mathbf{I}_{n_u}, \sigma_2 \mathbf{I}_{n_u}, \ldots, \sigma_{n_u} \mathbf{I}_{n_u})$ *is a block diagonal matrix.*

Proof The inequality (3.34) is equivalent to

$$z^T \left(A_{cl}^T P + P A_{cl} + C_y^T C_y\right) z + 2z^T (P B_w w + P B_l r_l) - \kappa^2 r_l^T r_l \leq 0 \tag{3.36}$$

for all (z, w, r_l) satisfying $|w_i| \leq \|\tilde{C}_{wi} z\|$. Using the S-procedure,

$$z^T \left(A_{cl}^T P + P A_{cl} + C_y^T C_y\right) z + 2z^T (P B_w w + P B_l r_l) - \kappa^2 r_l^T r_l$$
$$- \sum_{i=1}^{n_u} \sigma_i \left\{ w_i^T w_i - (\tilde{C}_{wi} z)^T (\tilde{C}_{wi} z) \right\} \leq 0.$$

Then, the above inequality condition is equivalent to LMI (3.35). □

Therefore, we can compute the $\mathscr{L}_2$ gain of the augmented error dynamics provable via quadratic functions by minimizing κ over the variables P and κ satisfying LMI (3.35). Specifically speaking, the convex optimization problem (COP) is written as follows:

$$\begin{aligned} &\text{minimize} \quad \kappa^2 \\ &\text{subject to} \quad P > 0, \qquad \Sigma \geq 0, \qquad \kappa^2 \geq 0, \quad \text{and} \quad \text{LMI (3.35)}. \end{aligned} \tag{3.37}$$

Example 3.4 ($\mathscr{L}_2$ gain of the locally Lipschitz uncertain system) Consider the following example introduced in [94]:

$$\begin{aligned} \dot{x}_1 &= x_2 + \Delta f_1(x_1), \\ \dot{x}_2 &= u \end{aligned} \tag{3.38}$$

where $\Delta f_1(x_1)$ is the non-Lipschitz uncertainty but bounded by a known C^1 function such that $|\Delta f_1(x_1)| \leq \rho_1(x_1) = x_1^2$ for all $x \in \mathscr{D} = \{x \in \Re^2 | |x_1| \leq 1\}$. The control objective is to make x_1 to track the desired value, $x_{1d}(t) = \sin t$.

If the DSC with nonlinear damping is applied, using Lemma 3.2, the closed-loop error dynamics is

$$\dot{z} = A_{cl} z + B_w w + B_{rd} r_d + B_{nd} n_d.$$

Since $f = 0$ in this example, the error dynamics is simplified to

$$\begin{aligned} \dot{z} &= A_{cl} z + B_w w + [B_e \ B_w] \begin{bmatrix} \ddot{x}_{1d} \\ \frac{S_1 \rho_1}{\varepsilon} \Phi d \end{bmatrix} + [B_u \ B_w] \begin{bmatrix} d \\ \frac{\rho_1^2}{2\varepsilon} d \end{bmatrix} \\ &= A_{cl} z + B_w w + B_l r_l \end{aligned}$$

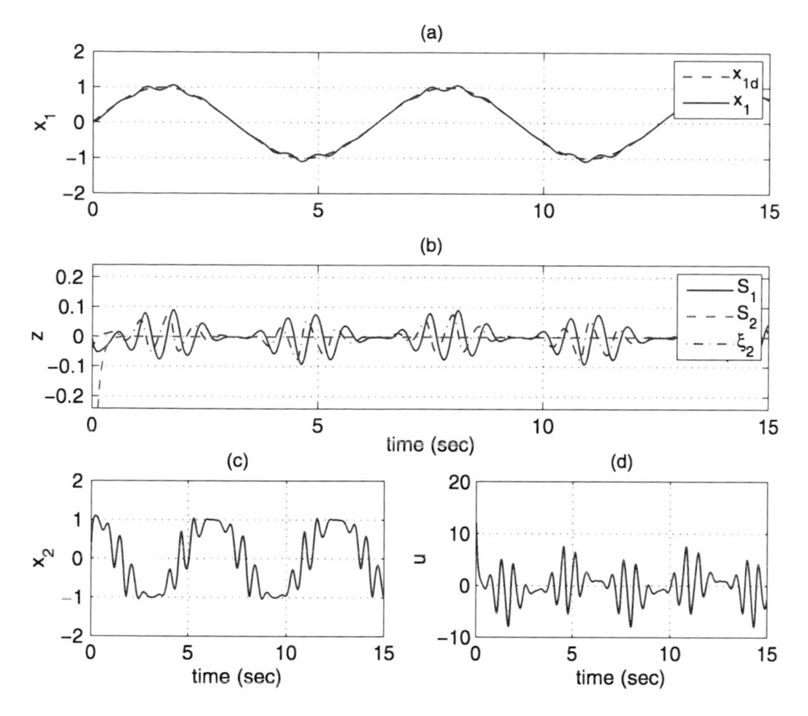

Fig. 3.11 Time responses of x, z, and u

where

$$w = \frac{\rho_1}{\varepsilon}\begin{bmatrix}\frac{\rho_1}{2} & 2S_1x_1 & 2x_1\dot{x}_{1d}\end{bmatrix}\begin{bmatrix}c_{z1}z \\ c_{z1}z \\ e_1^T z\end{bmatrix} = a^T C_w z,$$

$$B_l = [B_e \; B_w \; B_u \; B_w]$$

and all other vectors and matrices are defined in Example 3.3. Furthermore, since

$$\|a\| \le \left|\frac{\rho_1}{\varepsilon}\right|\sqrt{\rho_1^2/4 + 4x_1^2(S_1^2 + \dot{x}_{1d}^2)} \le (1/4+8)^{1/2}/\varepsilon := \mu$$

for all $x \in \mathscr{D}$ and $z \in \Omega = \{z \in \Re^3 \,|\, \|z\| \le 1\}$, the linear bound of w is written as $|w| \le \|\mu C_w z\|$.

For the given set of gains, $\{K_1, K_2, \tau_2\} = \{5, 10, 0.01\}$, and $\varepsilon = 0.2$, COP (3.37) is solved using MATLAB® and CVX as shown in MATLAB Program 3-2. The calculated $\mathscr{L}_2$ gain is 0.2412 and thus it is expected that the corresponding error dynamics has $\|H_{r_l \to y}\|_\infty \le 0.2412$. When the model uncertainty is given as $\Delta f_1 = x_1^2\sin(10x_1)$ for simulation, the time responses of x, z, and u are shown in Fig. 3.11.

MATLAB Program 3-2

```
%***** Define the dimension of matrices *****
n = size(Acl, 1);
m = size(Bw, 2);
nr = size(Br, 2);

%***** cvx version *****
cvx_begin sdp
   variable P(n,n) symmetric;
   variable kappa;
   variable sigma;
   minimize(kappa);
   kappa >= 0;
   sigma >= 0;
   P == semidefinite(n);
   -[Acl'*P + P*Acl + Cy'*Cy + sigma*Cw'*Cw P*Bw P*Br;
   Bw'*P -sigma*eye(m) zeros(m,nr);
   Br'*P zeros(nr,m) -kappa*eye(nr)] == semidefinite(n+m+nr);
cvx_end

%***** Calculation of L2 gain *****
L2 = sqrt(kappa);
```

Chapter 4
Observer-Based Dynamic Surface Control

While it is assumed that all information of the full state is measured via sensors in Chaps. 2 and 3, the entire state vector may not be measured in many cases, due to either cost considerations or sensor fidelity. Furthermore, if faults or complete failures occur in the sensor measurement, which will be discussed later in Chap. 7, only limited measurements will be available. Due to these reasons, observer-based nonlinear control or nonlinear compensator approaches have been studied for specific classes of nonlinear systems, e.g., high-gain observer-sliding control [24, 67], sliding observer-sliding control [2, 59], and backstepping observer-output feedback control [47, 55]. Although stability analysis has been performed for these approaches, the nonlinear compensator synthesis problem is typically very complex, if even tractable. This is because the observer and controller have a coupled structure and one cannot generally take advantage of the *separation principle* of linear systems.

Many observer design approaches such as the extended Kalman filter and pseudo linearization techniques can be found for the nonlinear systems in the literature (see in [62] and references therein). Among them, a geometric approach for observer design was proposed to construct exact observers for a general description of nonlinear systems [50, 53, 54]. However, the required conditions for the observers are extremely stringent and allow only a small applicable class [73]. To expand the class of nonlinear systems, extensions of the Luenberger observer design method have been studied for a class of Lipschitz nonlinear systems [73, 74, 97]. Moreover, sufficient conditions for existence and optimal design of the observer gain matrix were also performed via a linear matrix inequality (LMI) approach [3, 39, 87].

In Chap. 2, a systematic analysis method based on the closed-loop error dynamics including the filter error dynamics for DSC systems was proposed. It showed that the error dynamics can be described as linear error dynamics with a bounded perturbation term. Based upon the theory of quadratic stability, sufficient conditions for stability were derived. Therefore, since each error dynamics of the DSC and nonlinear observer have a well-defined structure for analysis, a nonlinear compensator design technique based on both methods will be investigated within the framework of convex optimization in this chapter. Furthermore, the separation principle for the

B. Song, J.K. Hedrick, *Dynamic Surface Control of Uncertain Nonlinear Systems*, Communications and Control Engineering,
DOI 10.1007/978-0-85729-632-0_4,

augmented closed-loop error dynamics of the observer-based DSC systems will be discussed to separate the observer design from the controller design.

The remainder of this chapter is divided as follows; Sect. 4.1 presents the class of nonlinear systems considered and the design procedure of the nonlinear observer. Next, based on the nonlinear observer, observer-based DSC (ODSC) is designed for the system and the corresponding augmented closed-loop error dynamics is derived in Sect. 4.2. Then, a separation principle for the error dynamics that makes the design methodology simpler is addressed. In Sect. 4.3, an LMI approach methodology to guarantee quadratic stability for the closed-error dynamics of ODSC is proposed. Finally, the design procedure of ODSC is proposed in Sect. 4.4 when uncertainty is considered.

4.1 Nonlinear Observer Design

4.1.1 Problem Statement

Consider the following class of nonlinear systems described in (2.5):

$$\begin{aligned} \dot{x} &= Ax + B_u u + f(x), \\ y &= Cx \end{aligned} \tag{4.1}$$

where the state $x \in \Re^n$, the control input $u \in \Re$, the measurement $y \in \Re^p$ where $p < n$, and $B_u = [0 \; \cdots \; 0 \; 1]^T \in \Re^{n_z}$. The assumptions for the system are summarized as follows:

A-1 The matrix A is a *lower-Hessenberg* matrix, that is, all elements of the first super-diagonal in the matrix A are not zero, i.e., $a_{i(i+1)} \neq 0$. Without loss of generality, suppose $a_{i(i+1)} = 1$ for all $i = 1, \ldots, n-1$.

A-2 The nonlinear function f and $[\partial f / \partial x]$ are continuous on $\mathscr{D} \in \Re^n$. Then, f is locally Lipschitz with a Lipschitz constant γ, i.e.,

$$\big\| f(x) - f(\hat{x}) \big\| \leq \gamma \|x - \hat{x}\|, \quad \forall (x, \hat{x}) \in \mathscr{D}. \tag{4.2}$$

If $[\partial f / \partial x]$ is bounded componentwise, i.e.,

$$\left\| \frac{\partial f_i}{\partial x} \right\| \leq \gamma_i$$

the positive constants γ_i can be used as a Lipschitz constant for f_i such that

$$\big| f_i(x) - f_i(\hat{x}) \big| \leq \gamma_i \|x - \hat{x}\|, \quad \forall (x, \hat{x}) \in \mathscr{D}. \tag{4.3}$$

A-3 The pair (A, C) is detectable.

4.1.2 Quadratic Stability of Observer

The observer to be considered for this class of systems will have the following form [73, 74, 97],

$$\dot{\hat{x}} = A\hat{x} + B_u u + f(\hat{x}) + L(y - C\hat{x}). \tag{4.4}$$

In order to study the convergence of the observer, the dynamics of the estimation error defined by $e := x - \hat{x}$ can be obtained as follows:

$$\begin{aligned}\dot{e} &= (A - LC)e + \{f(x) - f(\hat{x})\} \\ &= A_{ob}e + \phi\end{aligned} \tag{4.5}$$

where $A_{ob} = A - LC$ and $\phi = f(x) - f(\hat{x})$. Using the inequality constraints (4.2) or (4.3), the estimation error dynamics is summarized as

$$\begin{aligned}&\dot{e} = A_{ob}e + \phi, \\ &\|\phi\| \le \gamma \|e\| \quad \text{or} \quad \|\phi_i\| \le \gamma_i \|e\|\end{aligned} \tag{4.6}$$

where ϕ_i is the ith element of $\phi \in \Re^n$. It is of interest that the error dynamics can be considered as a linear system subject to the vanishing perturbation term, thus classified as a norm-bounded LDI (NLDI) in Sect. 2.5.2 or diagonal norm-bounded LDI (PNLDI) in Sect. 2.5.3 depending on the inequality constraint. Due to this analogy, Theorem 2.2 can be applied to (4.6) and the corresponding result was addressed originally by Thau [97] as follows

Theorem 4.1 *For the class of systems in* (4.1) *and observer forms in* (4.4), *the estimation error dynamics in* (4.6) *is derived for the given* L. *If* A_{ob} *is Hurwitz, i.e., there exist* $P > 0$ *and* $Q > 0$ *satisfying*

$$A_{ob}^T P + PA_{ob} = -Q \tag{4.7}$$

and

$$\gamma < \frac{\lambda_{\min}(Q)}{2\lambda_{\max}(P)}$$

for $\mathscr{D}_i = \{x \in \Re^n \mid \|J\| \le \gamma\} \subset \mathscr{D}$, *the origin in* (4.6) *is exponentially stable in* $\mathscr{D}_i$ *and thus* $\lim_{t\to\infty} e(t) = 0$.

Remark 4.1 As discussed in Remark 2.3, the ratio $\lambda_{\min}(Q)/2\lambda_{\max}(P)$ is maximized when $Q = I$. Therefore, if a gain matrix L is chosen such that

$$\gamma < \frac{1}{2\lambda_{\max}(P)}, \tag{4.8}$$

the estimation error converges to the origin asymptotically.

Remark 4.2 Using the result in Theorem 2.1, if there exist $P > 0$ and $\sigma \ge 0$ such that

$$\begin{bmatrix} A_{ob}^T P + PA_{ob} + \sigma\gamma^2 I & P \\ P & -\sigma I \end{bmatrix} < 0, \tag{4.9}$$

the origin in (4.6) is exponentially stable for the given L. Furthermore, using Algorithm 2.1, quadratic stability margin of estimation error dynamics can be computed as follows:

$$\begin{aligned} &\text{maximize} \quad \beta \\ &\text{subject to} \quad \tilde{P} > 0, \qquad \beta \geq 0, \\ &\qquad\qquad \begin{bmatrix} A_{ob}^T \tilde{P} + \tilde{P} A_{ob} + \beta I & \tilde{P} \\ \tilde{P} & -I \end{bmatrix} < 0. \end{aligned} \tag{4.10}$$

That is, for the given L, the maximum number of γ for quadratic stability of the error dynamics is $\sqrt{\beta}$, which is computed by solving COP (4.10).

The statement above can be described using the Schur complement as follows: suppose $X = X^T \in \Re^{n\times n}$, $Y = Y^T \in \Re^{m\times m}$, and $Z \in \Re^{n\times m}$.

$$\begin{bmatrix} X & Z \\ Z^T & Y \end{bmatrix} > 0 \implies Y > 0 \quad \text{and} \quad X - ZY^{-1}Z^T > 0.$$

Using this property, the Schur complement of LMI (4.9) is

$$A_{ob}^T P + P A_{ob} + \sigma \gamma^2 I - P(-\sigma I)^{-1} P < 0.$$

Let $\tilde{P} = P/\sigma$ and $\beta = \gamma^2$. Then

$$A_{ob}^T \tilde{P} + \tilde{P} A_{ob} + \beta I + PP < 0 \tag{4.11}$$

which is equivalent to the Schur complement of the matrix inequality condition in (4.10). Therefore, the maximum of β computed in COP (4.10) is the maximum of γ^2 which enables the solutions of LMI (4.9) to be feasible. These results are useful to check the stability of the observer once the gain matrix L is chosen to make A_{ob} Hurwitz.

Since the results in Theorem 4.1 and COP (4.10) provide only sufficient conditions for quadratic stability, a methodology using the benefit of coordinate transformations was proposed to reduce the Lipschitz constant and increase the distance to unobservability in the new coordinates [73, 74]. Suppose $z = Tx$ and $\hat{z} = T\hat{x}$ where T is invertible and called a transformation matrix. Then, the errors in the new coordinates are defined as $\tilde{e} = Te$. Equation (4.5) in the new coordinates is given by

$$\begin{aligned} \dot{\tilde{e}} &= T(A - LC)T^{-1}\tilde{e} + T\{f(T^{-1}z) - f(T^{-1}\hat{z})\} \\ &= \tilde{A}_{ob}\tilde{e} + \tilde{\phi} \end{aligned} \tag{4.12}$$

where $\tilde{A}_{ob} = T_c(A - LC)T_c^{-1}$ and $\tilde{\phi} = T\{f(T^{-1}z) - f(T^{-1}\hat{z})\}$. Since A_{ob} and $\tilde{A}_{ob}$ are similar, they have the same eigenvalues, but result in different values of $\lambda_{\max}(P)$ in (4.7) or β in COP (4.10). Therefore, there is no clear relation between $\lambda(A_{ob})$ and either $\lambda_{\max}(P)$ or between stability margin $\sqrt{\beta}$.

Example 4.1 (Quadratic stability margin and coordinate transformation) Consider the second order nonlinear system:

$$\begin{aligned} \dot{x}_1 &= x_2, \\ \dot{x}_2 &= u + l\sin(\omega x_1), \\ y &= x_1, \\ \Longrightarrow \quad \dot{x} &= Ax + Bu + f(x), \quad y = Cx \end{aligned}$$

where $x = [x_1\ x_2]^T \in \Re^2$, $u \in \Re$, and

$$A = \begin{bmatrix} 0 & 1 \\ 0 & 0 \end{bmatrix}, \qquad B = \begin{bmatrix} 0 \\ 1 \end{bmatrix}, \qquad f = \begin{bmatrix} 0 \\ l\sin(\omega x_1) \end{bmatrix}, \qquad C = [1\ 0].$$

Let us use an observer form in (4.4) with the gain matrix $L = [2\ 3]^T \in \Re^{2\times 1}$. It is noted that we have not followed any particular design method to choose L here, but simply made $(A - LC)$ Hurwitz, i.e.,

$$A_{ob} = A - LC = \begin{bmatrix} -2 & 1 \\ -3 & 0 \end{bmatrix}, \quad \lambda(A_{ob}) = -1 \pm \mathrm{i}\sqrt{2}.$$

In order to prove the stability of the error dynamics in (4.5), Theorem 4.1 and (4.10) are used and compared respectively. First, if a positive definite matrix P satisfying

$$A_{ob}^T P + P A_{ob} = -I$$

is computed using the command 'lyap' in MATLAB®, $\lambda_{\max}(P) = 1.3604$. Therefore, if $\gamma < 1/2.7208 = 0.3675$, the error dynamics is exponentially stable. That is, since the Lipschitz constant γ is $l\omega$, the estimation error converges to the origin if $l\omega < 0.3675$. Second, when COP (4.10) is solved as shown in MATLAB Program 4-1, $\beta = 0.5$ and thus the maximum of allowable γ is 0.7071. Therefore, if $l\omega < 0.7071$, the error goes to zero asymptotically. In this case, a larger value of allowable γ is computed by the second approach.

Next, consider the following coordinate transformation, $z = Tx$, with

$$T = \mathrm{diag}(4, 0.4).$$

The error dynamics in the new coordinates is given by

$$\dot{\tilde{e}} = T(A - LC)T^{-1}\tilde{e} + T\{f(T^{-1}z) - f(T^{-1}\hat{z})\}$$

where

$$Tf(T^{-1}z) = \begin{bmatrix} 0 \\ 0.4l\sin(\omega z_1/4) \end{bmatrix}.$$

It is seen that the new Lipschitz constant is $\tilde{\gamma} = 0.1 \times l\omega$. When the result of Theorem 4.1 is used, the ratio $1/2\lambda_{\max}(P) = 0.0561$. So the error dynamics becomes exponentially stable if $\tilde{\gamma} = 0.1 \times l\omega < 0.0561 \Rightarrow l\omega < 0.561$. When β is computed in COP (4.10), $\tilde{\gamma}^2 \le \beta = 0.0754 \Rightarrow 0.1 \times l\omega \le 0.2746$. Therefore, the condition for quadratic stability of the error dynamics is $l\omega \le 2.746$.

It is interesting to note that change of the Lipschitz constant via coordinate transformations does not have a direct relationship with either the ratio $1/2\lambda_{\max}(P)$ or the stability margin β. To prove this statement, let $T_1 = \text{diag}(4, 0.2)$. Then the Lipschitz constant becomes $\tilde{\gamma}_1 = 0.05 \times 1\omega$, which is a half of $\tilde{\gamma}$ above. However, the ratio and $\sqrt{\beta}$ are computed as 0.0147 and 0.1404 respectively. Therefore, the stability conditions are $l\omega \leq 0.294$ and $l\omega \leq 2.808$. In the case of the ratio computation, it becomes smaller even though the new Lipschitz constant decreases via coordinate transformation.

MATLAB Program 4-1

```
%% define matrices
A = [0 1;0 0];    B = [0 1]';
C = [1 0];    L = [2 3]';
Aob = A-L*C;

%% define the dimension of matrix
n = size(Aob, 1);

%% cvx version
cvx_begin
   variable P(n,n) symmetric;
   variable beta;
   maximize(beta);
   beta >= 0;
   P == semidefinite(n);
   -[Aob'*P + P*Aob + beta*eye(n) P; P -eye(n)] == semidefinite(n+n);
cvx_end
```

Instead of (4.8) and (4.9), a new condition for stability in the form of algebraic Riccati equation (ARE) has been proposed by Rajamani and Cho [74] and Aboky et al. [1] as follows:

Theorem 4.2 *If a gain matrix L is chosen such that for some* $P > 0$,

$$(A - LC)^T P + P(A - LC) + \gamma^2 PP + I < 0 \tag{4.13}$$

or

$$(A - LC)^T P + P(A - LC) + PP + \gamma^2 I < 0, \tag{4.14}$$

then the error dynamics (4.6) *is quadratically stabilized by the observer* (4.1).

Proof Consider the Lyapunov function candidate $V = e^T Pe$ where $P > 0$. Its derivative of V along the trajectory of (4.5) is

$$\dot{V} = e^T\left[(A - LC)^T P + P(A - LC)\right]e + 2e^T P\left[f(x) - f(\hat{x})\right]. \tag{4.15}$$

Using the inequality constraint (4.2),

$$2e^T P\left[f(x) - f(\hat{x})\right] \leq 2\gamma \|Pe\| \|e\| \leq \gamma^2 e^T P^T Pe + e^T e \tag{4.16}$$

or

$$2e^T P\left[f(x) - f(\hat{x})\right] \leq 2\gamma \|Pe\| \|e\| \leq e^T P^T Pe + \gamma^2 e^T e. \tag{4.17}$$

If the inequality (4.16) is used, (4.15) is bounded as

$$\dot{V} \le e^T\big[(A - LC)^T P + P(A - LC) + \gamma^2 PP + I\big]e.$$

Therefore, if the matrix inequality condition (4.13) is satisfied, $\dot{V} < 0$, thus the error dynamics in (4.5) becomes asymptotically stable. Similarly, if the inequality (4.17) is used and the matrix inequality condition (4.14) is satisfied, they are also stable. □

Remark 4.3 It is remarked that (4.13) is equal to (4.11), thus it is also equivalent to (4.9). After defining $\alpha = 1/\gamma^2$, the Schur complement of (4.14) is

$$\begin{bmatrix} A_{ob}^T P + P A_{ob} + I & P \\ P & -\alpha I \end{bmatrix} < 0. \tag{4.18}$$

Although the stability margin can be computed by solving COP (4.10), it can be also computed as follows

$$\begin{aligned} &\text{minimize} \quad \alpha \\ &\text{subject to} \quad P > 0, \qquad \alpha \ge 0, \quad \text{LMI (4.18)}. \end{aligned} \tag{4.19}$$

Then, the stability margin is calculated by $1/\sqrt{\alpha}$.

4.1.3 Design of Observer Gain Matrix

All results in the previous section provide a method to check the stability and to compute the stability margin for the given gain matrix L. However, they do not tell us how to design L to satisfy the stability condition. This observer design problem has been considered by several researchers in the literature. An algebraic Riccati equation for guaranteeing the quadratic stability was proposed by Raghavan and Hedrick [73]. The result can be summarized as follows.

For some small ε, if there exists a positive definite P such that

$$AP + PA^T + P\left(\gamma^2 I - \frac{1}{\varepsilon}C^T C\right)P + I + \varepsilon I = 0 \tag{4.20}$$

then the error dynamics (4.5) can be stabilized by $L = PC^T/2\varepsilon$.

Since the proof is in [73], we will investigate how this result is related with other results proposed in [74] and [1]. First, ARE (4.20) can be written as a following matrix inequality without loss of generality

$$AP + PA^T + P\left(\gamma^2 I - \frac{1}{\varepsilon}C^T C\right)P + I < 0. \tag{4.21}$$

Let $P = P_1^{-1}/\gamma^2$ where $P_1 > 0$. Then

$$A\frac{P_1^{-1}}{\gamma^2} + \frac{P_1^{-1}}{\gamma^2}A^T + \frac{P_1^{-1}}{\gamma^2}\left(\gamma^2 I - \frac{1}{\varepsilon}C^T C\right)\frac{P_1^{-1}}{\gamma^2} + I < 0.$$

Multiplying the above inequality on the left and right by $(\gamma P_1)^T$ and γP_1 respectively, we get

$$P_1 A + A^T P_1 + I - \frac{1}{\varepsilon\gamma^2} C^T C + \gamma^2 P_1 P_1 < 0. \tag{4.22}$$

Since the gain matrix L is

$$L = \frac{PC^T}{2\varepsilon} = \frac{P_1^{-1} C^T}{2\varepsilon\gamma^2}$$

substituting L in (4.22) gives that

$$(A - LC)^T P_1 + P_1(A - LC) + \gamma^2 P_1 P_1 + I < 0$$

which is equivalent to (4.13). Using the result of Theorem 4.1, the error dynamics (4.5) is quadratically stabilized by the observer (4.1). It is of interest that for $\varepsilon = 1$ the condition to design a gain matrix L is same with one proposed by Rajamani and Cho [74].

Similarly, let $P = P_2^{-1}$ where $P_2 > 0$. After substituting P with P_2^{-1} in (4.21) and multiplying the inequality on the left and right side by P_2^T and P_2,

$$P_2^T A + A^T P_2 + \gamma^2 I - \frac{1}{\varepsilon} C^T C + P_2 P_2 < 0. \tag{4.23}$$

Since the gain matrix L is

$$L = \frac{PC^T}{2\varepsilon} = \frac{P_2^{-1} C^T}{2\varepsilon}$$

substituting L in (4.23) gives that

$$(A - LC)^T P_2 + P_2(A - LC) + P_2 P_2 + \gamma^2 I < 0$$

which is equivalent to (4.14) in Theorem 4.1. When $\varepsilon = 1$, the inequality condition (4.23) is equal to one proposed by Aboky et al. [1].

Furthermore, if $P_2 = \gamma^2 P_1$, the matrix inequality (4.23) are equal to (4.22) and the same gain matrix L is calculated. Therefore, all results can be summarized as follows [87]:

Theorem 4.3 *For the class of systems in* (4.1) *and observer forms in* (4.4), *if there exist $P > 0$ and $\varepsilon > 0$ such that*

$$\begin{bmatrix} P^T A + A^T P + \gamma^2 I - \frac{1}{\varepsilon} C^T C & P \\ P & -I \end{bmatrix} < 0 \tag{4.24}$$

then the error dynamics (4.6) *can be stabilized quadratically by $L = P^{-1} C^T / 2\varepsilon$.*

For the design of the gain matrix $L = \frac{P^{-1} C^T}{2\varepsilon}$, the next question is under what conditions the positive definite matrix P and positive constant ε exist to satisfy (4.24). To address the problem, the idea of the distance to unobservability was proposed in [74] as follows

Definition 4.1 The distance to unobservability of the pair (A, C), $A \in \Re^{n\times n}$ and $C \in \Re^{p\times n}$, is defined as the magnitude of the smallest perturbation (E, F) that makes the pair $(A+E, C+F)$ unobservable. Specifically, this is the quantity

$$\Delta(A, C) = \inf_{(A+E, C+F)\in\mathscr{U}} \|[E, F]\|_2$$

where $\mathscr{U}$ is the set of the unobservable pair in $\Re^{n\times n} \times \Re^{p\times n}$.

Using the relation between the distance to unobservability and the Lipschitz constant of the nonlinear system, the following statement for stability was proposed in [74].

Theorem 4.4 *If A is stable and*

$$\gamma < \Delta(A, C) \tag{4.25}$$

there exists a matrix L such that the error dynamics (4.5) *is exponentially stable.*

However, there are still two limitations to use the result of the theorem: one is the condition that A is stable and the other is that the exact computation of $\Delta(A, C)$ is not available, but its upper bound can be computed. As discussed in [1], a number $\delta(A, B)$ is defined by Rajamani and Cho in [74]

$$\delta(A, C) = \min_{\omega\in\Re} \sigma_{\min} \begin{bmatrix} \mathrm{i}\omega - A \\ C \end{bmatrix}$$

where $\sigma_{\min}$ is the smallest singular value. It is the distance between the pair (A, C) and the set of pairs with an unobservable purely imaginary mode and there is no clear relation between $\Delta(A, C)$ and $\delta(A, C)$. In general, we have $\Delta(A, C) \le \delta(A, C)$. So $\gamma < \delta(A, C)$ can not imply $\gamma < \Delta(A, C)$ [1]. The condition to satisfy (4.25) was modified with an additional constraint by Aboky et al. as follows [1]:

If the Lispchitz constant γ satisfies

$$\gamma < \delta(A, C) \quad \text{and} \quad \gamma < \delta\left(A, \frac{\gamma}{\|C\|_2} C\right)$$

then there exists a gain matrix L which stabilizes the error dynamics (4.6) quadratically.

While a bisection algorithm is used for computing δ numerically to guarantee the existence of a quadratically stabilizing observer gain matrix [1, 11], the solution of the ARE proposed in [1, 73, 74] does not provide a specific level of performance for the observer [39]. Therefore, the last question in this section is how to optimize the observer's performance as well as guarantee the existence of a quadratically stabilizing observer gain matrix. To provide a possible solution, the question will be formulated in a form of convex optimization problems. First, the existence problem of the positive definite matrix P and positive constant ε to satisfy (4.24) can be formulated in the LMI problem as follows:

For a fixed $\varepsilon \in [a,\ b]$,

$$\begin{aligned} &\text{maximize} \quad \beta \\ &\text{subject to} \quad P > 0, \qquad \beta > 0, \\ &\qquad \begin{bmatrix} P^T A + A^T P + \beta I - \frac{1}{\varepsilon} C^T C & P \\ P & -I \end{bmatrix} < 0. \end{aligned} \tag{4.26}$$

If the computed stability margin $\alpha = \sqrt{\beta}$ is greater than γ, it guarantees the existence of P and ε in (4.24).

Second, to optimize the performance of an observer, we need to define a desirable property for the observer. Among many desirable properties such as magnitude of elements of the gain matrix, decay rate, and $\mathcal{L}_2$ gain, the magnitude of elements of the gain matrix is only considered to reduce the amplification of sensor measurement noise. However, consideration of other desirable properties can be referred to [12, 39]. Since the observer gain matrix depends only upon the inverse of P in Theorem 4.3, the minimum eigenvalue of P should be maximized to reduce the maximum singular value of the observer gain matrix elements. Therefore, it can be formulated as follows: for the given γ,

$$\begin{aligned} &\text{maximize} \quad \lambda_{\min}(P) \\ &\text{subject to} \quad P > 0, \\ &\qquad \begin{bmatrix} P^T A + A^T P + \gamma^2 I - \frac{1}{\varepsilon} C^T C & P \\ P & -I \end{bmatrix} < 0 \end{aligned} \tag{4.27}$$

where ε is one of solutions for COP (4.26).

Finally, the design procedure for the observer gain matrix is summarized as follows [87]:

Algorithm 4.1 Procedure to design the observer gain

Step 1. Solve COP (4.26) iteratively with logarithmic spacing of $\varepsilon_k \in [10^{-n},\ 10^n]$. If $\beta_k < \gamma^2$ for all k, go to Step 2. Otherwise, go to Step 3.

Step 2. Use a coordinate transformation as suggested in (4.12) to reduce the Lipschitz constant. Go to Step 1 with $\tilde{A} = TAT^{-1}$ and $\tilde{C} = CT^{-1}$.

Step 3. If there is a positive integer k such that $\beta_k \geq \gamma^2$, let the gain matrix L be

$$L = \frac{P^{-1} C^T}{2\varepsilon_k}$$

where P is a solution of COP (4.27).

Example 4.2 (Design of Nonlinear Observers) The following example is considered to illustrate the proposed design technique for a fourth-order model. The model represents a flexible joint robotic arm and was found in the literature [2, 73, 90].

$$\begin{aligned}
\dot{\theta}_m &= \omega_m, \\
\dot{\omega}_m &= \frac{k}{J_m}(\theta_l - \theta_m) - \frac{B}{J_m}\omega_m + \frac{K_\tau}{J_m}u, \\
\dot{\theta}_l &= \omega_1, \\
\dot{\omega}_l &= -\frac{k}{J_l}(\theta_l - \theta_m) - \frac{mgh}{J_l}\sin(\theta_1)
\end{aligned} \tag{4.28}$$

where θ_m and θ_1 are the angular position of the motor and link, respectively, and ω_m and ω_1 are the angular velocity of both. Moreover, J_m and J_l represent the inertia of the motor and controlled link, m stands for the pointer mass, and k is the torsional spring constant. B represents the viscous friction and K_τ is the amplifier gain. Then, the system model (4.28) can be described by the following equation:

$$\begin{aligned}
\dot{x} &= Ax + B_u u + B_f f(x), \\
y &= Cx
\end{aligned} \tag{4.29}$$

where $x = [\theta_m \ \omega_m \ \theta_l \ \omega_l]^T \in \Re^4$, and the matrices are

$$A = \begin{bmatrix} 0 & 1 & 0 & 0 \\ -48.6 & -1.25 & 48.6 & 0 \\ 0 & 0 & 0 & 1 \\ 19.5 & 0 & -19.5 & 0 \end{bmatrix}, \qquad C = \begin{bmatrix} 1 & 0 & 0 & 0 \\ 0 & 1 & 0 & 0 \end{bmatrix},$$

$$B_u = [0\ 21.6\ 0\ 0]^T, \qquad B_f = [0\ 0\ 0\ 1]^T, \qquad f(x) = -3.33\sin(\theta_l).$$

Furthermore, the pair (A, C) is controllable and there exists a positive constant γ such that

$$\left\| f(x) - f(\hat{x}) \right\| \le \gamma \|e\|.$$

Then, we can let $\gamma = 3.33$ in this example.

Step 1. Calculation of Stability Margin Equation (4.4) is used for the observer and the corresponding error dynamics is derived as in (4.5). After solving COP (4.26) iteratively with logarithmic spacing of $\varepsilon_k \in [10^{-1},\ 10]$ for the given $\gamma = 3.33$, Fig. 4.1 shows the stability margin with respect to ε_k. Since $\alpha_k = \sqrt{\beta_k} < \gamma$ for all k, we need to go to Step 2 to reduce the Lipschitz constant via coordinate transformation.

Step 2. Coordinate Transformation Suppose $T_c = \text{diag}(1, 1, 1, 0.1)$ which is used in [74]. With this coordinate transformation, the system (4.29) becomes

$$\begin{aligned}
\dot{\tilde{x}} &= T_c A T_c^{-1}\tilde{x} + T_c B_u u + T_c f(x), \\
y &= C T_c^{-1}\tilde{x}
\end{aligned} \tag{4.30}$$

with the inequality constraint

$$\left\| T_c^{-1} f(x) - T_c^{-1} f(\hat{x}) \right\| < 0.1\gamma \|e\| = \tilde{\gamma} \|e\|.$$

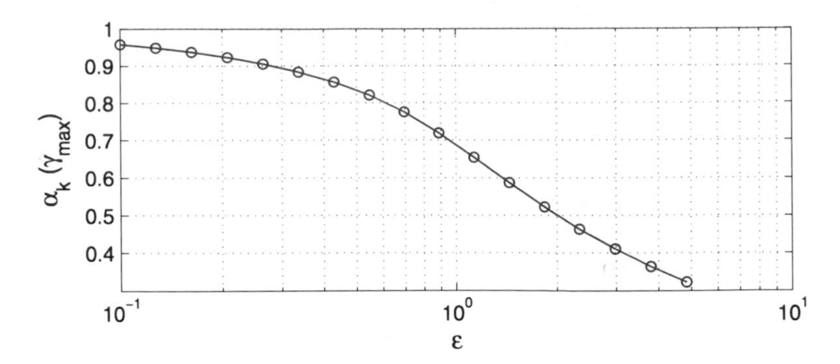

Fig. 4.1 Stability margin with respect to ε_k for the given $\gamma = 3.33$

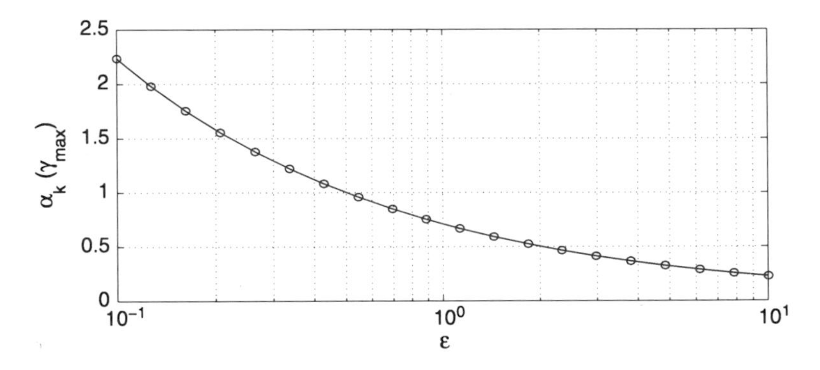

Fig. 4.2 Stability margin with respect to ε_k for the given $\gamma = 0.333$

Then, we have $\tilde{A} = T_c A T_c^{-1}$, $\tilde{C} = C T_c^{-1}$, and the new Lipschitz constant $\tilde{\gamma} = 0.333$. Figure 4.2 shows the stability margin (α_k) with respect to ε_k for the given $\tilde{\gamma}$. Since $\alpha_k = 0.3630$ for the $\varepsilon_k = 4.8329$ in Fig. 4.2, it is summarized that $\alpha_k \geq \gamma$ for all k satisfying $\varepsilon_k \leq 4.8329$. Therefore, it is implied that there exist a positive definite matrix P and a positive constant ε satisfying (4.24).

Step 3. Design of Observer Gain Matrix L If COP (4.27) is solved numerically using CVX for the given $\varepsilon_k = 1$, the gain matrix L is calculated as

$$L = \begin{bmatrix} 0.8155 & 0.4422 \\ 0.4422 & 6.1759 \\ 0.8080 & 1.2933 \\ 0.6979 & 2.5587 \end{bmatrix}. \tag{4.31}$$

When a sinusoidal input with frequency of 1 Hz is used to drive the system dynamics, Fig. 4.3 shows the state estimates for the given gain matrix in (4.31).

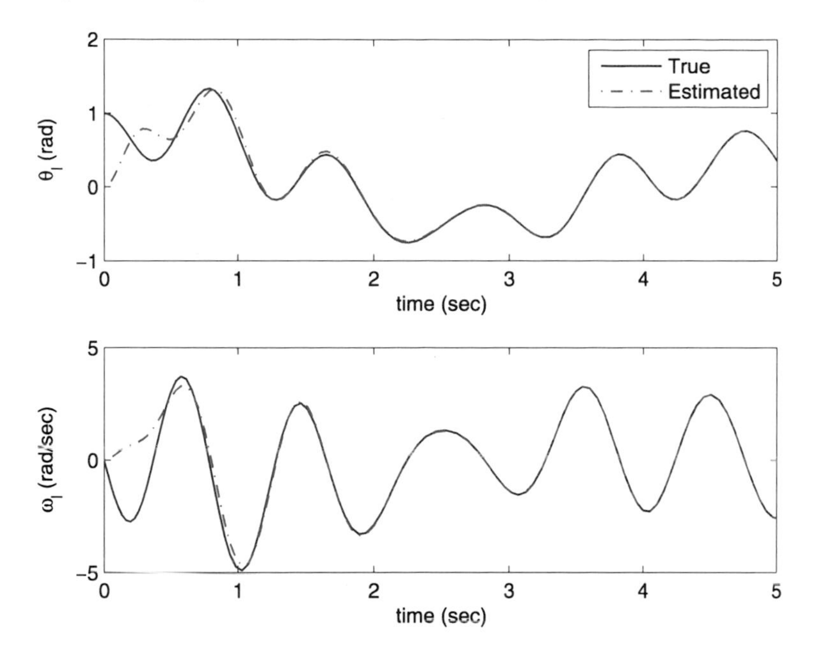

Fig. 4.3 Estimation of the link angle and angular velocity

4.2 A Separation Principle

4.2.1 Preliminary Design of ODSC

The standard design procedure of DSC which stabilizes the given system is described in Sect. 2.3. The only difference here is that the estimated state based on the nonlinear observer is used instead of the measured state information. After the first super-diagonal elements of A are decoupled, the system (4.1) and observer dynamics can be rewritten componentwise, for $1 \le i \le n-1$,

$$\begin{cases} \dot{x}_i = x_{i+1} + \bar{A}_i x + f_i(x), \\ \dot{x}_n = u + \bar{A}_n x + f_n(x), \end{cases} \tag{4.32}$$

$$\begin{cases} \dot{\hat{x}}_i = \hat{x}_{i+1} + \bar{A}_i \hat{x} + f_i(\hat{x}) + L_i(y - C\hat{x}), \\ \dot{\hat{x}}_n = u + \bar{A}_n \hat{x} + f_n(\hat{x}) + L_n(y - C\hat{x}) \end{cases} \tag{4.33}$$

where $\bar{A} := A - U$ where U is a square matrix whose first super-diagonal elements are one and elsewhere zero, i.e., $U = \text{diag}([1, \ldots, 1], 1)$, and $\bar{A}_i$ and L_i are the ith row of $\bar{A}$ and L, respectively.

As derived in Sect. 2.3, the sliding surface and synthetic input are calculated based on the estimated state information instead of the full state measurement. That is, (2.10) is replaced by

$$\bar{x}_{i+1} := -\bar{A}_i \hat{x} - f_i(\hat{x}_1, \ldots, \hat{x}_i) + \dot{x}_{id} - K_i \hat{S}_i \tag{4.34}$$

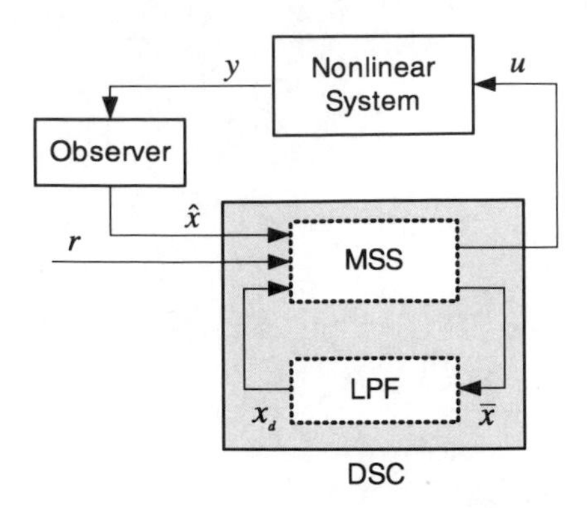

Fig. 4.4 Schematic framework of observer-based DSC

where $\hat{S}_i := \hat{x}_i - x_{id}$. Finally, after defining $\hat{S}_n = \hat{x}_n - x_{nd}$, the control input u is derived as

$$u = -\bar{A}_n\hat{x} - f_n(\hat{x}_1, \ldots, \hat{x}_n) + \dot{x}_{nd} - K_n\hat{S}_n. \tag{4.35}$$

Based on the graphical interpretation shown in Fig. 4.4, the design procedure can be summarized as the following iteration logic:

For $i = 1$ to n,

$$\left.\begin{aligned} \hat{S}_i &= \hat{x}_i - x_{id} \\ \bar{x}_{i+1} &= -\bar{A}_i\hat{x} - f_i(\hat{x}_1, \ldots, \hat{x}_i) + \dot{x}_{id} - K_i\hat{S}_i \end{aligned}\right\} \quad \text{MSS}$$

$$\left.\tau_{i+1}\dot{x}_{(i+1)d} + x_{(i+1)d} = \bar{x}_{i+1}\right\} \quad \text{LPF}$$

If $i = n$,

$$\left.\begin{aligned} \hat{S}_n &= \hat{x}_n - x_{nd} \\ u &= -\bar{A}_n\hat{x} - f_n(\hat{x}_1, \ldots, \hat{x}_n) + \dot{x}_{nd} - K_n\hat{S}_n \end{aligned}\right\} \quad \text{MSS}$$

End

End

4.2.2 Augmented Error Dynamics

As developed for the DSC design methodology in Chap. 2, we can describe the closed-loop system by a set of augmented error dynamics including the observer error dynamics [83]. First, after adding and subtracting $\hat{x}_{i+1}$, $x_{(i+1)d}$, and $\bar{x}_{i+1}$ to (4.32),

$$\begin{aligned} \dot{x}_i &= (x_{i+1} - \hat{x}_{i+1}) + (\hat{x}_{i+1} - x_{(i+1)d}) \\ &\quad + (x_{(i+1)d} - \bar{x}_{i+1}) + \bar{x}_{i+1} + \bar{A}_i x + f_i. \end{aligned} \tag{4.36}$$

Then, substituting the definition of $\bar{x}_{i+1}$ from (4.34) in the above equation and using the definitions of e and $\hat{S}$, we can rewrite (4.36) as follows:

$$\dot{x}_i = e_{i+1} + \hat{S}_{i+1} + \xi_{(i+1)} + \bar{A}_i e + \dot{x}_{id} - K_i \hat{S}_i + f_i(x) - f_i(\hat{x})$$

where the low pass filter error is defined as $\xi_{i+1} := x_{(i+1)d} - \bar{x}_{i+1}$. Moreover, since $\dot{S}_i = \dot{x}_i - \dot{x}_{id} = \dot{e}_i + \dot{\hat{S}}_i$, the system dynamics (4.1) with the nonlinear observer and DSC can be described as a function of the surface error, observer error, and low pass filter error, i.e.,

$$\begin{cases} \dot{e}_i + \dot{\hat{S}}_i = e_{i+1} + \hat{S}_{i+1} + \xi_{i+1} + \bar{A}_i e - K_i \hat{S}_i + f_i(x) - f_i(\hat{x}), \\ \dot{e}_n + \dot{\hat{S}}_n = \bar{A}_n e - K_n \hat{S}_n + f_n(x) - f_n(\hat{x}) \end{cases} \tag{4.37}$$

for $1 \leq i \leq n-1$. It can be written in matrix form as follows:

$$\frac{d}{dt}(\hat{S} + e) = \mathbf{A}_{11}\hat{S} + \mathbf{A}_{12}e + \xi + \phi \tag{4.38}$$

where $\hat{S} = \hat{x} - x_d \in \Re^{n_S}$, $e = x - \hat{x} \in \Re^{n_e}$, $\xi = x_d - \bar{x} \in \Re^{n_\xi}$, and

$$\begin{aligned} &\mathbf{A}_{11} := U - \mathrm{diag}(K_1, \ldots, K_n), \qquad \mathbf{A}_{12} := A, \\ &\phi := f(x) - f(\hat{x}). \end{aligned}$$

Next, we investigate the filter error dynamics which is included for DSC design. Differentiating the filter error (ξ), we obtain

$$\begin{cases} \dot{\xi}_2 = \dot{x}_{2d} - \dot{\bar{x}}_2 = \dot{x}_{2d} + \bar{A}_1 \dot{\hat{x}} + \dot{f}_1(\hat{x}) - \ddot{x}_{1d} + K_1 \dot{\hat{S}}_1, \\ \dot{\xi}_{i+1} = \dot{x}_{(i+1)d} - \dot{\bar{x}}_{i+1} = -\frac{\xi_{i+1}}{\tau_{i+1}} + \bar{A}_i \dot{\hat{x}} + \dot{f}_i(\hat{x}) - \ddot{x}_{id} + K_i \dot{\hat{S}}_i \end{cases}$$

for $2 \leq i \leq n-1$. This can be rewritten as

$$\begin{cases} -K_1 \dot{\hat{S}}_1 + \dot{\xi}_2 = -\frac{\xi_2}{\tau_2} + \bar{A}_1 \dot{\hat{x}} + \dot{f}_1(\hat{x}) - \ddot{x}_{1d}, \\ -K_i \dot{\hat{S}}_i + \dot{\xi}_{i+1} - \frac{\dot{\xi}_i}{\tau_i} = -\frac{\xi_{i+1}}{\tau_{i+1}} + \bar{A}_i \dot{\hat{x}} + \dot{f}_i(\hat{x}). \end{cases} \tag{4.39}$$

Furthermore, the observer dynamics in (4.4) can be written as

$$\begin{cases} \dot{\hat{x}}_i = (\hat{x}_{i+1} - x_{(i+1)d}) + (x_{(i+1)d} - \bar{x}_{i+1}) + \bar{x}_{i+1} + \bar{A}_i \hat{x} + f_i(\hat{x}) + L_i C e, \\ \dot{\hat{x}}_n = u + \bar{A}_n \hat{x} + f_n(\hat{x}) + L_n(y - C\hat{x}). \end{cases}$$

If $\bar{x}_{i+1}$ and u in (4.34) and (4.35) are used, respectively,

$$\begin{aligned} &\begin{cases} \dot{\hat{x}}_1 = \hat{S}_2 + \xi_2 - K_1 \hat{S}_1 + \dot{x}_{1d} + L_1 C e, \\ \dot{\hat{x}}_i = \hat{S}_{i+1} + \xi_{i+1} - K_i \hat{S}_i + \dot{x}_{id} + L_i C e, \quad \text{for } i = 2, \ldots, n \end{cases} \\ &\Longrightarrow \quad \dot{\hat{x}} = C_z z + b_r \dot{x}_{1d} + LCe \end{aligned} \tag{4.40}$$

where $z = [\hat{S}^T \ \xi^T]^T \in \Re^{n_S + n_\xi}$, $C_z = [\underline{A}_{11} \ T_\xi]$ and $b_r = [1 \ 0 \ \cdots \ 0]^T$ are defined in Lemma 2.1. Therefore, the filter error dynamics in (4.39) can be written in the matrix form as

$$\begin{aligned}\frac{d}{dt}(-K\hat{S}+T_\xi\xi) &= A_{22}\xi+\bar{A}_{1:n-1}(C_z z+b_r\dot{x}_{1d}+LCe)+\dot{f}(\hat{x})-b_r\ddot{x}_{1d}\\ &= \mathbf{A}_{31}\hat{S}+\mathbf{A}_{32}e+\mathbf{A}_{33}\xi+\psi+\bar{A}_{1:n-1}b_r\dot{x}_{1d}-b_r\ddot{x}_{1d}\end{aligned} \quad (4.41)$$

where $A_{22} := -\operatorname{diag}(1/\tau_2,\dots,1/\tau_n)$ which is defined in Lemma 2.1, and $\bar{A}_{1:n-1}$ is the matrix consisting of the first to $(n-1)$th row of $\bar{A}$, and $\psi := \frac{\partial f(\hat{x})}{\partial \hat{x}}\dot{\hat{x}}$ such that $\psi_{ij} = \frac{\partial f_i}{\partial \hat{x}_j}\dot{\hat{x}}_j$. Furthermore, the submatrices denote

$$\mathbf{A}_{31} := \bar{A}_{1:n-1}\underline{A}_{11}, \qquad \mathbf{A}_{32} := \bar{A}_{1:n-1}LC, \qquad \mathbf{A}_{33} := A_{22}+\bar{A}_{1:n-1}T_\xi.$$

Finally, combining estimation error dynamics (4.5) with surface error dynamics (4.38) and filter error dynamics (4.41), we have the augmented error dynamics as follows:

$$\begin{aligned}\frac{d}{dt}\left[\begin{array}{cc|c}\mathbf{I} & \mathbf{I} & \mathbf{0}\\ \mathbf{0} & \mathbf{I} & \mathbf{0}\\ \hline -K & \mathbf{0} & T_\xi\end{array}\right]\left[\begin{array}{c}\hat{S}\\ e\\ \hline \xi\end{array}\right] &= \left[\begin{array}{cc|c}\mathbf{A}_{11} & \mathbf{A}_{12} & \mathbf{I}\\ \mathbf{0} & \mathbf{A}_{22} & \mathbf{0}\\ \hline \mathbf{A}_{31} & \mathbf{A}_{32} & \mathbf{A}_{33}\end{array}\right]\left[\begin{array}{c}\hat{S}\\ e\\ \hline \xi\end{array}\right]+\left[\begin{array}{cc}\mathbf{I} & \mathbf{0}\\ \mathbf{I} & \mathbf{0}\\ \hline \mathbf{0} & \mathbf{I}\end{array}\right]\left[\begin{array}{c}\phi\\ \psi\end{array}\right]\\ &\quad+\left[\begin{array}{cc}0_{n_S} & 0_{n_S}\\ 0_{n_e} & 0_{n_e}\\ \hline \bar{A}_{1:n-1}b_r & -b_r\end{array}\right]\left[\begin{array}{c}\dot{x}_{1d}\\ \ddot{x}_{1d}\end{array}\right]\end{aligned} \quad (4.42)$$

where $\mathbf{A}_{22} := A_{ob} = A - LC$. Since the matrix on the left hand side of the above equation is invertible with an inverse given by:

$$T^{-1} = \left[\begin{array}{cc|c}\mathbf{I} & \mathbf{I} & \mathbf{0}\\ \mathbf{0} & \mathbf{I} & \mathbf{0}\\ \hline -K & \mathbf{0} & T_\xi\end{array}\right]^{-1} = \left[\begin{array}{cc|c}\mathbf{I} & -\mathbf{I} & \mathbf{0}\\ \mathbf{0} & \mathbf{I} & \mathbf{0}\\ \hline T_\xi^{-1}K & -T_\xi^{-1}K & T_\xi^{-1}\end{array}\right], \quad (4.43)$$

the augmented closed-loop error dynamics for the observer-based DSC can be reformulated as [83]

$$\dot{z}_e = A_e z_e + B_w g + B_h h \quad (4.44)$$

where

$$z_e := \left[\begin{array}{c}\hat{S}\\ e\\ \xi\end{array}\right] \in \Re^{3n-1} := \Re^{n_z}, \qquad g := \left[\begin{array}{c}\phi\\ \psi\end{array}\right] \in \Re^{n_w}, \qquad h := \left[\begin{array}{c}\dot{x}_{1d}\\ \ddot{x}_{1d}\end{array}\right] \in \Re^2$$

and the matrices are

$$A_e = \left[\begin{array}{cc|c}\mathbf{A}_{11} & LC & \mathbf{I}\\ \mathbf{0} & \mathbf{A}_{22} & \mathbf{0}\\ \hline T_\xi^{-1}(K\mathbf{A}_{11}+\mathbf{A}_{31}) & T_\xi^{-1}(KLC+\mathbf{A}_{32}) & T_\xi^{-1}(K\mathbf{I}+\mathbf{A}_{33})\end{array}\right],$$

$$B_w = \left[\begin{array}{cc}\mathbf{0} & \mathbf{0}\\ \mathbf{I} & \mathbf{0}\\ \hline \mathbf{0} & T_\xi^{-1}\end{array}\right], \qquad B_h = \left[\begin{array}{cc}0_{n_S} & 0_{n_S}\\ 0_{n_e} & 0_{n_e}\\ \hline T_\xi^{-1}\bar{A}_{1:n-1}b_r & -T_\xi^{-1}b_r\end{array}\right].$$

It turns out that the closed-loop error dynamics can be expressed as a linear system subject to a nonlinear term g and an external input h. Furthermore, using the assumptions for the system (4.1) in Sect. 4.1.1, the error dynamics in (4.44) can be written as a linear system subject to both a vanishing and a nonvanishing perturbation. The result can be summarized in the following lemma.

Lemma 4.1 *For the given class of nonlinear system* (4.1), *the augmented closed-loop error dynamics with observer-based DSC is*

$$\dot{z}_e = A_e z_e + B_w w + B_r r \tag{4.45}$$

and there exists C_{zi} *such that*

$$|w_i| \leq \|C_{zi} z_e\|$$

where

$$\psi = \frac{\partial f(\hat{x})}{\partial \hat{x}}\dot{\hat{x}} = \frac{\partial f(\hat{x})}{\partial \hat{x}}(C_{ze}z_e + b_r \dot{x}_{1d}) = JC_{ze}z_e + J_1\dot{x}_{1d} := p + q,$$
$$C_{ze} = [\underline{A}_{11}\ LC\ T_\xi],$$

$$w = \begin{bmatrix} \phi \\ p \end{bmatrix}, \quad r = \begin{bmatrix} q \\ \dot{x}_{1d} \\ \ddot{x}_{1d} \end{bmatrix}, \quad B_r = \left[\begin{array}{c|cc} \mathbf{0} & 0_{ns} & 0_{ns} \\ \mathbf{0} & 0_{n_e} & 0_{n_e} \\ \hline T_\xi^{-1} & T_\xi^{-1}\bar{A}_{1:n-1}b_r & -T_\xi^{-1}b_r \end{array}\right],$$

and other matrices are defined in Lemma 2.1.

Proof For the augmented error dynamics in (4.44), the nonlinear term g needs to be decomposed into vanishing and nonvanishing terms. First, using the assumption (4.3), the componentwise upper bound of ϕ in (4.44) is

$$|\phi_i| = \left|f_i(x) - f_i(\hat{x})\right| \leq \gamma_i\|x - \hat{x}\| = \gamma_i\|e\| := \|C_{zi}z_e\| \tag{4.46}$$

where $C_{zi} = [\mathbf{0}\ \gamma_i\mathbf{I}\ \mathbf{0}]$ for $i = 1, \ldots, n_\phi$. Next, the component of ψ is written as

$$\psi_i = \frac{\partial f_i(\hat{x})}{\partial \hat{x}}\dot{\hat{x}} = \sum_{j=1}^{n}\frac{\partial f_i}{\partial \hat{x}_j}\dot{\hat{x}}_j = \bar{J}_i\dot{\hat{x}}$$

where $\bar{J}_i$ is the ith row of Jacobian matrix J. Using (4.40), ψ_i for $i = 1, \ldots, n_\psi$ can be decomposed as

$$\begin{aligned}\psi_i &= \bar{J}_i[\underline{A}_{11}\ LC\ T_\xi]z_e + \bar{J}_i b_r\dot{x}_{1d} = \bar{J}_i C_{ze}z_e + J_{i1}\dot{x}_{1d} := p_i + q_i, \\ |p_i| &= |\bar{J}_i C_{ze}z_e| \leq \gamma_i\|C_{ze}z_e\| := \|C_{zi}z_e\|.\end{aligned} \tag{4.47}$$

Therefore, combining (4.46) with (4.47), there exists C_{zi} such that

$$|w_i| \leq \|C_{zi}z_e\|$$

where

$$C_{zi} = \begin{cases} \gamma_i[\mathbf{0}\ \mathbf{I}\ \mathbf{0}], & \text{for } i = 1, \ldots, n_\phi, \\ \gamma_i[\underline{A}_{11}\ LC\ T_\xi], & \text{for } i = n_\phi + 1, \ldots, n_\phi + n_\psi. \end{cases}$$

Moreover, since $\psi = p + q$, the error dynamics (4.44) are rewritten as

$$\dot{z}_e = A_e z_e + B_w w + B_r r, \quad |w_i| \leq \|C_{zi}z_e\|.$$

□

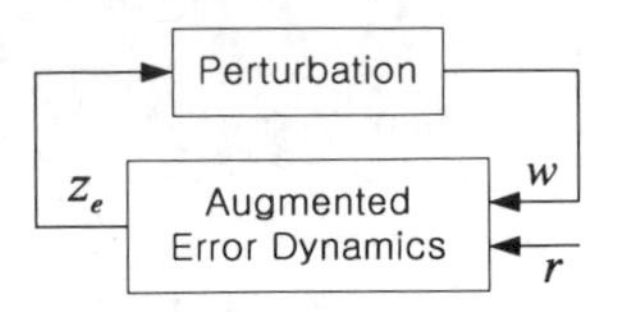

Fig. 4.5 Error dynamics of the closed-loop system controlled via ODSC

Remark 4.4 It is remarked that the augmented error dynamics is quite similar with those in Lemma 2.1 in the sense that they are considered as a linear system subject to a vanishing perturbation w and a nonvanishing perturbation r (see also in Fig. 4.5). Therefore, if the stabilization or regulation problem is considered, $r = 0$ due to $\dot{x}_{1d} = \ddot{x}_{1d} = 0$, and the augmented error dynamics becomes

$$\begin{aligned} \dot{z}_e &= A_e z_e + B_w w, \\ |w_i| &\leq \|C_{zi} z_e\|. \end{aligned} \tag{4.48}$$

Furthermore, if x_{1d} is not zero or constant, but a feasible trajectory, as discussed in Sect. 2.6 or 3.3, r is bounded on the domain $\mathscr{D}$ such as

$$\|r\| = \left\| \begin{bmatrix} J_1 & 0_{n_q} \\ 1 & 0 \\ 0 & 1 \end{bmatrix} \begin{bmatrix} \dot{x}_{1d} \\ \ddot{x}_{1d} \end{bmatrix} \right\| \leq \left\| \begin{bmatrix} J_1 & 0_{n_q} \\ 1 & 0 \\ 0 & 1 \end{bmatrix} \right\| \left\| \begin{bmatrix} \dot{x}_{1d} \\ \ddot{x}_{1d} \end{bmatrix} \right\| \leq mc := r_0$$

where m is the matrix norm of the first matrix and c is the vector norm of $[\dot{x}_{1d}\ \ddot{x}_{1d}]^T$.

4.2.3 Separation Principle of Error Dynamics

Since r is an external input and w is bounded by a state-dependent linear function in the augmented error dynamics (4.45), the matrix A_e is crucial to represent the characteristics of the closed-loop systems. Therefore, $(3n - 1)$ eigenvalues of the system are obtained from the characteristic equation,

$$\begin{aligned} &\det(\lambda \mathbf{I} - A_e) = 0 \\ \iff &\det\left(\left[\begin{array}{cc|c} \lambda\mathbf{I} - \mathbf{A}_{11} & -LC & -\mathbf{I} \\ \mathbf{0} & \lambda\mathbf{I} - \mathbf{A}_{22} & \mathbf{0} \\ \hline -T_\xi^{-1}(K\mathbf{A}_{11} + \mathbf{A}_{33}) & -T_\xi^{-1}(KLC + \mathbf{A}_{32}) & \lambda\mathbf{I} - T_\xi^{-1}(K\mathbf{I} + \mathbf{A}_{33}) \end{array}\right]\right) = 0 \\ \iff &\det(\lambda\mathbf{I} - \mathbf{A}_{22}) \cdot \det\left(\begin{bmatrix} \lambda\mathbf{I} - \mathbf{A}_{11} & -\mathbf{I} \\ -T_\xi^{-1}(K\mathbf{A}_{11} + \mathbf{A}_{31}) & \lambda\mathbf{I} - T_\xi^{-1}(K\mathbf{I} + \mathbf{A}_{33}) \end{bmatrix}\right) = 0. \end{aligned}$$

Therefore, each eigenvalue satisfies either

$$\det(\lambda\mathbf{I} - \mathbf{A}_{22}) = 0 \quad \text{or} \tag{4.49}$$

$$\det\left(\begin{bmatrix} \lambda\mathbf{I} - \mathbf{A}_{11} & -\mathbf{I} \\ -T_\xi^{-1}(K\mathbf{A}_{11} + \mathbf{A}_{31}) & \lambda\mathbf{I} - T_\xi^{-1}(K\mathbf{I} + \mathbf{A}_{33}) \end{bmatrix}\right) = 0. \tag{4.50}$$

Remark 4.5 It turns out that (4.49) is the characteristic equation of the observer error dynamics (4.12), and (4.50) is the characteristic equation of the DSC error dynamics. If $\bar{A} = 0$ in (4.32), i.e., there is no linear function in $f(x)$ in (4.1), the submatrices become

$$\mathbf{A}_{11} = A_{11}, \qquad \mathbf{A}_{31} = \mathbf{0}, \qquad \mathbf{A}_{33} = A_{22}$$

where A_{11} and A_{22} are defined in Lemma 2.1. Then, (4.50) is equal to the characteristic equation derived in Chap. 2. This is quite similar to the well-known separation principle for observer state feedback control in linear system theory, in the sense that the control modes can be set independently of the observer modes. Therefore, the K_i and τ_i for DSC and L for the observer can be designed separately [83].

4.3 Quadratic Stabilization and Tracking

A sufficient condition for quadratic stabilization and tracking via observer-based DSC will be presented in this section. Since the separation principle holds in the augmented error dynamics, either the nonlinear observer or the DSC can be designed independently and each sufficient condition for either the observer or DSC design can be found in the literature and Chap. 2. Furthermore, due to the similarity of the augmented error dynamics, (4.45), the theorems in Chap. 2 can be used for the augmented error dynamics with the observer-based DSC. Therefore, theorems for observer-based DSC will be stated without proof.

Remark 4.6 $V(t) = z_e(t)^T P z_e(t)$ in Definition 2.4 is called a *quadratic Lyapunov-like* function in the sense that $\dot{V} < 0$ holds only outside a set $\varepsilon_P = \{z_e \in \Re^{n_z} | z_e^T P z_e \leq 1\}$. Moreover, the set is a *controlled invariant* and *reachable set* [12, 88].

The quadratic Lyapunov function for the regulation problem needs to be obtained to test feasibility of the observer gain and the DSC gain set stabilizing the closed-loop systems, while the quadratic Lyapunov-like function for the tracking problem depends on the characteristics of the reference signal. For this reason, two theorems for the augmented error dynamics will be presented respectively as follows:

Theorem 4.5 *Suppose that the closed-loop error dynamics* (4.48) *is given for the set of controller gains, Θ, and observer gain matrix, L for all x in a domain $\mathscr{D}$. The matrices A_e and B_w are given. The nonlinear system* (4.1) *is quadratically stabilizable via observer-based DSC if there exist $P > 0$ and $\Sigma = \mathrm{diag}(\sigma_1, \ldots, \sigma_{n_w}) \geq 0$ such that*

$$\begin{bmatrix} A_e^T P + P A_e + C_z^T \Sigma_B C_z & P B_w \\ B_w^T P & -\Sigma \end{bmatrix} < 0 \tag{4.51}$$

where $C_z = \mathrm{diag}(C_{z1}, \ldots, C_{zn_w})$ and $\Sigma_B = \mathrm{diag}(\sigma_1 I, \ldots, \sigma_{n_w} I)$ are block diagonal matrices.

For the tracking problem, formulation of a convex optimization problem depends on the external input r as discussed in Sect. 2.6. If it is assumed that the reference input has bounded peaks, i.e. $r^T r \le r_0^2$ where r_0 is the bound on the peak, the following theorem can be stated for the quadratic tracking.

Theorem 4.6 *Suppose that the closed-loop error dynamics* (4.48) *is given for the set of controller gains, Θ, and observer gain matrix, L for all x in a domain $\mathcal{D}$. The matrices A_e and B_w are given. The nonlinear system* (4.1) *is quadratically trackable via observer-based DSC if there exist $P > 0$, $\Sigma = \mathrm{diag}(\sigma_1, \ldots, \sigma_{n_w}) \ge 0$, and $\alpha \ge 0$ such that*

$$\begin{bmatrix} A_e^T P + P A_e + \alpha P + C_z^T \Sigma_B C_z & P B_w & P \tilde{B}_r \\ B_w^T P & -\Sigma & 0 \\ \tilde{B}_r^T P & 0 & -\alpha I \end{bmatrix} \le 0 \tag{4.52}$$

where $\tilde{B}_r = r_0 B_r$.

Example 4.3 (Design of Observer-based DSC) Consider the system introduced in Example 4.2 and suppose the objective of a controller is to make θ_m track $\theta_d = 0.7 - \cos(3t)$. The same example is introduced in [2].

Since the nonlinear observer is designed in Example 4.2, let us design the observer-based DSC (ODSC). Define the first error surface as $S_1 := \theta_m - \theta_d$. After differentiating S_1 along the trajectory of (4.28), let the synthetic input $\bar{\omega}_m$ be

$$\begin{aligned} \dot{S}_1 &= \dot{\theta}_m - \dot{\theta}_d = \omega_m - \dot{\theta}_d, \\ \bar{\omega}_m &:= \dot{\theta}_d - K_1 \hat{S}_1 \end{aligned}$$

where $\hat{S}_1 = \hat{\theta}_m - \theta_d$. Next, after the synthetic input passes through the low-pass filter as

$$\tau_2 \dot{\omega}_{md} + \omega_{md} = \bar{\omega}_m, \quad \omega_{md}(0) := \bar{\omega}_m(0),$$

define the second surface as $S_2 = \omega_m - \omega_{md}$. Similarly differentiating it, we get the control input u as follows

$$\begin{aligned} \dot{S}_2 &= A_2 x + \frac{K_\tau}{J_m} u - \dot{\omega}_{md}, \\ u &= \frac{J_m}{K_\tau} \left[-A_2 \hat{x} + \dot{\omega}_{md} - K_2 \hat{S}_2 \right] \end{aligned}$$

where A_2 is the second row of the matrix A and $\hat{S}_2 = \hat{\omega}_m - \omega_{md}$.

Using the result of Lemma 4.1, we can derive the augmented closed-loop error dynamics as presented in (4.44):

$$\dot{z}_e = A_e z_e + B_w w + B_r r \tag{4.53}$$

where $z_e = [\hat{S}_1 \ \hat{S}_2 \ e_1 \ e_2 \ e_3 \ e_4 \ \xi_2]^T \in \Re^7$, $w = \sin(\theta_1) - \sin(\hat{\theta}_1)$, $r = \ddot{\theta}_d$, and the matrices are as follows:

$$A_e = \begin{bmatrix} A_{11} & LCT_c^{-1} & A_{13} \\ \mathbf{0} & T_c(A - LC)T_c^{-1} & \mathbf{0} \\ KA_{11} & KLCT_c^{-1} & K_1 - \frac{1}{\tau_2} \end{bmatrix} \in \Re^{7\times 7},$$

$$A_{11} = \begin{bmatrix} -K_1 & 1 \\ 0 & -K_2 \end{bmatrix}, \qquad A_{13} = \begin{bmatrix} 1 \\ 0 \end{bmatrix}, \qquad K = [K_1 \ 0],$$

$$B_w = \begin{bmatrix} 0 \ 0 \,\vdots\, 0 \ 0 \ 0 \ -0.333 \,\vdots\, 0 \end{bmatrix}^T \in \Re^{7\times 1}, \qquad B_r = \begin{bmatrix} 0 \ 0 \,\vdots\, 0 \ 0 \ 0 \ 0 \,\vdots\, -1 \end{bmatrix}^T \in \Re^{7\times 1}.$$

When the observer gain matrix in (4.31) is used and the DSC gain set is $\{K_1, K_2, \tau_2\} = \{10, 40, 0.02\}$, the eigenvalues of A_e are

$$\lambda(A_e) = \{\underbrace{-13.8197, -36.1803, -40}, \underbrace{-2.3263 \pm i9.58466, -0.9573, -2.6316}\}$$

where the eigenvalues in the first under-brace come from DSC characteristic equation and the second set from the observer characteristic equation.

Figure 4.6 presents numerical solution of the following semi-positive definite problem to find the "smallest" ellipsoidal tracking error bound along line search of α: for a fixed α,

$$\begin{aligned} &\text{maximize} && \lambda_{\min}(P) \\ &\text{subject to} && P > 0, \qquad \Sigma \geq 0, \quad \text{LMI (4.52).} \end{aligned} \tag{4.54}$$

It is noted that the problem is solved numerically using CVX and the objective function in the problem can be changed, depending on the definition of "smallest" such as the volume and the largest semi-axis [12]. Once the ellipsoidal bound satisfies the performance specification, the chosen observer gain matrix and DSC gain set will be one of the candidates for the certain nonlinear system as long as there are no more requirements.

Figures 4.7 and 4.8 (see dash-dot lines in the figures) show simulation results in terms of the observer error and DSC tracking error responses without considering the model uncertainty. For the initial condition given as $x(0) = [1.2, 0, 2, 0]$, it is shown that all the observer errors converge to zero and the tracking error stays within the pre-calculated ellipsoidal error bound after a certain time.

When an uncertainty is considered, e.g., $\Delta f_1 = c\theta_m \sin(\theta_m)$ in the first row of (4.29), the uncertain nonlinear system is

$$\dot{x} = Ax + B_u u + f(x) + \Delta f(x) \tag{4.55}$$

where $\Delta f = [\Delta f_1 \ 0 \ 0 \ 0]^T \in \Re^4$. Figures 4.7 and 4.8 (see solid lines in the figures) show simulation results for $c = 0.3$ in terms of the observer error and DSC tracking error responses when the model uncertainty is considered. Comparing Fig. 4.7(a) with Fig. 4.8(c), we can see that the observer error is directly correlated with the tracking error. That is, the estimation error (e) does not converge to zero asymptotically, thus resulting in a larger error. It is motivated to consider the uncertainty and investigate how it affects the augmented error dynamics with observer-based DSC.

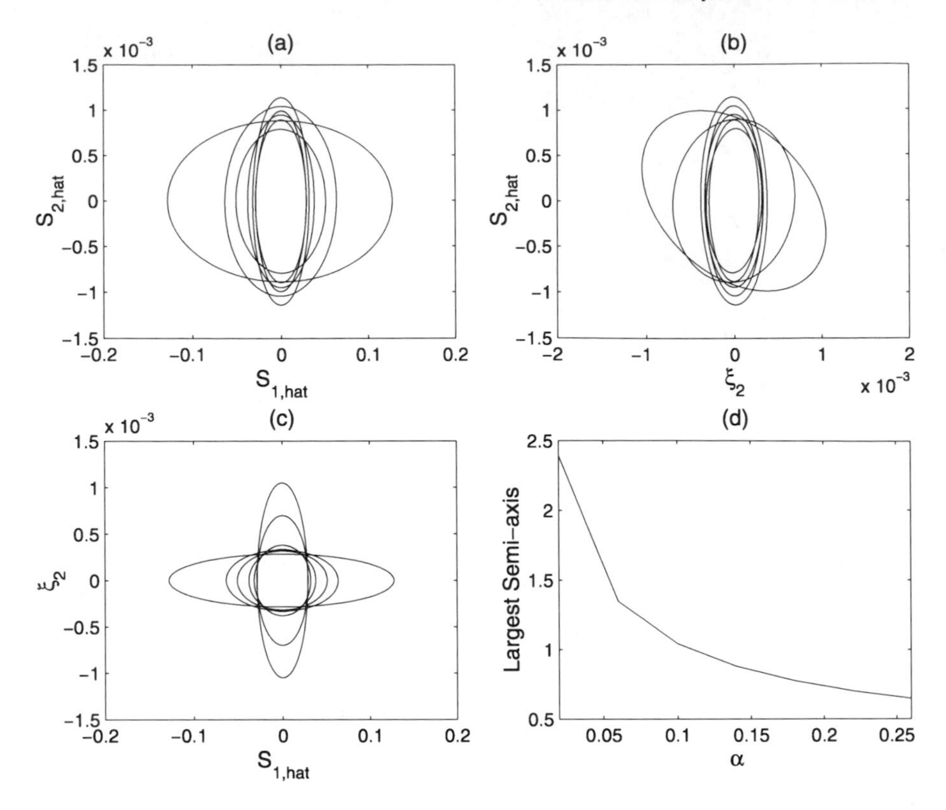

Fig. 4.6 Quadratic Lyapunov-like function (**a**)–(**c**) and corresponding largest semi-axis (**d**)

4.4 Consideration of Uncertainty

Consider uncertainties in the nonlinear system (4.1) as follows

$$\begin{aligned} \dot{x} &= Ax + B_u u + f(x) + \Delta f(x), \\ y &= Cx. \end{aligned} \tag{4.56}$$

If the nonlinear observer in (4.4) is used, the estimation error dynamics is

$$\dot{e} = (A - LC)e + f(x) - f(\hat{x}) + \Delta f = A_{ob}e + \phi + \Delta f. \tag{4.57}$$

The surface error dynamics in (4.38) is rewritten with the addition of uncertainties as follows

$$\frac{d}{dt}(\hat{S} + e) = \mathbf{A}_{11}\hat{S} + \mathbf{A}_{12}e + \xi + \phi + \Delta f. \tag{4.58}$$

While the filter error dynamics in Chap. 3 includes the uncertainty, the filter error dynamics for the observer-based DSC does not, thus the same equation in (4.41) is derived.

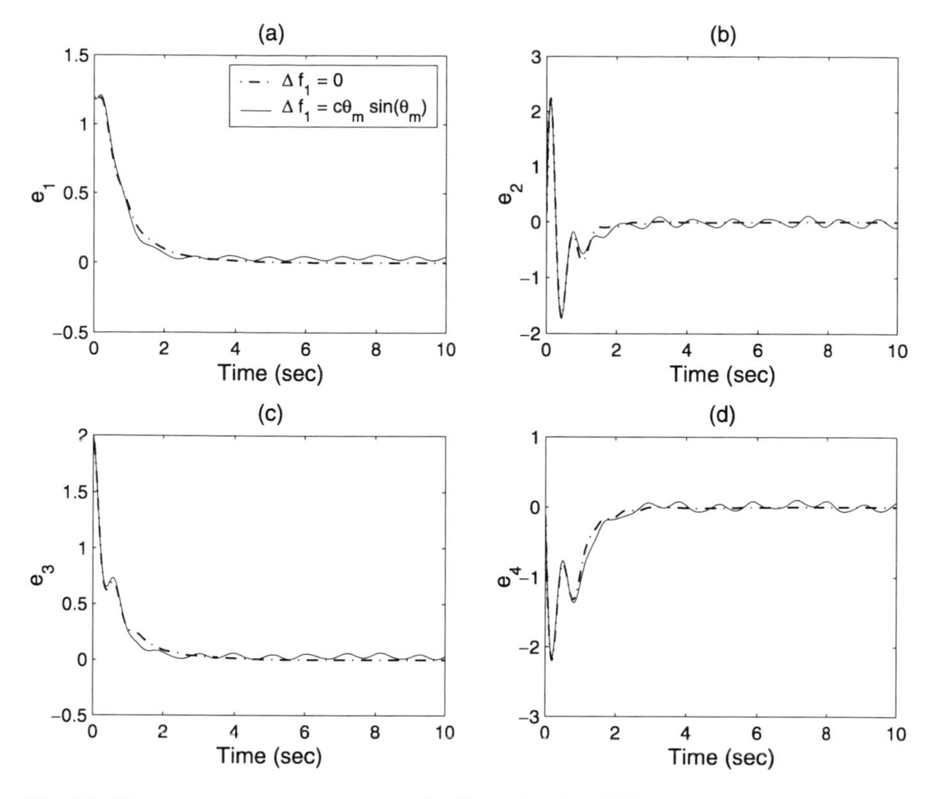

Fig. 4.7 Observer errors responses—*dash-dot line*: $\Delta f = 0$, *solid line*: $\Delta f_1 = c\theta_m \sin(\theta_m)$

Therefore, the augmented error dynamics is

$$\frac{d}{dt}\left[\begin{array}{cc|c} \mathbf{I} & \mathbf{I} & \mathbf{0} \\ \mathbf{0} & \mathbf{I} & \mathbf{0} \\ \hline -K & \mathbf{0} & T_\xi \end{array}\right]\left[\begin{array}{c} \hat{S} \\ e \\ \hline \xi \end{array}\right] = \left[\begin{array}{cc|c} \mathbf{A}_{11} & \mathbf{A}_{12} & \mathbf{I} \\ \mathbf{0} & \mathbf{A}_{22} & \mathbf{0} \\ \hline \mathbf{A}_{31} & \mathbf{A}_{32} & \mathbf{A}_{33} \end{array}\right]\left[\begin{array}{c} \hat{S} \\ e \\ \hline \xi \end{array}\right] + \left[\begin{array}{cc} \mathbf{I} & \mathbf{0} \\ \mathbf{I} & \mathbf{0} \\ \hline \mathbf{0} & \mathbf{I} \end{array}\right]\left[\begin{array}{c} \phi \\ \psi \end{array}\right]$$

$$+ \left[\begin{array}{cc} 0_{nS} & 0_{nS} \\ 0_{n_e} & 0_{n_e} \\ \hline A_{1:n-1} b_r & -b_r \end{array}\right]\left[\begin{array}{c} \dot{x}_{1d} \\ \ddot{x}_{1d} \end{array}\right] + \left[\begin{array}{c} \mathbf{I} \\ \mathbf{I} \\ \hline \mathbf{0} \end{array}\right]\Delta f$$

where $\mathbf{A}_{12} = U$, $\mathbf{A}_{31} = \mathbf{A}_{32} = \mathbf{0}$. After multiplying T^{-1}, we can have the augmented error dynamics as follows:

$$\dot{z}_e = A_e z_e + B_w g + B_h h + B_\Delta \Delta f.$$

Then, using Lemma 4.1, the augmented error dynamics with uncertainties is written as

$$\begin{aligned} \dot{z}_e &= A_e z_e + B_w w + B_r r + B_\Delta \Delta f, \\ |w_i| &\leq \|C_{zi} z_e\| \end{aligned} \tag{4.59}$$

where $B_\Delta = [\mathbf{0}\ \mathbf{I}\ \vdots\ \mathbf{0}]^T$.

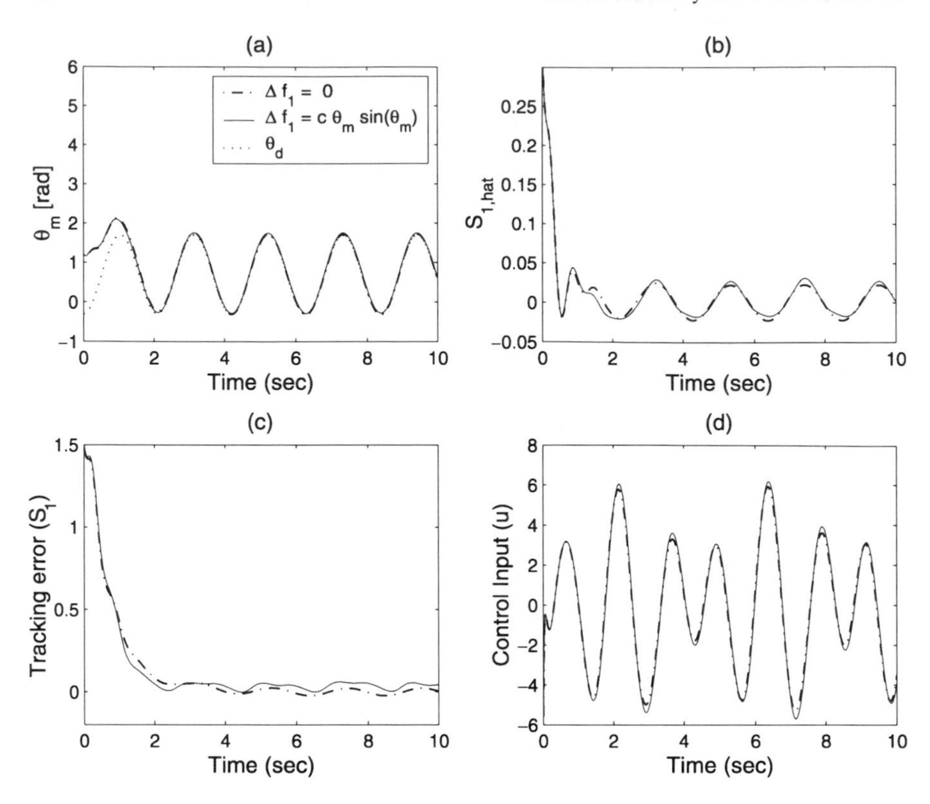

Fig. 4.8 θ_m and tracking error (**a**), (**c**), surface error $\hat{S}_1$ (**b**), and control input (**d**)—*Dash-dot line*: $\Delta f = 0$, *solid line*: $\Delta f_1 = c\theta_m \sin(\theta_m)$

Remark 4.7 It is interesting to note that the first and third block matrices of the matrix B_Δ in the closed error dynamics (4.59) are a zero matrix. This implies that any model uncertainty (Δf) is not directly related with the DSC error dynamics, i.e. S and ξ subspaces. Consequently, as long as there exists a robust nonlinear observer to overcome the uncertain nonlinearity Δf, the nominal DSC design methodology can be used without considering the uncertainty.

4.4.1 Redesign of Nonlinear Observer

Based on the separation principle in Sect. 4.2.3, we need to redesign the observer gain matrix when the uncertainty is considered. According to (4.59), the estimation error dynamics is

$$\dot{e} = (A - LC)e + f(x) - f(\hat{x}) + \Delta f(x) = A_{ob}e + \phi + \Delta f(x).$$

If Δf is considered to be an unknown exogenous input, a desirable property of an observer is that the state estimates should be insensitive to disturbances and uncertainties. To consider this property, the induced $\mathcal{L}_2$ gain between the uncertainty Δf

and the estimation error e, signified as $\|H_{\Delta\to e}\|_\infty$, can be minimized by redesigning the observer gain matrix L.

Theorem 4.7 *For the given nonlinear system in* (4.56), *suppose that the nonlinear observer is given as* (4.4) *on a domain* $\mathscr{D}$. *The observer error dynamics in* (4.57) *has* $\|H_{\Delta\to e}\|_\infty \le \kappa$ *if there exist* $P > 0$, $\varepsilon \ge 0$, *and* $\kappa \ge 0$ *such that*

$$\begin{bmatrix} P^T A + A^T P + (1+\gamma^2)I - \frac{1}{\varepsilon}C^T C & P & P \\ P & -I & 0 \\ P & 0 & -\kappa^2 I \end{bmatrix} < 0 \tag{4.60}$$

and the resulting observer gain matrix is $L = \frac{P^{-1}C^T}{2\varepsilon}$.

Proof Suppose there exist $V(e) = e^T P e$, $P > 0$, and $\kappa \ge 0$ such that

$$\dot{V} + e^T e - \kappa^2 \Delta f^T \Delta f \le 0. \tag{4.61}$$

After integrating the left side of (4.61) from 0 to T with the assumption that $e(0) = 0$,

$$V(T) + \int_0^T \left(e^T e - \kappa^2 \Delta f^T \Delta f\right) dt \le 0.$$

Since $V(T) \ge 0$, this implies that $\|H_{\Delta\to e}\|_\infty \le \kappa$ by definition [12]. The inequality (4.61) is equivalent to

$$e^T \left(A_{ob}^T P + P A_{ob} + I\right)e + 2e^T P(\phi + \Delta f) - \kappa^2 \Delta f^T \Delta f \le 0 \tag{4.62}$$

for all $(e, \phi, \Delta f)$ satisfying $\|\phi\| \le \gamma\|e\|$. Using the inequality condition in (4.23), the inequality condition in (4.62) holds if

$$e^T\left\{A_{ob}^T P + P A_{ob} + PP + \left(1+\gamma^2\right)I\right\}e + 2e^T P\Delta f - \kappa^2 \Delta f^T \Delta f < 0.$$

If $L = \frac{P^{-1}C^T}{2\varepsilon}$ is used, the above inequality becomes

$$e^T\left\{A^T P + PA + PP + \left(1+\gamma^2\right)I - \frac{1}{\varepsilon}C^T C\right\}e + 2e^T P\Delta f - \kappa^2 \Delta f^T \Delta f < 0$$

$$\Longleftrightarrow \quad \begin{bmatrix} A^T P + PA + PP + (1+\gamma^2)I - \frac{1}{\varepsilon}C^T C & P \\ P & -\kappa^2 I \end{bmatrix} < 0.$$

Finally, using the Schur complement, the above inequality condition is equivalent to an LMI (4.60). □

4.4.2 Design Procedure of ODSC

The design procedure for ODSC is summarized as follows:

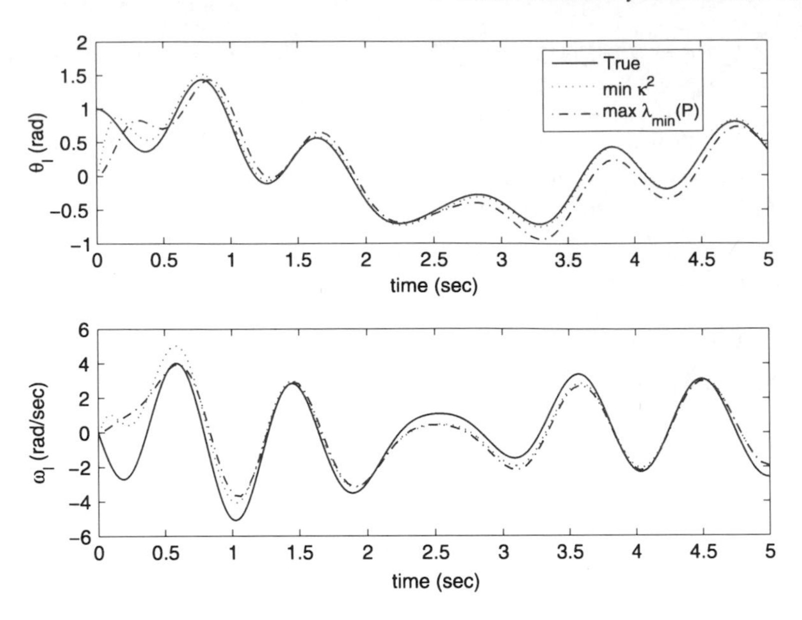

Fig. 4.9 Estimation of the link angle and angular velocity

Algorithm 4.2 Procedure to design ODSC

Step 1. Use Algorithm 4.1 to design the observer gain matrix L. If the uncertainty is considered, the following convex optimization problem is solved instead of COP (4.26): for the given γ and ε,

$$\begin{aligned} &\text{minimize} \quad \kappa^2 \\ &\text{subject to} \quad P > 0, \quad \text{LMI (4.60)}. \end{aligned} \tag{4.63}$$

Step 2. Use the design method suggested in Chap. 2 under the assumption that all states are measured. Using the separation principle in Sect. 4.2.3, a set of controller parameters are designed independently.

Step 3. Once the design of both the observer gain matrix and controller parameters is completed, depending on the characteristics of Δf either quadratic boundedness or input–output stability in the term of an induced $\mathscr{L}_2$ gain can be investigated using the augmented error dynamics (4.59). For example, (4.57) can be rewritten as

$$\begin{aligned} \dot{z}_e &= A_e z_e + B_w w + [B_r \; B_\Delta] \begin{bmatrix} r \\ \Delta f \end{bmatrix} := A_e z_e + B_w w + B_{re} r_e, \\ |w_i| &\le \|C_{zi} z_e\|. \end{aligned} \tag{4.64}$$

If $r_e^T r_e \le r_0$ as discussed in Sect. 4.3, the performance of the nonlinear compensator (or ODSC) in the term of quadratic stability and tracking can be predicted via convex optimization similarly using Theorems 4.5 or 4.6.

Example 4.4 Let us consider Example 4.2 with the uncertainty introduced in Example 4.3. To decouple the effect of DSC completely, a sinusoidal function is used for u as done in Example 4.2. After solving COP (4.63) for the $\gamma = 0.333$ and $\varepsilon = 0.1$, the new observer gain matrix is used for estimation and the simulation result for $c = 0.3$ is shown in Fig. 4.9. With a larger magnitude of the elements of L, the performance is slightly improved, especially for the estimation of θ_l. Consequently, the overall performance of ODSC is also improved due to more accurate estimation.

Chapter 5
Constrained Stabilization

In most of applications, it is natural to assume that actuators have limited performance, e.g., limited magnitude and/or response speed of an output value generated by the actuator, nonlinearities due to friction or hysteresis, and measurement noise of sensors embedded in the actuator. This fact motivates the stabilization problem of linear systems subject to actuator saturation. For example, based on quadratic stabilization and invariant set theory, semi-global stabilization of the linear system subject to actuator saturation has been studied in [38, 41]. Related research can be found in a survey paper [8] and its references. While there are many analysis and synthesis methods for uncertain linear systems considering actuator saturation in the literature [8, 9, 56], a few studies have been found for a nonlinear system subject to actuator saturation which guarantees the stability of the closed-loop system via a nonlinear controller.

This chapter is arranged in three sections. In Sect. 5.1, a nonlinear system is considered and a preliminary DSC design procedure is presented. Section 5.2 includes the main results and it will be shown how to estimate the initial condition set which guarantees quadratic stability and satisfies constrained conditions of the control input via an LMI approach. In Sect. 5.3, it will be shown how the system with model uncertainties is stabilized by DSC, even though there is input saturation and a region of attraction is estimated by an ellipsoidal invariant set.

5.1 Problem Statement and Preliminaries

As introduced in (2.3) in Chap. 2, a nonlinear system in strict-feedback form is considered as follows:

$$\begin{aligned}\dot{x}_i &= x_{i+1} + f_i(x_1, \dots, x_i) \quad \text{for } i = 1, \dots, n-1,\\ \dot{x}_n &= u\end{aligned} \tag{5.1}$$

where $f(x)$ and $[\partial f(x)/\partial x]$ are continuous on $\mathscr{D} \in \Re^n$ and the control input u is bounded, i.e., $|u| \le u_0$ where u_0 is a strictly positive constant. Furthermore, suppose that the control objective is $x_1(t) \to 0$.

B. Song, J.K. Hedrick, *Dynamic Surface Control of Uncertain Nonlinear Systems*, Communications and Control Engineering,
DOI 10.1007/978-0-85729-632-0_5,

Although the constrained DSC design methodology is quite similar to unconstrained DSC design discussed in Chap. 2, the difference is that the control input ($u \in \Re$) based on the desired input is applied to the system with a magnitude constraint, which is described by a saturation function. Mathematically, the control input can be written as

$$u = \begin{cases} u_d, & \text{if } |u_d| \le u_0, \\ u_0 \cdot \text{sgn}(u_d), & \text{otherwise} \end{cases} \tag{5.2}$$

where u_d is derived in (2.12) of Chap. 2. Then, the error dynamics of the nonlinear system (5.1) is written as

$$\begin{aligned} \dot{S}_1 &= -K_1 S_1 + S_2 + (x_{2d} - \bar{x}_2), \\ &\vdots \\ \dot{S}_{n-1} &= -K_{n-1} S_{n-1} + S_n + (x_{nd} - \bar{x}_n), \\ \dot{S}_n &= -\dot{x}_n + u = \xi_n / \tau_n + u. \end{aligned} \tag{5.3}$$

It is noted that the control input (u) is separated from the error dynamics to describe input saturation, i.e., u_d in (2.12) is not yet applied to (5.1); thus the equation in the last line of (5.3) is different from the one in (2.14).

Finally, the overall equation of the augmented error dynamics including (5.3) and $(n-1)$th order filter error dynamics in (2.16) can be written using the result of Lemma 2.1 as follows:

$$\begin{aligned} \dot{z} &= A_{op} z + B_u u + B_w w, \\ \|w\| &\le \gamma \|C_z z\| \end{aligned} \tag{5.4}$$

where $z = [S_1 \;\cdots\; S_n \;\xi_2 \;\cdots\; \xi_n]^T \in \Re^{n_z}$, $u \in \Re$,

$$w = [\dot{f}_1 \;\cdots\; \dot{f}_{n-1}]^T = \frac{\partial f}{\partial x}\dot{x} = J(x) C_z z \in \Re^{n_w}, \quad \text{and} \quad \|J(x)\| = \left\| \frac{\partial f}{\partial x} \right\| \le \gamma$$

where γ is a positive Lipschitz constant and C_z is defined in Lemma 2.1. Furthermore, the matrices A_{op} and B_u are

$$A_{op} = \begin{bmatrix} \tilde{A}_{11} & \tilde{A}_{12} \\ T_\xi^{-1} K \tilde{A}_{11} & T_\xi^{-1}(K \tilde{A}_{12} + A_{22}) \end{bmatrix} \in \Re^{n_z \times n_z},$$

$$B_u = [0 \cdots 0 \; 1 \,\vdots\, 0 \cdots 0]^T \in \Re^{n_z},$$

the corresponding sub-block matrices are

$$\tilde{A}_{11} = \begin{bmatrix} -K_1 & 1 & \cdots & 0 \\ 0 & -K_2 & \ddots & \vdots \\ \vdots & \vdots & \ddots & 1 \\ 0 & 0 & \cdots & 0 \end{bmatrix}, \qquad \tilde{A}_{12} = \begin{bmatrix} 1 & \cdots & 0 \\ \vdots & \ddots & \vdots \\ 0 & \cdots & 1 \\ 0 & \cdots & 1/\tau_n \end{bmatrix},$$

and A_{22}, K, T_ξ, and B_w are defined in Lemma 2.1.

Definition 5.1 The set $\mathscr{I} \subset \Re^{n_z}$ is called *controlled invariant* for the system (5.4) if there exists a feedback control law $\phi : \mathscr{D} \rightarrow \Re$

$$u = \phi\{z(t)\}$$

such that for all $z(0) \in \mathscr{I}$ the solution $z(t) \in \mathscr{I}$ for $t > 0$. That is, $\mathscr{I}$ is positively invariant for the closed-loop system.

Given a continuous function $\Psi : \Re^{n_z} \rightarrow \Re$ and $\rho \in (0, \infty)$ we define the compact and convex set as

$$\mathscr{E}[\Psi, \rho] \triangleq \left\{ z \in \Re^{n_z} : \Psi(z) \leq \frac{1}{\rho} \right\}.$$

If Ψ is a gauge function,[1] then the set $\mathscr{E}[\Psi, \rho]$ is a convex and compact set containing the origin [8].

Definition 5.2 For the given error dynamics (5.4), the convex and compact set $\Omega = \mathscr{E}[\Psi, \rho] \subset \Re^{n_z}$ is said to be a region of attraction if there exists $\beta > 0$ and a control law $u = \phi(z)$ such that for every $z(0) \in \Omega$

$$\Psi\{z(t)\} \leq e^{-\beta t} \Psi\{z(0)\}$$

for all $z \in \Omega$.

Remark 5.1 $z(t)$ goes to the origin ($z = 0 \in \Re^{n_z}$) as $\Psi\{z(t)\} \rightarrow 0$ for $t \rightarrow \infty$ in the set Ω, so $\Omega \subseteq \mathscr{I}$. Consequently, $x(t) \in \mathscr{D}$ also converges to the origin ($x = 0 \in \Re^n$) as $z(x) : \mathscr{D} \rightarrow \Omega \in \Re^{n_z}$ does.

Definition 5.3 For the given error dynamics (5.4), a nonlinear system (5.1) is quadratically stabilizable via DSC if there exist a positive definite matrix P and a control law $u = \phi(z)$ such that

$$(A_{op}z + B_u u + B_w w)^T P z + z^T P (A_{op} z + B_u u + B_w w) < 0 \tag{5.5}$$

holds for all $z \in \Omega$. Furthermore, it is locally quadratically stabilizable via DSC if the above holds for all z in a subset Ω_i of Ω.

From the hypothesis that the control input u is bounded, the input can be described mathematically by use of a saturation function: for instance, see [23, 41] for a linear system. More specifically, the single control input is

$$u = \text{sat}[\dot{x}_{nd} - K_n S_n] = \text{sat}\left[c^T z\right] := \begin{cases} c^T z, & \text{if } |c^T z| \leq u_0, \\ u_0 \cdot \text{sign}(c^T z), & \text{otherwise} \end{cases} \tag{5.6}$$

where $c = [0 \ \cdots \ -K_n \ 0 \ \cdots \ -1/\tau_n]^T \in \Re^{n_z}$ and u_0 is the maximum input of u. Then, the equation for the error dynamics in (5.4) can be rewritten as follows:

$$\begin{aligned} \dot{z} &= \left[A_{op} + \lambda(z) B_u c^T\right] z + B_w w := A_{sat}(\lambda) z + B_w w, \\ \|w\| &\leq \gamma \|C_z z\| \end{aligned} \tag{5.7}$$

[1] $\Psi(-z) = -\Psi(z)$ if and only if it is a norm.

where

$$\lambda(z) = \begin{cases} 1, & \text{if } |c^T z| \le u_0 \\ \frac{|u_0|}{|c^T z|} \operatorname{sign}(c^T z), & \text{otherwise.} \end{cases} \tag{5.8}$$

Remark 5.2 If there is no saturation, $\lambda(z) = 1$, the error dynamics in (5.7) is equivalent to (2.23) in Sect. 2.5. It is also interesting to note that if some degree of saturation is allowed, i.e., $\lambda(z) = \lambda_1 \in (0, 1)$ for the given initial condition, the error dynamics (5.7) is not a single equation, but a linear differential inclusion (LDI) given by

$$\dot{z} = A_{sat}(\lambda) z + B_w w \tag{5.9}$$

where the set $A_{sat}(\lambda) \triangleq \{X \in \Re^{n_z \times n_z} | X = A_{sat}(\lambda^*) \ \forall \lambda^* \text{ satisfying } \lambda_1 \le \lambda^* \le 1\}$.

Let a constraint set $\mathscr{C} \triangleq \{z \in \Re^{n_z} || c^T z| \le u_0\}$, which is compact and convex. Given the nonlinear system (5.1) with DSC and a constrained input, we will analyze and design DSC which guarantees quadratic stabilization of the system (5.7) in a *conservative* sense that a region of attraction $\Omega \subset \mathscr{C}$. Moreover, we will maximize the region where quadratic stability holds for the system although there is some allowable degree of saturation in the control input.

5.2 Local Regulation and Regions of Attraction

Among the positively invariant set candidates in the literature, there are two kinds of families which have had a great deal of attention in control theory: ellipsoidal sets and polyhedral sets. The polyhedral sets are more "flexible" than the ellipsoidal ones, but require a more complex mathematical representation [8]. For simplicity and powerful tools such as the Lyapunov equation and LMI software, ellipsoidal sets will be used to estimate the region of attraction in this chapter.

Let a continuous function $\Psi : \mathscr{D} \to \Re$ be a quadratic function as

$$\Psi(z) := z^T \bar{P} z, \quad \bar{P} > 0.$$

Then, the ellipsoidal sets are defined as

$$\mathscr{E}[\Psi, \rho] = \left\{ z \in \mathscr{D} \subset \Re^{n_z} | z^T \bar{P} z \le \frac{1}{\rho}, \ \forall \rho \in (0, \infty) \right\}. \tag{5.10}$$

The set can be rewritten as

$$\mathscr{E}[P] \triangleq \left\{ z \in \Re^{n_z} | z^T P z \le 1, \ \forall P := \rho \bar{P} > 0 \right\}. \tag{5.11}$$

Moreover, in the domain, $\lambda(z) \in (0, 1]$ is bounded and its lower bound is defined by

$$\lambda_0 := \min_{\forall z \in \mathscr{E}[P]} \lambda(z) \tag{5.12}$$

such that $\lambda_0 \le \lambda(z) \le 1$. The *extended* constraint set $\mathscr{C}(\lambda_0) \subset \mathscr{D}$, in the sense that some degree of saturation is allowed, is defined as follows:

$$\mathscr{C}(\lambda_0) = \{ z \in \Re^{n_z} | \, |c^T z| \le u_0/\lambda_0 \}. \tag{5.13}$$

This is a polyhedral set and equivalent to the constraint set $\mathscr{C}$ when $\lambda_0 = 1$.

The constrained stabilization problem can be cast into the calculation of the largest invariant ellipsoid contained in the polyhedron $\mathscr{C}(\lambda_0)$ where the quadratic stability condition (5.5) holds for the norm-bounded closed-loop system in (5.7). All these conditions can be expressed in terms of LMIs as follows [84]:

Theorem 5.1 *Suppose the augmented error dynamics is given in* (5.7) *for the given set of controller gains,* $\Theta = \{K_1, \dots, K_n, \tau_2, \dots, \tau_n\}$. *The origin is exponentially stable in Lyapunov level set* $\mathscr{E}[P]$ *if there exist* $P > 0$, $\sigma \ge 0$, *and* $z(0)$ *such that*

$$\begin{bmatrix} A_0^T P + P A_0 + \sigma \gamma^2 C_z^T C_z & P B_w \\ B_w^T P & -\sigma I \end{bmatrix} < 0, \tag{5.14}$$

$$\begin{bmatrix} P & \lambda_0 c \\ \lambda_0 c^T & u_0^2 \end{bmatrix} \ge 0, \tag{5.15}$$

$$\begin{bmatrix} 1 & z^T(0) \\ z(0) & P^{-1} \end{bmatrix} \ge 0 \tag{5.16}$$

where $A_0 := A_{op} + \lambda_0 B_u c^T$ *for* $\lambda_0 = 1$.

Proof It is proven that the norm-bounded linear differential inclusion (NLDI) is quadratically stable as long as LMI (5.14) holds (see in Theorem 2.1) within $\mathscr{E}[P] \subset \mathscr{C}(\lambda_0)$. This inclusion can also be written as a matrix inequality (refer to [12, Sect. 7.2.3] and [38])

$$\begin{bmatrix} P^{-1} & (c^T P^{-1})^T \\ c^T P^{-1} & (u_0/\lambda_0)^2 \end{bmatrix} \ge 0.$$

By use of the Schur complement, it is equivalent to LMI (5.15). Finally, the initial state $z(0)$ should be within $\mathscr{E}[P]$, i.e., $z^T(0) P z(0) \le 1$. This is equivalent to the inequality (5.16). □

Remark 5.3 A conservative region of attraction to avoid any saturation of the system can be obtained by solving LMI (5.14) and (5.15) via convex optimization. Furthermore, if there is no input constraint, $\lambda_0 = 1$ for all $z \in \Omega$, and LMI (5.15) and (5.16) are not necessary in Theorem 5.1 to design DSC. Then, Theorem 5.1 is equivalent to Theorem 2.1 in Chap. 2.

Theorem 5.2 *For the given hypotheses in Theorem* 5.1, *the set* $\mathscr{E}[P]$ *is a region of attraction if there exist* $P > 0$, $\sigma \ge 0$, $z(0)$, *and* $\varepsilon > 0$ *such that*

$$\begin{bmatrix} A_0^T P + P A_0 + 2\varepsilon P + \sigma \gamma^2 C_z^T C_z & P B_w \\ B_w^T P & -\sigma I \end{bmatrix} \le 0, \tag{5.17}$$

LMI (5.15) *and* (5.16).

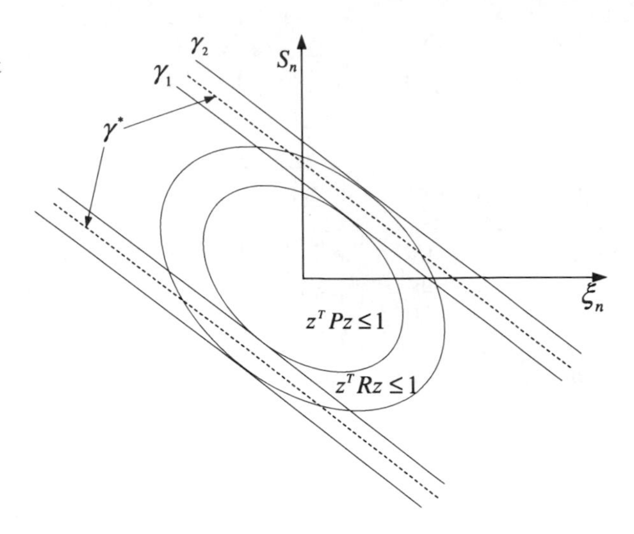

Fig. 5.1 Ellipsoidal initial condition bounds for different λs

Using the fact that the conditions for the quadratic Lyapunov function ($V(z) = z^T Pz$, $dV(z)/dt \leq -2\varepsilon V(z)$) for all trajectories are equivalent to (see in [12])

$$V\{z(t)\} \leq e^{-2\varepsilon t} V\{z(0)\},$$

so that $\|z(t)\| \leq e^{-\varepsilon t}\kappa(P)^{1/2}\|z(0)\|$ for all trajectories and therefore the decay rate of (5.7) is at least ε. we can derive the above conditions as done in Theorem 5.1.

Remark 5.4 If there is an additional constraint such that a specific decay rate ε^* is given, then we need to put another condition $\varepsilon \geq \varepsilon^* > 0$ in the corollary. Furthermore, the decay rate can be maximized by solving the following problem:

$$\begin{array}{ll} \text{maximize} & \varepsilon \\ \text{subject to} & P > 0, \qquad \sigma \geq 0, \qquad \varepsilon > 0, \\ & \text{LMI (5.15)} \quad \text{and} \quad (5.17). \end{array}$$

Next, we will investigate the existence of the quadratic Lyapunov function for the error dynamics, LDI (5.9), when there is saturation. As a first step, the lower bound of $\lambda(z)$, λ_0, will be given sequentially, i.e., a set $\lambda_0 = \{\lambda_0(1), \ldots, \lambda_0(k)\}$ where $\lambda_0(1) = 1$, $\lambda_0(i+1) < \lambda_0(i)$ for $1 \leq i \leq k$, and $\lambda_0(k) > 0$. That is, if there exists a quadratic Lyapunov function for all $\lambda \in [\lambda_0(2), \lambda_0(1)] \subset (0, 1]$, then we will increase a degree of saturation and try to obtain the Lyapunov function for another $\lambda \in [\lambda_0(3), \lambda_0(2)]$. Then, this procedure will be iterated as long as there exists a quadratic Lyapunov function for all $\lambda \in [\lambda_0(k), \lambda_0(k-1)]$. Now an interesting question is what conditions guarantee the existence of a quadratic Lyapunov function for all $\lambda \in [\lambda_2, \lambda_1] = [\lambda_0(i+1), \lambda_0(i)] \subset (0, 1]$, under an assumption that there exist solutions P_1 and P_2 satisfying Theorem 5.1 for λ_1 and λ_2, respectively (see in Fig. 5.1).

Finally, if the existence of the Lyapunov function for all $\lambda \in [\lambda_0(i+1), \lambda_0(i)]$ is guaranteed, then $\mathscr{E}(R) \supset \mathscr{E}(P)$ (see two ellipsoids in Fig. 5.1) is an *enlarged* positively invariant initial condition set in the sense that saturation is considered. The conditions for the existence are considered in the following theorem [84].

Theorem 5.3 *Suppose there exist* $\{P_1, \sigma_1, \lambda_1\}$ *and* $\{P_2, \sigma_2, \lambda_2\}$ *satisfying all LMIs in Theorem* 5.1 *where* $\lambda_1 > \lambda_2$ *(see also in Fig.* 5.1*). Furthermore, if* P_1, σ_1, λ_1, *and* λ_2 *satisfy*

$$\begin{bmatrix} A_0^T P + P A_0 + (\lambda_1 - \lambda_2) c c^T + \sigma \gamma^2 C_z^T C_z & P B_u & P B_w \\ B_u^T P & -\frac{1}{\lambda_1 - \lambda_2} & 0 \\ B_w^T P & 0 & -\sigma I \end{bmatrix} < 0 \tag{5.18}$$

where $A_0 := A_{op} + \lambda_1 B_u c^T$. *Then, there exist* $\tilde{P}(\lambda)$ *and* $\tilde{\sigma}(\lambda)$ *satisfying Theorem* 5.1 *for any* $A_0(\lambda) = A_{op} + \lambda B_u c^T$, $\lambda \in [\lambda_2, \lambda_1]$.

The proof is given in Appendix A.4. It is interesting to note that a solution of the condition (5.14) will be close to that of the condition (5.18) when an interval of the sequence $\lambda_0(k)$ is small enough. Based on the results in this section, the algorithm to estimate the positively invariant initial condition set is proposed as follows:

Algorithm 5.1 Calculation of region of attraction

Step 1. Assign controller gains and filter time constants such that there exists a solution for Theorem 5.1 as well as A_0 is Hurwitz for $\lambda_0 = 1$.
Step 2. Find the smallest $\lambda_0(k)$ for $[\lambda_0(k), \lambda_0(k-1)] \in (0, 1]$ such that

$$\begin{array}{ll} \text{minimize} & \operatorname{Tr}(P) \\ \text{subject to} & P > 0, \quad \Sigma \geq 0, \quad \text{LMI (5.14)}, \quad (5.15), \quad \text{and} \quad (5.18). \end{array} \tag{5.19}$$

Example 5.1 Consider the following example with an input constraint

$$\begin{aligned} \dot{x}_1 &= x_2 + x_1^2, \\ \dot{x}_2 &= u \end{aligned} \tag{5.20}$$

where $|u| \leq u_0$ and $u_0 = 10$. Suppose the objective of a controller is to drive $x \to 0 \in \Re^2$ in $\mathscr{D} = \{x \in \Re^2 | \|x\| \leq 1\}$.

As discussed in Sect. 5.1, the equation of the error dynamics with DSC is

$$\begin{aligned} \dot{z} &= \left[A_{op} + \lambda(z) B_u c^T\right] z + B_w w \\ &= A_{sat} z + B_w w \end{aligned} \tag{5.21}$$

where $z = [S_1 \ S_2 \ \xi_2]^T \in \Re^3$, $w = \dot{f}_1 \in \Re$, and matrices A_{op}, B_w, B_u, and C_u are derived as follows:

$$A_{op} = \left[\begin{array}{cc|c} -K_1 & 1 & 1 \\ 0 & 0 & 1/\tau_2 \\ \hline -K_1^2 & K_1 & K_1 - 1/\tau_2 \end{array}\right],$$

$$B_w = \left[\begin{array}{c} 0 \\ 0 \\ \hline 1 \end{array}\right], \qquad B_u = \left[\begin{array}{c} 0 \\ 1 \\ \hline 0 \end{array}\right], \qquad c = \left[\begin{array}{c} 0 \\ -K_2 \\ \hline -1/\tau_2 \end{array}\right],$$

$$A_{sat} = \left[\begin{array}{cc|c} -K_1 & 1 & 1 \\ 0 & -\lambda(z)K_2 & \frac{1-\lambda(z)}{\tau_2} \\ \hline -K_1^2 & K_1 & K_1 - 1/\tau_2 \end{array}\right].$$

Furthermore, w is bounded by

$$\|w\| \leq \gamma \|C_z z\|$$

where $C_z = [-K_1 \ 1 \ 1]$ and γ is considered to be 2, since

$$|J| = \left|\frac{\partial x_1^2}{\partial x_1}\right| = |2x_1| \leq 2$$

for all $x \in \mathscr{D}$.

When the controller gains and filter time is assigned as $\{K_1, K_2, \tau_2\} = \{2, 10, 0.05\}$, the eigenvalues of $A_{op} + B_u c^T$ are $\{-2.254, -10, -17.746\}$. If the following convex problem is solved numerically (refer to MATLAB Program 5-1)

$$\begin{array}{ll} \text{minimize} & \text{Tr}(P) \\ \text{subject to} & P > 0, \quad \sigma \geq 0, \quad \text{LMI (5.14) and (5.15)}, \end{array} \tag{5.22}$$

the positive definite matrix P_1 is

$$P_1 = \begin{bmatrix} 1.3199 & -0.1681 & 0.4043 \\ -0.1681 & 1.2569 & 1.8531 \\ 0.4043 & 1.8531 & 4.1625 \end{bmatrix}. \tag{5.23}$$

MATLAB Program 5-1

```
%***** Define parameters and the dimension of matrices *****
gamma = 2;
lambda0 = 1;
n = size(A0, 1);
m = size(Bw, 2);

%***** cvx version *****
cvx_begin
  variable P(n,n) symmetric;
  variable sigma;
  minimize(trace(P));
  sigma >= 0;
  P == semidefinite(n);
  -[A0'*P + P*A0 + sigma*gamma*gamma*Cz'*Cz P*Bw;
  Bw'*P -sigma*eye(m)] == semidefinite(n+m);
  [P lambda0*c; lambda0*c' u0*u0] == semidefinite(n+1);
cvx_end
```

When an initial condition is given as $x(0) = [-0.7071, \ 0.7071]^T \in \mathscr{D}$, time responses of x and u are shown in Fig. 5.2(d) and the corresponding z is projected onto 2-D space as shown in Figs. 5.2(a)–(c). Furthermore, it is validated graphically that the given initial condition satisfies $z^T(0)P_1 z(0) \leq 1$. In Fig. 5.2(c), two solid lines represent $|c^T z| = u_0$ where the control input u reaches input saturation and it is also shown that there is no saturation for the given initial condition. Therefore, by the result of Theorem 5.1, the calculated ellipsoidal invariant set

$$\mathscr{E}[P_1] = \left\{ z \in \Re^{n_z} | z^T P_1 z \leq 1 \right\}$$

where P_1 is given in (5.23) is a region of attraction in the z coordinate.

When COP (5.19) is solved iteratively with inverse order of logarithmic spacing of $\lambda_0(k) \in [10^{-1},\ 1]$ for $k = 1, \ldots, 40$ (refer to MATLAB Program 5-2), there exists a solution of COP (5.22) up to $\lambda_0(k) \in [0.1914,\ 0.1805]$ for $k = 29$ and the solution P_2 of COP (5.22) for $\lambda_0 = 0.1805$ is

$$P_2 = \begin{bmatrix} 0.6658 & -0.1842 & -0.0862 \\ -0.1842 & 0.0957 & 0.0616 \\ -0.0862 & 0.0616 & 0.2028 \end{bmatrix}. \tag{5.24}$$

When the initial condition $x(0) = [0.6\ 0.1]^T$ is given, time responses of x and u are shown in Fig. 5.3(d). It is noted that there is input saturation in u during the time less than 0.1 (s). Figures 5.3(a)–(c) show that a larger region of attraction is estimated by $\mathscr{E}[P_2]$ (compare them with those in Fig. 5.2). Furthermore, since $z(0) \in \mathscr{E}[P_2]$, it is expected that the error goes to the origin and this is validated via simulation in Fig. 5.3.

MATLAB Program 5-2

```
%***** Define parameters and the dimension of matrices *****
gamma = 2;
lambda0 = 1;
n = size(A0, 1);
m = size(Bw, 2);

%***** cvx version *****
cvx_begin
   variable P(n,n) symmetric;
   variable sigma;
   minimize(trace(P));
   sigma >= 0;
   P == semidefinite(n);
   -[A0'*P + P*A0 + sigma*gamma*gamma*Cz'*Cz P*Bw;
   Bw'*P -sigma*eye(m)] == semidefinite(n+m);
   -[A0'*P + P*A0+(lambda1-lambda2)*c*c'+sigma*gamma*gamma*Cz'*Cz P*Bu
   P*Bw;
   Bu'*P -1/(lambda1-lambda2) 0;Bw'*P 0 -sigma*eye(m)] == semidefinite(n+m+1);
   [P lambda0*c; lambda0*c' u0*u0] == semidefinite(n+1);
cvx_end
```

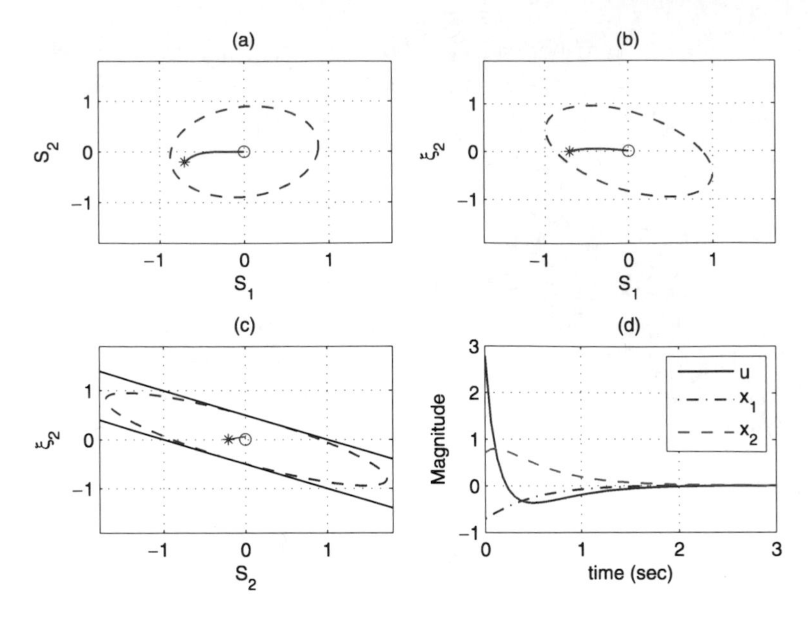

Fig. 5.2 Estimate of a region of attraction and time responses of x and u for $x(0) = [-0.7071,\ 0.7071]^T$

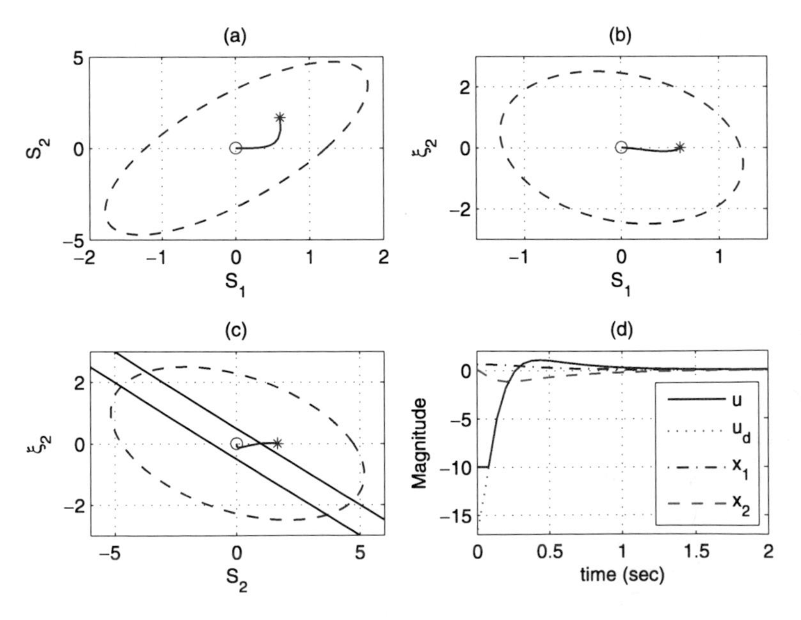

Fig. 5.3 Estimate of a region of attraction and time responses of x and u for $x(0) = [0.6, 0.1]^T$

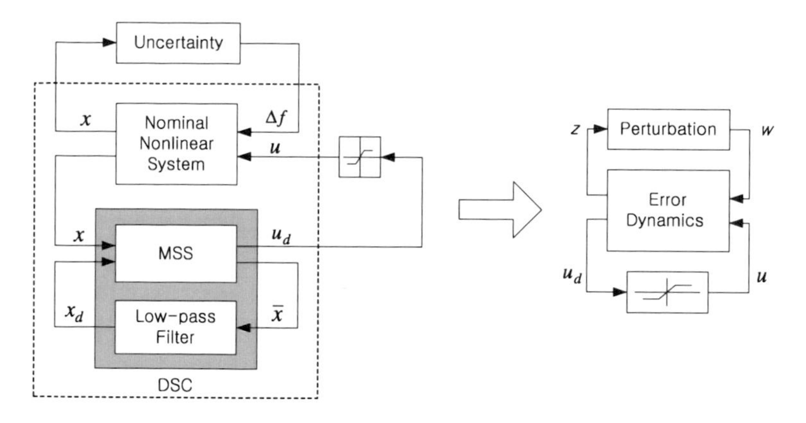

Fig. 5.4 Schematic of DSC with an input constraint and closed-loop error dynamics

5.3 Robust Constrained Stabilization

As introduced in (3.1) in Chap. 3, the strict-feedback nonlinear system with unmatched nonlinearities is considered as follows:

$$\begin{aligned} \dot{x}_i &= x_{i+1} + f_i(x_1, \ldots, x_i) + \Delta f_i(x_1, \ldots, x_i) \quad \text{for } i = 1, \ldots, n-1, \\ \dot{x}_n &= u \end{aligned} \tag{5.25}$$

where the nonlinear function f_i and model uncertainty Δf_i are assumed to be globally Lipschitz in $\mathscr{D}$ as assumed in Chap. 3. Moreover, the additional assumption is that control input u is bounded, i.e., $|u| \le u_0$ where u_0 is a strictly positive constant.

A graphical representation of the structural DSC design procedure for the nonlinear systems with input saturation is shown in Fig. 5.4. As derived in (5.4), the overall equations of the error dynamics including the nonlinear system (5.25) and $n-1$ low-pass filter equations can be written as follows:

$$\begin{aligned} \dot{z} &= A_{op} z + B_u u + B_w w + B_\Delta \Delta f = A_{op} z + B_u u + [B_\Delta \ B_w] \begin{bmatrix} \Delta f \\ w \end{bmatrix} \\ &= A_{op} z + B_u u + B_{\Delta w} w_u \end{aligned} \tag{5.26}$$

where $f_\Delta = [\Delta f_1 \ \cdots \ \Delta f_{n-1}]^T \in \Re^{n_\Delta}$ and

$$B_\Delta = \begin{bmatrix} I_{nn_\Delta} \\ T_\xi^{-1} K \end{bmatrix} \in \Re^{n_z \times n_\Delta}.$$

Since the model uncertainty is assumed to be globally Lipschitz, by the result of Lemma 3.1, there exists a matrix $\tilde{C}_i$ such that

$$|w_{ui}| \le \left\| \tilde{C}_i z \right\| \tag{5.27}$$

for $i = 1, \ldots, n_{wu}$. This result is also addressed in (3.12) in Remark 3.1. Therefore, the augmented error dynamics in (5.26) with componentwise inequality constraints (5.27) can be considered as a diagonal norm-bounded linear differential inclusion (DNLDI). As discussed in Chap. 3, all results in Sect. 5.2 can be extended by replacing $\sigma\gamma^2 C_z^T C_z$ and $-\sigma I$ by $\tilde{C}_i^T \Sigma_B \tilde{C}_i$ and $-\Sigma$ where $\Sigma = \mathrm{diag}(\sigma_1, \ldots, \sigma_{n_{wu}})$ is a diagonal matrix and $\Sigma_B = \mathrm{diag}(\sigma_1 I, \ldots, \sigma_{n_{wu}} I)$ is a block diagonal matrix.

Example 5.2 Consider the nonlinear system with uncertainty as well as an input constraint in Example 5.1:

$$\begin{aligned}\dot{x}_1 &= x_2 + (1+\Delta)x_1^2 = x_2 + f_1(x_1) + \Delta f_1(x_1),\\ \dot{x}_2 &= u\end{aligned}$$

where $f_1 = x_1^2$ and Δ represents a parametric uncertainty satisfying $|\Delta| < 0.5$. Suppose the objective of a controller is to drive $x \to 0 \in \Re^2$ in $\mathscr{D} = \{x \in \Re^2 | \|x\| \le 1\}$.

As discussed in Sect. 5.1, the equation of error dynamics with DSC is

$$\begin{aligned}\dot{z} &= \left[A_{op} + \lambda(z) B_u K\right] z + B_w w + B_\Delta \Delta f_1\\ &= A_{sat} z + B_{\Delta w} w_u\end{aligned}$$

where $w_u = [\Delta f_1 \ \dot{f}_1]^T \in \Re^2$,

$$B_{\Delta w} = [B_\Delta \ B_w] = \begin{bmatrix} 1 & 0 \\ 0 & 0 \\ \hdashline K_1 & 1 \end{bmatrix},$$

and the other matrices and vectors are defined in Example 5.1. Furthermore, the componentwise upper bound of w_u is derived as

$$|w_u| = \left|\begin{bmatrix} \Delta f_1 \\ \dot{f}_1 \end{bmatrix}\right| = \left|\begin{bmatrix} \Delta x_1^2 \\ 2x_1\dot{x}_1 \end{bmatrix}\right| \le \left|\begin{bmatrix} 0.5x_1 \\ 2\dot{x}_1 \end{bmatrix}\right| = \left|\begin{bmatrix} \tilde{C}_1 z \\ \tilde{C}_2 z \end{bmatrix}\right| := \left|\tilde{C} z\right| \tag{5.28}$$

for all $x \in \mathscr{D}$ where

$$\begin{aligned}\tilde{C}_1 &= 0.5[1\ 0\ 0] \quad \text{and} \quad \tilde{C}_2 = 2C_2 = 2\sqrt{2}\begin{bmatrix} \tilde{C}_1 \\ C_z \end{bmatrix} \quad \text{where } C_z = [-K_1\ 1\ 1],\\ \dot{x}_1 &= (x_2 - x_{2d}) + (x_{2d} - \bar{x}_2) + \bar{x}_2 + f_1 + \Delta f_1 = S_2 + \xi_2 - K_1 S_1 + \Delta f_1\\ &:= C_z z + \Delta f_1,\\ |\dot{x}_1| &\le |C_z z| + |\Delta f_1| \le |C_z z| + \left|\tilde{C}_1 z\right| = \left\| \begin{bmatrix} \tilde{C}_1 \\ C_z \end{bmatrix} z \right\|_1 \le \|C_2 z\|_2.\end{aligned}$$

It is noted that the inequality in (5.28) is defined componentwise and the matrix C_2 is calculated by use of the equivalence of norms as

$$C_2 := \sqrt{2}\begin{bmatrix} \tilde{C}_1 \\ C_z \end{bmatrix}.$$

Suppose a set of controller gains and a filter time constant is assigned as

$$\{K_1, K_2, \tau_2\} = \{2, 10, 0.05\}.$$

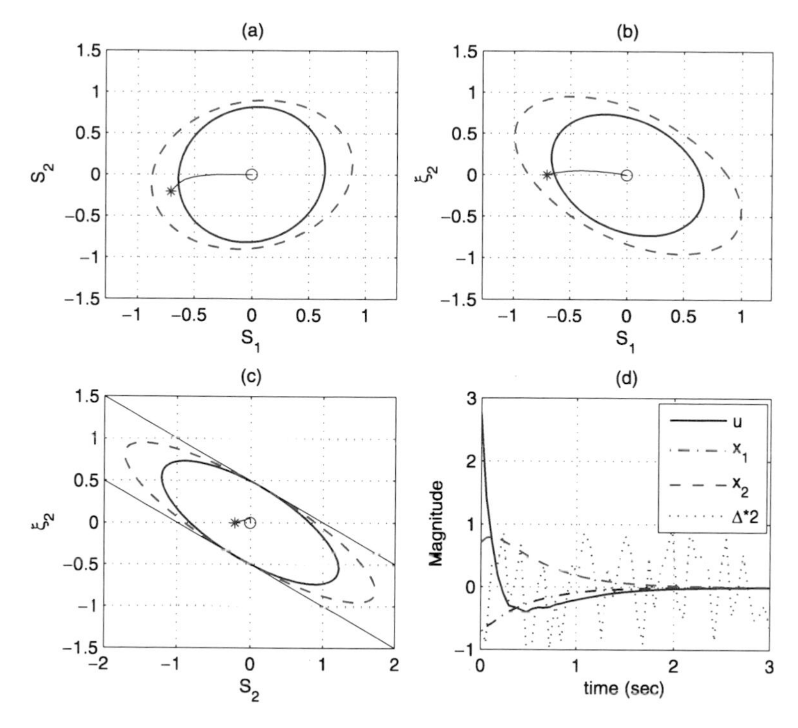

Fig. 5.5 Estimate of a region of attraction and time responses of states and a control input for $x(0) = [-0.7071, 0.7071]^T$

If the following convex problem is solved numerically for $\lambda_0 = 1$

$$\begin{array}{ll} \text{minimize} & \mathrm{Tr}(P) \\ \text{subject to} & P > 0, \qquad \Sigma \geq 0, \quad \text{LMI (5.15)}, \quad \text{and} \\ & \begin{bmatrix} A_0^T P + P A_0 + \tilde{C}^T \Sigma_B \tilde{C} & P B_w \\ B_w^T P & -\Sigma \end{bmatrix} < 0 \end{array} \tag{5.29}$$

where $\Sigma = \mathrm{diag}(\sigma_1,\ \sigma_2) \in \Re^{2\times 2}$ and $\Sigma_B = \mathrm{diag}(\sigma_1,\ \sigma_2 I) \in \Re^{3\times 3}$, the positive definite matrix P_3 is

$$P_3 = \begin{bmatrix} 2.4382 & -0.1497 & 0.4570 \\ -0.1497 & 1.5029 & 1.7906 \\ 0.4570 & 1.7906 & 4.1523 \end{bmatrix}. \tag{5.30}$$

Figures 5.5(a)–(c) show the projections of $\mathscr{E}(P_3)$ in (5.30) (see a solid line of ellipsoids in the figure) and they are compared with ones of $\mathscr{E}(P_1)$ calculated in Example 5.1 (see a dashed line of ellipsoids in the figure). Since the uncertainty Δf_1 is considered, thus a larger upper bound of w_u is used, it is expected that a smaller $\mathscr{E}(P)$ to guarantee the quadratic stability is calculated. Therefore, when $\{x_1(0), x_2(0)\} = \{-0.7071, 0.7071\} \in \mathscr{D}$, the corresponding $z(0)$ is in $\mathscr{E}(P_1)$, not in $\mathscr{E}(P_3)$ while the $z(0)$ (see a mark $*$ in Fig. 5.5) is in the input constraint set

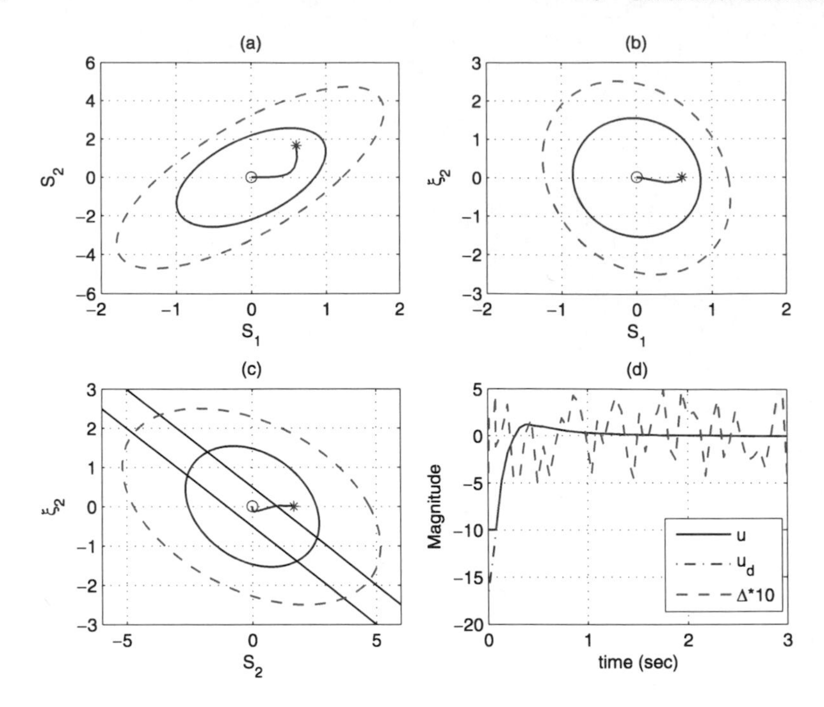

Fig. 5.6 Estimate of a region of attraction and time responses of states and a control input for $x(0) = [0.6, 0.1]^T$

$\{z \in \Re^3 | |c^T z| \leq 10\}$ which graphically represents an area between two solid straight lines in Fig. 5.5(c). However, it is shown in Fig. 5.5(d) that the closed-loop system is stabilized without saturation even if the parametric uncertainty Δ is changed as shown in Fig. 5.5(d).

The above result motivates us to expand $\mathscr{E}(P)$ by allowing a degree of saturation. If the following convex optimization problem is solved iteratively with inverse order of logarithmic spacing of $\lambda_0(k) \in [0.1, \ 1]$ for $k = 1, \ldots, 40$ (refer to Example 5.1):

$$\begin{aligned} &\text{minimize} \quad \mathrm{Tr}(P) \\ &\text{subject to} \quad P > 0, \quad \Sigma \geq 0, \quad \text{LMI (5.15) and (5.29), and} \\ &\begin{bmatrix} A_0^T P + P A_0 + (\lambda_1 - \lambda_2) c c^T + C_z^T \Sigma_B C_z & P B_u & P B_w \\ B_u^T P & -\frac{1}{\lambda_1 - \lambda_2} & 0 \\ B_w^T P & 0 & -\Sigma \end{bmatrix} < 0, \end{aligned} \tag{5.31}$$

there exists a solution of COP (5.31) up to $\lambda_0(k) \in [\lambda_2(k), \lambda_1(k)] = [0.2894, 0.3070]$ for $k = 21$ and the solution P_4 of COP (5.29) for $\lambda_0 = 0.2894$ is

$$P_4 = \begin{bmatrix} 1.3862 & -0.2821 & -0.0931 \\ -0.2821 & 0.2096 & 0.0943 \\ -0.0931 & 0.0943 & 0.4656 \end{bmatrix}. \tag{5.32}$$

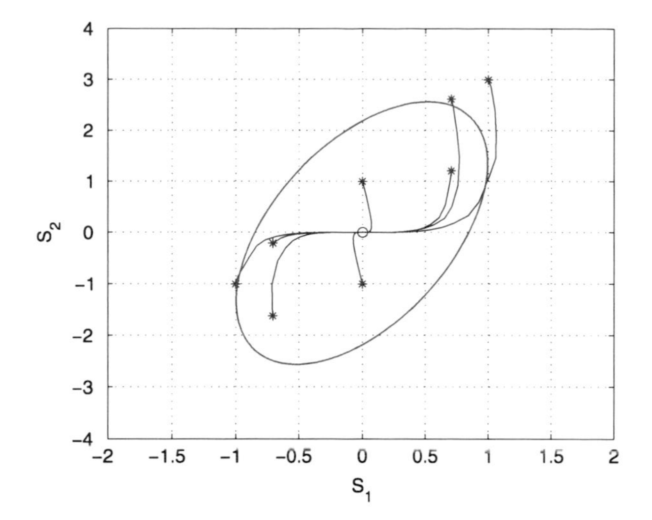

Fig. 5.7 Phase portrait for a set of initial conditions $x(0) \in \mathscr{D}$

Figure 5.6 shows the projections of $\mathscr{E}(P_4)$ in the shape of an ellipse and they are compared with ones of $\mathscr{E}(P_2)$ which is given in (5.24) of Example 5.1. As mentioned above, it is shown that a smaller ellipsoid is calculated due to the inclusion of the uncertainty. However, it is large enough so that $z(0)$ corresponding to $x(0) = [-0.7107,\ 0.7107]^T$ is in $\mathscr{E}(P_4)$. Furthermore, when $\{x_1(0), x_2(0)\} = \{0.6, 0.1\} \in \mathscr{D}$ and Δ is given in Fig. 5.6(d), it is shown that the system is stabilized although there is saturation initially about 0.2 second—refer to u and u_d in Fig. 5.6(d).

Finally, a set of eight different initial conditions on $\mathscr{D}$ are assigned as

$$x(0) = [\pm 1, 0]^T, \qquad [0, \pm 1]^T, \quad \text{and} \quad \left[\pm 1/\sqrt{2}, \pm 1/\sqrt{2}\right]^T.$$

The corresponding trajectories of the state x are shown in Fig. 5.7. Three initial conditions, $[\pm 1, 0]^T]$ and $[1/\sqrt{2}, 1/\sqrt{2}]^T$, are placed outside $\mathscr{E}(P_4)$ while the system is stabilized for all given initial conditions. It is implied that a *conservative* region of attraction $\mathscr{E}(P)$ is calculated in the sense that a worst case scenario is considered, thus the upper bound of w_u is used.

Chapter 6
Multi-Input Multi-Output Mechanical Systems

Interest in the development of autonomous systems such as biped robots, autonomous vehicles, and unmanned aerial vehicles is increasing and more effort has been devoted on the control problem of interconnected mechanical systems. In general, interconnected mechanical systems have been modeled as a class of nonlinear systems with multi-inputs and multi-outputs [69, 72]. In this chapter, it will be discussed how the design methodology of DSC for a class of single-input single-output (SISO) nonlinear systems developed in Chap. 2 can be extended to a multi-input multi-output (MIMO) nonlinear system. Later the results of this chapter will be applied to a biped robot well known as an interconnected mechanical system in Chap. 9. Only a few nonlinear control techniques with applications to biped walking can be found in the literature due to the mathematical complexity of the dynamic model and none of the control approaches has proven stable walking of the nonlinear controller based on biped robot models. Consequently, the complexity of nonlinear control for biped walking challenges us to come up with systematic analysis and design procedures to meet both stability and performance.

If the given system is a large-scale complex nonlinear system, the corresponding controller is in general complicated, thus resulting in difficulties in analysis and design of the controller. Another control approach to deal with this problem is a decentralized control approach which is motivated to design a set of local controllers to reduce the burden of computation in the controller and information communication among subsystems in the 1970s (e.g., see in [78, 96] and references therein). More specifically, while decentralized adaptive control has been developed for the linear system subject to nonlinear interactions among subsystems [27, 31, 42], the decentralized adaptive output feedback design has been proposed in the literature for a system of which all states are not measured [46, 48, 114]. Significant progress in the area of decentralized control has been made during the last three decades, and the decentralized approaches for a class of nonlinear systems were proposed in the 1990s. For instance, decentralized adaptive control techniques for interconnected nonlinear subsystems have been found in the literature [45, 105]. However, as addressed in Chap. 1, most of the work is based on integrator backstepping technique whose complexity is due to an "explosion of terms" that may result in difficulties for control synthesis.

B. Song, J.K. Hedrick, *Dynamic Surface Control of Uncertain Nonlinear Systems*, Communications and Control Engineering,
DOI 10.1007/978-0-85729-632-0_6,

The remainder of this chapter is organized into four sections. MIMO mechanical systems are introduced in Sect. 6.1 and a synthesis method of DSC gains for a fully-actuated holonomic system of N particles is provided in Sect. 6.2. Among the feasible sets of controller gains satisfying quadratic stabilization or input-to-output properties, a specific set of gains are determined by minimizing their magnitude via convex optimization. Then, the design methodology of DSC developed in Sect. 6.2 is extended to the rigid body dynamics in Sect. 6.3. Finally, a decentralized approach of DSC is introduced for a class of nonlinear systems in Sect. 6.4.

6.1 Fully-Actuated Mechanical System

Consider an n degrees of freedom dynamical system with generalized coordinates $q \in \Re^n$ and external forces $Q \in \Re^n$. The standard form of Lagrange's equation for the system is written as

$$\frac{d}{dt}\left(\frac{\partial \mathbf{L}}{\partial \dot{q}_i}(q, \dot{q})\right) - \frac{\partial \mathbf{L}}{\partial q_i}(q, \dot{q}) = Q \tag{6.1}$$

where

$$\mathbf{L}(q, \dot{q}) = \mathbb{T}(q, \dot{q}) - \mathbb{V}(q)$$

is the Lagrangian function, $\mathbf{T}(q, \dot{q})$ is the kinetic energy function which is assumed to be of the form

$$\mathbf{T}(q, \dot{q}) = \frac{1}{2}\dot{q}^T D(q)\dot{q},$$

where $D(q) = D^T(q) \in \Re^{n \times n}$ is the generalized inertia matrix, and $\mathbf{V}(q)$ is the potential function. If three types of external forces are considered such that

$$Q = -\frac{\partial F(\dot{q})}{\partial \dot{q}} + Q_d + Mu$$

where $F(\dot{q})$ is the Rayleigh dissipation function which by definition satisfies

$$\dot{q}^T \frac{\partial F(\dot{q})}{\partial \dot{q}} \geq 0,$$

Q_d is the exogenous input due to the effect of disturbances, and Mu represent the action of control, the equivalent equation of motion of (6.1) is derived in [68] as follows:

$$D(q)\ddot{q} + H(q, \dot{q})\dot{q} + g(q) + \frac{\partial F(\dot{q})}{\partial \dot{q}} = Mu + Q_d \tag{6.2}$$

where $H(q, \dot{q})$ is related to centrifugal and Coriolis forces and $g(q)$ is defined by

$$g(q) = \frac{\partial \mathbf{V}(q)}{\partial q}.$$

If it is assumed that the system is a fully-actuated system with no internal damping and exogenous input, (6.2) can be simplified as

$$D(q)\ddot{q} + H(q,\dot{q})\dot{q} + g(q) = u. \tag{6.3}$$

Then, the equation of motion in (6.3) can be written in state-space form as follows:

$$\begin{cases} \dot{x}_1 = x_2, \\ \dot{x}_2 = D^{-1}(x_1)\{u - H(x_1,x_2)x_2 - g(x_1)\} \end{cases} \tag{6.4}$$

where $x_1 = q \in \Re^n$ and $x_2 = \dot{q} \in \Re^n$ and D is always invertible because $D(x_1)$ is a positive definite matrix.

Furthermore, if a holonomic system of n particles is considered, the Lagrange's equation in the standard form can be written in a more explicit form of the equation of motion as follows [34]:

$$\ddot{q}_i + f_i(q,\dot{q}) = u_i \quad (i = 1,2,\ldots,n)$$

where q, $\dot{q} \in \Re^n$, $f_i : \Re^n \times \Re^n \to \Re$ is the nonlinear function and $u_i \in \Re$ is the generalized force including a control input. The resulting equations of motion for the MIMO nonlinear system are of the state-space form

$$\frac{d}{dt}\begin{bmatrix} x_{1i} \\ x_{2i} \end{bmatrix} = \begin{bmatrix} x_{2i} \\ -f_i(x_1,x_2) \end{bmatrix} + \begin{bmatrix} 0 \\ 1 \end{bmatrix} u_i \tag{6.5}$$

where $x_{1i} = q_i$ and $x_{2i} = \dot{q}_i$.

6.2 Synthesis of Dynamic Surface Control

For the clarity of the mathematical derivation, the simpler system shown in (6.5) is considered for synthesis of a DSC technique in this section [86]. Then, the proposed analysis and design method will be extended to a holonomic system in (6.4) composed of n rigid bodies in Sect. 6.3.

6.2.1 Error Dynamics

Suppose the reference input, $\mathbf{x}_d = [x_{d1} \ \cdots \ x_{dn}]^T \in \Re^n$, is given for the system in (6.5). Let us use the design procedure introduced in Sect. 2.3 for the ith subsystem. First, define the first error surface as $S_{1i} := x_{1i} - x_{di}$ for $1 \le i \le n$. After differentiating the error surface and using (6.5), we get

$$\dot{S}_{1i} = \dot{x}_{1i} - \dot{x}_{di} = x_{2i} - \dot{x}_{di}.$$

Sequentially, the synthetic input x_{2i} and the first-order lowpass filter is derived as follows:

$$\bar{x}_{2i} = \dot{x}_{di} - K_{1i}S_{1i}, \tag{6.6}$$

$$\tau_{2i}\dot{x}_{2i,d} + x_{2i,d} = \bar{x}_{2i}, \quad x_{2i,d}(0) := \bar{x}_{2i}(0) \tag{6.7}$$

where $K_{1i} > 0$ and $\tau_{2i} > 0$ are the controller gains chosen later to guarantee quadratic stability and boundedness.

Next, after defining $S_{2i} := x_{2i} - x_{2i,d}$ and differentiating it

$$\dot{S}_{2i} = \dot{x}_{2i} - \dot{x}_{2i,d} = -f_i + u_i - \dot{x}_{2i,d},$$

the desired control input is chosen as

$$u_i = f_i + \dot{x}_{2i,d} - K_{2i} S_{2i} \tag{6.8}$$

where $\dot{x}_{2i,d}$ is calculated as $\dot{x}_{2i,d} = (\bar{x}_{2i} - x_{2i,d})/\tau_{2i}$ using (6.7).

As explained in Sect. 2.4, a set of augmented error dynamics for the closed-loop system is derived as follow:

$$\begin{cases} \dot{x}_{1i} = (x_{2i} - x_{2i,d}) + (x_{2i,d} - \bar{x}_{2i}) + \bar{x}_{2i} = S_{2i} + \xi_{2i} + \dot{x}_{di} - K_{1i} S_{1i}, \\ \dot{x}_{2i} = \dot{x}_{2i,d} - K_{2i} S_{2i}, \\ \dot{\xi}_{2i} = \frac{d}{dt}(x_{2i,d} - \bar{x}_{2i}) = -\frac{1}{\tau_{2i}} \xi_{2i} - \ddot{x}_{di} + K_{1i} \dot{S}_{1i}. \end{cases} \tag{6.9}$$

Then, we have the augmented error dynamics in the form of block matrices as follows:

$$T_i \dot{\mathbf{z}}_i = \bar{A}_i \mathbf{z}_i + B_i \ddot{x}_{di} \tag{6.10}$$

where $\mathbf{z}_i := [S_{1i}, S_{2i}, \xi_{2i}]^T \in \Re^3$ and

$$T_i = \begin{bmatrix} 1 & 0 & 0 \\ 0 & 1 & 0 \\ -K_{1i} & 0 & 1 \end{bmatrix},$$

$$\bar{A}_i = \begin{bmatrix} -K_{1i} & 1 & 1 \\ 0 & -K_{2i} & 0 \\ 0 & 0 & -1/\tau_{2i} \end{bmatrix}, \quad \text{and} \quad B_i = \begin{bmatrix} 0 \\ 0 \\ -1 \end{bmatrix}.$$

Therefore, the augmented error dynamics in (6.10) is rewritten as

$$\dot{\mathbf{z}}_i = A_i \mathbf{z}_i + B_i \ddot{x}_{di} \quad (i = 1, 2, \ldots, n) \tag{6.11}$$

where

$$A_i = T_i^{-1} \bar{A}_i = \begin{bmatrix} 1 & 0 & 0 \\ 0 & 1 & 0 \\ K_{1i} & 0 & 1 \end{bmatrix} \bar{A}_i = \begin{bmatrix} -K_{1i} & 1 & 1 \\ 0 & -K_{2i} & 0 \\ -K_{1i}^2 & K_{1i} & K_{1i} - 1/\tau_{2i} \end{bmatrix}.$$

Remark 6.1 Using terminologies in [12], the error dynamics (6.11) can be considered as a polytopic linear differential inclusion (PLDI) which is a set of linear time-invariant systems subject to an exogenous input. This fact motivates us to analyze stability and design a set of controller gains, $\{K_{1i}, K_{2i}, \tau_{2i}\}$, via convex optimization.

6.2.2 Synthesis for Stabilization

6.2.2.1 Stability Analysis

If either a stabilization problem or a set-point regulation problem is considered with $\ddot{x}_{di} = 0$, the error dynamics in (6.11) becomes

$$\dot{\mathbf{z}}_i = A_i \mathbf{z}_i. \tag{6.12}$$

Suppose the Lyapunov function candidate is

$$V(\mathbf{z}) = \mathbf{z}^T \mathbf{P} \mathbf{z} = \begin{bmatrix} \mathbf{z}_1^T & \cdots & \mathbf{z}_n^T \end{bmatrix} \operatorname{diag}(P_1, \ldots, P_n) \begin{bmatrix} \mathbf{z}_1 \\ \vdots \\ \mathbf{z}_n \end{bmatrix} = \sum_{i=1}^{n} \mathbf{z}_i^T P_i \mathbf{z}_i$$

where $\mathbf{z} \in \Re^{3n}$, $\mathbf{z}_i \in \Re^3$, and $P_i \in \Re^{3\times 3}$ is a positive definite matrix. Since the derivative of $V(\mathbf{z})$ along trajectories of (6.12)

$$\frac{d}{dt} V(\mathbf{z}) = \sum_{i=1}^{n} \mathbf{z}_i^T \left(A_i^T P_i + P_i A_i\right) \mathbf{z}_i,$$

a sufficient condition for quadratic stability of the error dynamics (6.12) is the existence of a positive definite matrix P_i such that

$$A_i^T P_i + P_i A_i < 0 \quad \text{for } i = 1, \ldots, n.$$

Or equivalently we have a positive definite matrix Q_i such that

$$A_i Q_i + Q_i A_i^T < 0 \quad \text{for } i = 1, \ldots, n. \tag{6.13}$$

Remark 6.2 The existence of Q_i satisfying (6.13) is equivalent to A_i being Hurwitz. Using the result in Example 2.2, A_i is Hurwitz as long as K_{1i}, K_{2i}, and τ_{2i} are all positive.

If measurement noise is considered such that

$$\tilde{\mathbf{x}}_i = \mathbf{x}_i + \mathbf{v}_i \tag{6.14}$$

where $\mathbf{v}_i = [v_{1i} \; v_{2i}]^T \in \Re^2$, both (6.6) and (6.8) are rewritten as

$$\begin{aligned} \bar{x}_{2i} &= \dot{x}_{di} - K_{1i} S_{1i} - K_{1i} v_{1i}, \\ u_i &= f_i(\tilde{\mathbf{x}}) + \dot{x}_{2i,d} - K_{2i} S_{2i} - K_{2i} v_{2i}. \end{aligned}$$

The overall system in (6.9) is written with $\dot{x}_{di} = \ddot{x}_{di} = 0$ as

$$\begin{cases} \dot{x}_{1i} = S_{2i} + \xi_{2i} - K_{1i} S_{1i} - K_{1i} v_{1i}, \\ \dot{x}_{2i} = \dot{x}_{2i,d} - K_{2i} S_{2i} + \{f_i(\tilde{\mathbf{x}}) - f_i(\mathbf{x})\} - K_{2i} v_{2i}, \\ \dot{\xi}_{2i} = -\frac{1}{\tau_{2i}} \xi_{2i} + K_{1i} \dot{S}_{1i} - K_{1i} \dot{v}_{1i} \end{cases}$$

and equivalently can be written in matrix form as

$$T_i\dot{\mathbf{z}}_i = \bar{A}_i\mathbf{z}_i + \begin{bmatrix} -K_{1i}v_{1i} \\ \Delta f_i - K_{2i}v_{2i} \\ -K_{1i}\dot{v}_{1i} \end{bmatrix} = \bar{A}_i\mathbf{z}_i + \bar{\mathbf{w}}_i \tag{6.15}$$

where $\Delta f_i = f_i(\tilde{\mathbf{x}}) - f_i(\mathbf{x})$ and T_i, $\bar{A}_i$ are defined in (6.10). Then, the augmented error dynamics is written as follows:

$$\begin{cases} \dot{\mathbf{z}}_i = A_i\mathbf{z}_i + \mathbf{w}_i, \\ \mathbf{y}_i = C_i\mathbf{z}_i \end{cases} \tag{6.16}$$

where $\mathbf{w}_i = T_i^{-1}\bar{\mathbf{w}}_i$ and $\mathbf{y}_i$ is the output of the error dynamics. For example, if the control objective is to make S_{1i} go to zero, the output is defined by $\mathbf{y}_i := S_{1i}$, thus $C_i = [1\ 0\ 0]$.

A desirable property of DSC is to make the control objective $\mathbf{y}$ insensitive to $\mathbf{w}$ representing disturbances due to measurement noise. To consider this property, the induced $\mathscr{L}_2$ gain between the exogenous input $\mathbf{w}$ and the output of state error $\mathbf{y}$, defined in Definition 3.1, is minimized by redesigning the matrix A_i in (6.16). When the induced $\mathscr{L}_2$ gain of the augmented error dynamics described in (6.16) is defined as

$$\kappa^2 = \|H_{\mathbf{w}\to\mathbf{y}}\|_\infty^2 = \sup_{\|\mathbf{w}\|_2\neq 0} \frac{\|\mathbf{y}\|_2^2}{\|\mathbf{w}\|_2^2} = \sup_{\|\mathbf{w}_i\|_2\neq 0} \frac{\sum_{i=1}^n \|\mathbf{y}_i\|_2^2}{\sum_{i=1}^n \|\mathbf{w}_i\|_2^2}. \tag{6.17}$$

For the given matrices A_i and C_i in (6.16), the calculation of the induced $\mathscr{L}_2$ gain can be summarized in LMI form as follows:

Theorem 6.1 *For the given nonlinear system in* (6.5), *the augmented error dynamics in* (6.16) *has* $\|H_{\mathbf{w}\to\mathbf{y}}\|_\infty \le \gamma$ *if there exist* $Q_i > 0$ *and* $\kappa \ge 0$ *such that*

$$\begin{bmatrix} A_iQ_i + Q_iA_i^T + I & Q_iC_i^T \\ C_iQ_i & -\kappa^2 I \end{bmatrix} \le 0, \quad i = 1,\dots,n. \tag{6.18}$$

Proof Suppose there exist $V(\mathbf{z}) = \sum_{i=1}^n \mathbf{z}_i^T P_i\mathbf{z}_i$, $P_i > 0$, and $\gamma \ge 0$ such that

$$\dot{V} + \mathbf{y}^T\mathbf{y} - \kappa^2\mathbf{w}^T\mathbf{w} \le 0. \tag{6.19}$$

After integrating the left side of (6.19) from 0 to T with the assumption that $\mathbf{z}(0) = 0$,

$$V(T) + \int_0^T \sum_{i=1}^n \left(\mathbf{y}_i^T\mathbf{y}_i - \kappa^2\mathbf{w}_i^T\mathbf{w}_i\right) dt \le 0.$$

Since $V(T) \ge 0$, this implies that $\|H_{\mathbf{w}\to\mathbf{y}}\|_\infty^2 \le \kappa^2$ from (6.17).

The inequality (6.19) is equivalent to

$$\sum_{i=1}^n \left\{\mathbf{z}_i^T\left(A_i^TP_i + P_iA_i\right)\mathbf{z}_i + 2\mathbf{z}_i^TP_i\mathbf{w}_i + \mathbf{y}_i^T\mathbf{y}_i - \kappa^2\mathbf{w}_i^T\mathbf{w}_i\right\} \le 0. \tag{6.20}$$

The inequality condition in (6.20) holds if

$$\mathbf{z}_i^T\left(A_i^T P_i + P_i A_i + C_i^T C_i\right)\mathbf{z}_i + 2\mathbf{z}_i^T P_i \mathbf{w}_i - \kappa^2 \mathbf{w}_i^T \mathbf{w}_i \leq 0.$$

This is written in matrix form as follows:

$$\begin{bmatrix} A_i^T P_i + P_i A_i + C_i^T C_i & P_i \\ P_i & -\kappa^2 I \end{bmatrix} \leq 0.$$

Using the Schur complement, the above linear matrix inequality (LMI) condition is equivalent to

$$A_i^T P_i + P_i A_i + C_i^T C_i + \frac{1}{\kappa^2} P_i P_i \leq 0.$$

After multiplying the inequality on the left and right side by P_i^{-1} and defining $Q_i = \kappa^2 P_i^{-1}$, the above inequality becomes

$$A_i Q_i + Q_i A_i^T + \frac{1}{\kappa^2}(C_i Q_i)^T C_i Q_i + I \leq 0 \tag{6.21}$$

which is equivalent to (6.18) by the Schur complement. □

6.2.2.2 Decomposition of Controller Gains

The controller gain matrix A_i in (6.16) can be decomposed into two matrices: one is a function of K_{1i} and the other is a function of K_{2i} and τ_{2i}. That is,

$$A_i = \tilde{A}_i(K_{1i}) + F\Theta_i(K_{2i}, \tau_{2i}) \tag{6.22}$$

where

$$\tilde{A}_i = \begin{bmatrix} -K_{1i} & 1 & 1 \\ 0 & 0 & 0 \\ -K_{1i}^2 & K_{1i} & K_{1i} \end{bmatrix}, \qquad F = \begin{bmatrix} 0 & 0 \\ 1 & 0 \\ 0 & 1 \end{bmatrix},$$
$$\Theta_i = \begin{bmatrix} 0 & -K_{2i} & 0 \\ 0 & 0 & -1/\tau_{2i} \end{bmatrix}. \tag{6.23}$$

Then, using (6.22), the matrix inequality in (6.13) is written as

$$\left(\tilde{A}_i + F\Theta_i\right)Q_i + Q_i\left(\tilde{A}_i + F\Theta_i\right)^T < 0. \tag{6.24}$$

After defining $Y = \Theta_i Q_i$ and substituting $\Theta_i = Y Q_i^{-1}$ into (6.24), the inequality condition is written as

$$\tilde{A}_i Q_i + Q_i \tilde{A}_i^T + FY + Y^T F^T < 0.$$

An alternate equivalent condition for quadratic stabilization can be derived using Finsler's lemma as follows [12]: There exist $Q_i > 0$ and $\sigma > 0$ such that

$$\tilde{A}_i Q_i + Q_i \tilde{A}_i^T - \sigma F F^T < 0$$

with $Y = -(\sigma/2)F^T$. Since the above LMI is homogeneous in Q_i and σ, we can take $\sigma = 1$ without loss of generality. Furthermore, it is necessary to consider a specific structure of Θ_i defined in (6.23), i.e., all off-diagonal elements of $F\Theta_i$ should be zero. Therefore, if there exists a diagonal positive definite matrix $Q_i > 0$ such that

$$\tilde{A}_i Q_i + Q_i \tilde{A}_i^T - FF^T < 0, \tag{6.25}$$

a stabilizing gain matrix Θ_i is given by $\Theta_i = -(1/2)F^T Q_i^{-1}$ and all off-diagonal elements of $F\Theta_i$ are zero.

Next, if (6.22) and the elimination of matrix variables is used, the matrix inequality in (6.21) can be written as

$$\tilde{A}_i Q_i + Q_i \tilde{A}_i^T - \sigma FF^T + \frac{1}{\kappa^2}(C_i Q_i)^T C_i Q_i + I \le 0.$$

The result can be summarized as follows:

Theorem 6.2 *The augmented error dynamics in* (6.16) *has* $\|H_{\mathbf{w}\to\mathbf{y}}\|_\infty \le \gamma$ *if there exists a diagonal matrix* $Q_i > 0$, $\sigma > 0$, *and* $\kappa \ge 0$ *such that*

$$\begin{bmatrix} \tilde{A}_i Q_i + Q_i \tilde{A}_i^T - \sigma FF^T + I & Q_i C_i^T \\ C_i Q_i & -\kappa^2 I \end{bmatrix} \le 0, \quad i = 1, \ldots, n \tag{6.26}$$

and the resulting gain matrix is $\Theta_i = -(\sigma/2)F^T Q_i^{-1}$.

Using the result of Theorem 6.2, the induced $\mathscr{L}_2$ gain can be calculated by minimizing κ via convex optimization as follows:

$$\begin{array}{lll} \text{minimize} & \kappa^2 & \\ \text{subject to} & Q_i > 0, \quad Q_i \text{ diagonal}, & \sigma > 0, \\ & \kappa^2 \ge 0, \quad \text{and} \quad \text{LMI (6.26)}. & \end{array} \tag{6.27}$$

6.2.2.3 Optimal Design of DSC

If the convex optimization problem in (6.27) is solved numerically, roughly speaking, a smaller κ is calculated as the magnitudes of controller gains increase. On the other hand, if you look at $\mathbf{w}_i$ in (6.15), the amplification of sensor measurement noise is proportional to the magnitude of controller gains, K_{1i} and K_{2i}. Thus, it is preferable to minimize their magnitude to reduce the amplification of measurement noise. Furthermore, combining (6.7) with (6.6) and (6.14), the filter equation (6.7) is written as

$$\begin{aligned} \tau_{2i}\dot{x}_{2i,d} + x_{2i,d} &= \bar{x}_{2i}(\tilde{x}_{1i}, \dot{x}_{di}) \\ &= \dot{x}_{di} - K_{1i}S_{1i} - K_{1i}v_{1i}. \end{aligned}$$

It is well known that $1/\tau_{2i}$ in the lowpass filter is a cutoff frequency. If $1/\tau_{2i}$ increases, the noise v_{1i} may not be attenuated but $x_{2i,d}$ approaches $\bar{x}_{2i}$ more quickly, thus favorable with respect to stability.

Therefore, the design of controller gains can be interpreted as a trade-off between disturbance rejection and noise attenuation. More specifically speaking, we

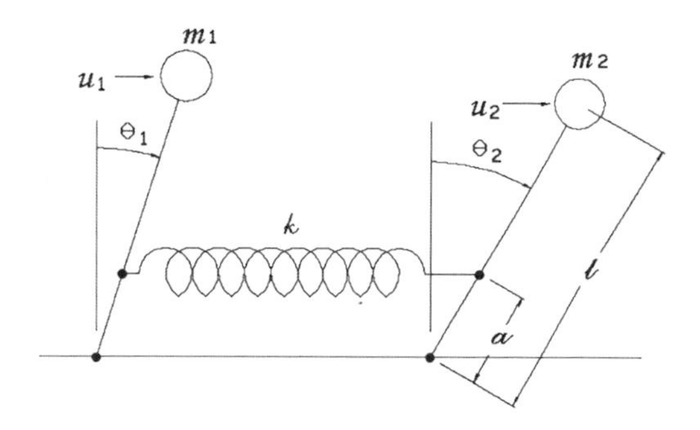

Fig. 6.1 Schematic of coupled inverted pendulums

will compute a set of controller gains, $\{K_{1i}, K_{2i}, 1/\tau_{2i}\}$, by minimizing its magnitude as well as by making the induced $\mathscr{L}_2$ gain of the closed-loop system less than a certain κ^* which is determined a priori based on performance specifications.

The design procedure of the controller gains consists of two steps: the first step is to determine K_{1i} to make sure $\kappa \le \kappa^*$ by solving COP (6.27) and the second step is to determine both K_{2i} and τ_{2i} by minimizing their magnitude among feasible solutions satisfying LMI (6.26) and $0 \le \kappa \le \kappa^*$. Since the magnitude of Θ_i is dependent on σ as well as the inverse of Q, we need to minimize σ and maximize $\lambda_{\min}(Q_i)$. This is called a multi-objective optimization problem. Using scalarization for finding Pareto optimal points (refer to [12]), this step can be formulated in a form of a convex optimization problem as follows: for a given $\delta \in [0, 1]$,

$$\begin{aligned} &\text{minimize} \quad \delta\sigma - (1-\delta)\sum_{i=1}^{n} \lambda_{\min}(Q_i) \\ &\text{subject to} \quad Q_i > 0, \quad Q_i \text{ diagonal}, \quad \sigma > 0, \\ &\qquad\qquad\quad 0 \le \kappa \le \kappa^*, \quad \text{and} \quad \text{LMI (6.26)}. \end{aligned} \tag{6.28}$$

Algorithm 6.1 The synthesis procedure of DSC for a stabilization problem is summarized as follows:

Step 1. Solve COP (6.27) iteratively with logarithmic spacing of $K_{1i}[j] \in [10^{-n}, 10^{n}]$ for $j = 1, \dots, N$. If there exists an integer j^* such that $\kappa[j^*] \le \kappa^*$, go to Step 2 with $K_{1i}^* = K_{1i}[j^*]$.
Step 2. Solve COP (6.28) with given K_{1i}^* and κ^*. Then, the resulting gain matrix $\Theta_i = -(\sigma/2)F^T Q_i^{-1}$, thus $K_{2i}^* = \Theta_i(1, 2)$ and $\tau_{2i}^* = 1/\Theta_i(2, 3)$ where $\Theta_i(j, k)$ is the $(j \times k)$th element of Θ_i.

Example 6.1 (Stabilization of coupled inverted double pendulums) The stabilization problem of coupled inverted double pendulums shown in Fig. 6.1 is considered in [27, 31]. They can be modeled as a fourth-order differential equation of motion using Lagrange's equation as follows:

$$\ddot{\theta}_1 = \frac{g}{l}\sin\theta_1 + u_1 + \frac{ka^2}{m_1 l^2}(\sin\theta_2\cos\theta_2 - \sin\theta_1\cos\theta_1),$$

$$\ddot{\theta}_2 = \frac{g}{l}\sin\theta_2 + u_2 + \frac{ka^2}{m_2 l^2}(\sin\theta_1\cos\theta_1 - \sin\theta_2\cos\theta_2)$$

where m_i is a mass of the ith particle, θ_i is an angle of the ith massless rod, and l is the length of the massless rods, k is a spring stiffness constant, and g is a gravity constant.

The model can be written in the form of (6.5) as follows:

$$\begin{aligned}\dot{\mathbf{x}}_i &= \begin{bmatrix} x_{2i} \\ \frac{g}{l}\sin x_{1i} + \frac{ka^2}{m_i l^2}(\sin x_{1j}\cos x_{1j} - \sin x_{1i}\cos x_{1i}) \end{bmatrix} + \begin{bmatrix} 0 \\ 1 \end{bmatrix} u_i \\ &= \begin{bmatrix} x_{2i} \\ -f_i(\mathbf{x}) \end{bmatrix} + \begin{bmatrix} 0 \\ 1 \end{bmatrix} u_i \end{aligned} \tag{6.29}$$

where

$$\mathbf{x} = \begin{bmatrix} \mathbf{x}_1 \\ \mathbf{x}_2 \end{bmatrix} \in \Re^4, \qquad \mathbf{x}_i = \begin{bmatrix} \theta_i \\ \dot{\theta}_i \end{bmatrix} = \begin{bmatrix} x_{1i} \\ x_{2i} \end{bmatrix} \in \Re^2, \quad j = \begin{cases} 2, & \text{if } i = 1, \\ 1, & \text{if } i = 2, \end{cases}$$

for $i = 1, 2$, and x_{ki} is the kth element of $\mathbf{x}_i$. Then, using (6.11), the augmented error dynamics is

$$\dot{\mathbf{z}}_i = A_i \mathbf{z}_i \quad (i = 1,\ 2)$$

where A_i is given in (6.11). Suppose that $a = \frac{g}{l} = 1$ and

$$b_i = \frac{ka^2}{m_i l^2} = \begin{cases} 0.25, & \text{for } i = 1, \\ 0.5, & \text{for } i = 2. \end{cases}$$

If the control objective is $x_{1i} \to 0$ and measurement noise is considered, the augmented error dynamics is written in a form of (6.16) where $C_i = [1\ 0\ 0]$.

Step 1. When COP (6.27) is solved iteratively with logarithmic spacing of $K_{1i}[j] \in [0.1, 10]$ for $j = 1, \ldots, 20$, Fig. 6.2 shows the calculated κ with respect to K_{1i}. When κ^* is given by 0.5, $\kappa[j] = 0.4281$ for $j = 14$ and the corresponding $K_{1i}[j] = 2.3357$. It is noted that COP (6.27) is solved numerically by CVX [32], refer to MATLAB Program 6-1.

Step 2. For the given $K_{1i}^* = 2.3357$, $\kappa^* = 0.5$, and $\delta = 0.5$, the calculated Θ_i by solving COP (6.28) is

$$\Theta_i = \begin{bmatrix} 0 & -14.3766 & 0 \\ 0 & 0 & -2.6806 \end{bmatrix} \quad \text{for } i = 1, 2.$$

Thus, $K_{2i}^* = 14.3766$ and $\tau_{2i}^* = 1/2.6806 = 0.373$. It is noted that COP (6.28) is solved by modifying the objective function and adding the inequality constraint $\kappa \le \kappa^*$ as shown in MATLAB Program 6-2.

MATLAB Program 6-1

```
%***** Define the dimension of matrices *****
n = size(A1, 1);
m = size(C1, 2);

%***** cvx version *****
cvx_begin sdp
   variable Q1(n,n) diagonal;
   variable Q2(n,n) diagonal;
   variables gamma sigma;
   minimize(gamma);
   gamma >= 0;    sigma >= 0;
   Q1 == semidefinite(n);    Q2 == semidefinite(n);
   -[Q1*A1' + A1*Q1 - sigma*F*F' + eye(n) Q1*C1'; C1*Q1 -gamma*eye(m)]
      == semidefinite(n+m);
   -[Q2*A2' + A2*Q2 - sigma*F*F' + eye(n) Q2*C2'; C2*Q2 -gamma*eye(m)]
      == semidefinite(n+m);
cvx_end

%***** Calculation of L2 gain *****
L2 = sqrt(gamma);
```

MATLAB Program 6-2

```
%***** cvx version *****
cvx_begin sdp
   :
   minimize(0.5*sigma - 0.5*lambda_min(Q1) -0.5*lambda_min(Q2));
   0.25 - gamma >= 0
   :
cvx_end

%***** Calculation of Theta *****
Theta1 = -sigma/2*F'*inv(Q1);
Theta2 = -sigma/2*F'*inv(Q2);
```

For the given set of controller gains, $\{K_{1i}^*, K_{2i}^*, \tau_{21}^*\} = \{2.3357, 14.3766, 0.373\}$ calculated above, it is validated that the calculated induced $\mathscr{L}_2$ using the result of Theorem 6.1 is 0.4292, which is less than κ^*. Furthermore, when a set of initial conditions are given as

$$\{\theta_1(0),\ \dot{\theta}_1(0),\ \theta_2(0),\ \dot{\theta}_2(0)\} = \{-\pi/12,\ 0,\ \pi/6,\ 0\},$$

the time responses of $\mathbf{x}_i$ and u_i are shown in Fig. 6.3 and $\mathbf{x}_i$ converges to the origin asymptotically.

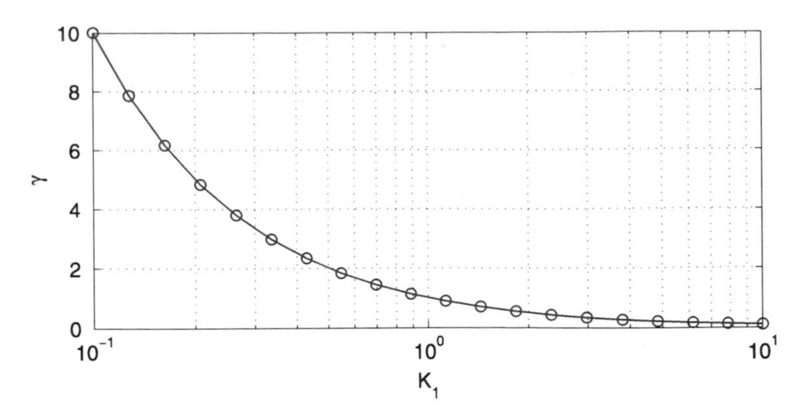

Fig. 6.2 Induced $\mathscr{L}_2$ gain with respect to K_{1i}

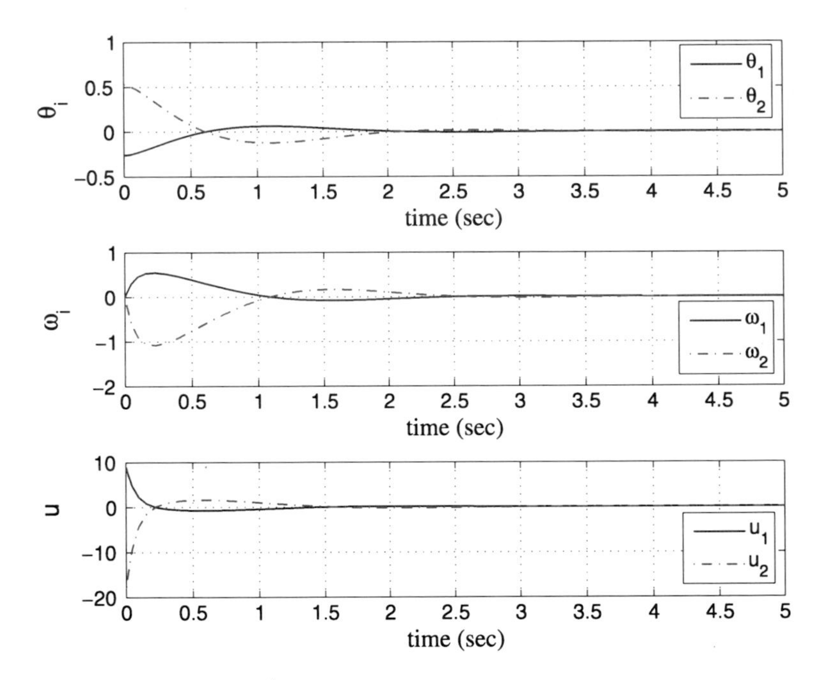

Fig. 6.3 Time responses of θ_1, $\dot{\theta}_i$, and u_i

When the sensor measurement noise with a relatively high frequency and small magnitude is considered only in the first inverted pendulum, i.e., $v_{1i} = v_{2i} = 0.05\sin(50t)$ for $i = 1$ in (6.15) for simulation, the performance comparison between the first and second inverted pendulums is shown in Fig. 6.4 for a set of initial conditions given by

$$\{\theta_1(0),\ \dot{\theta}_1(0),\ \theta_2(0),\ \dot{\theta}_2(0)\} = \{-\pi/12,\ 0,\ \pi/12,\ 0\}.$$

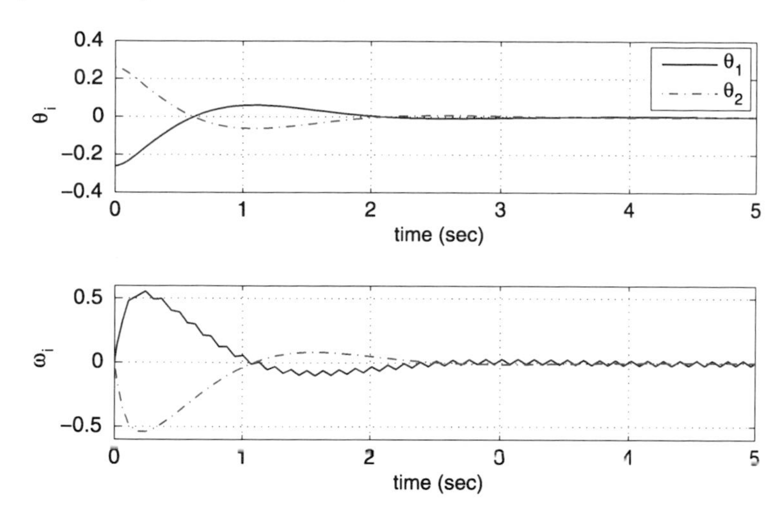

Fig. 6.4 Time responses of θ_1 and $\dot{\theta}_i$ for $\tau_{2i} = 0.373$

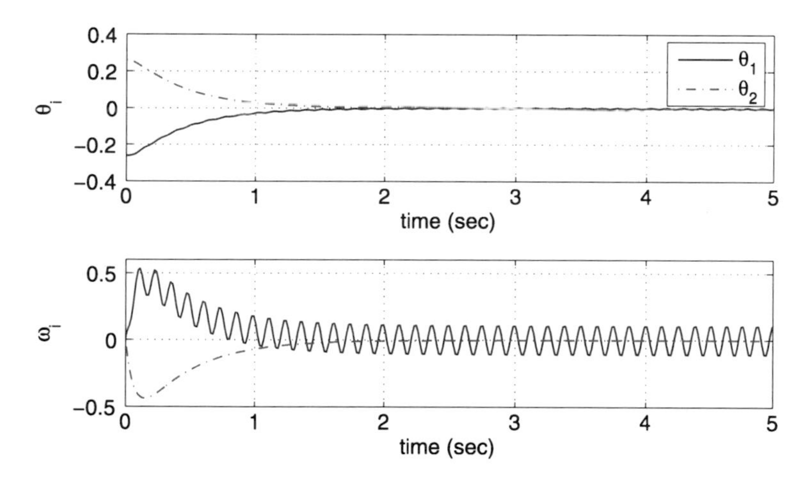

Fig. 6.5 Time responses of θ_1 and $\dot{\theta}_i$ for $\tau_{2i} = 0.01$

It is shown that the time response of $\mathbf{x}_1$ with measurement noise is close to that of $\mathbf{x}_2$ without measurement noise. It implies that the optimal choice of Θ_1 results in noise attenuation.

If τ_{2i} is chosen with an arbitrary small magnitude such as $\tau_{2i} = 0.01$, it is shown in Fig. 6.5 that good performance of noise attenuation cannot be achieved. This result implies that a choice of τ_{2i} is a trade-off between quadratic stability and noise attenuation.

6.2.3 Synthesis for Quadratic Tracking

If a tracking problem is considered, i.e., $\ddot{x}_{di} \neq 0$, the error dynamics with DSC is derived in (6.11). Furthermore, it is assumed that $\ddot{x}_{di}$ is bounded such that $|\ddot{x}_{di}| \leq d_0$ for some $d_0 > 0$. Then the error dynamics is written as

$$\begin{cases} \dot{\mathbf{z}}_i = A_i \mathbf{z}_i + B_{ri} r_i, & |r_i| \leq 1, \\ \mathbf{y}_i = C_i \mathbf{z}_i \end{cases} \tag{6.30}$$

where $r_i = \ddot{x}_{di}/d_0$ and $B_{ri} = d_0 B_i = [0\ 0\ \ -d_0]^T$. The reachable sets of the augmented error dynamics in (6.30) are defined as

$$\mathscr{R}_i = \left\{ \mathbf{z}_i | \mathbf{z}_i, r_i \text{ satisfy (6.30)},\ \mathbf{z}_i = 0,\ T \geq 0 \right\}.$$

Then, the ellipsoid $\mathscr{E}(P_i) = \{\mathbf{z}_i | \mathbf{z}_i^T P_i \mathbf{z}_i \leq 1\}$ contains the reachable set $\mathscr{R}_i$ if there exists $P_i > 0$ such that

$$(A_i \mathbf{z}_i + B_{ri} r_i)^T P_i \mathbf{z}_i + \mathbf{z}_i^T P_i (A_i \mathbf{z}_i + B_{ri} r_i) < 0 \tag{6.31}$$

for all $\mathbf{z}_i$ and r_i satisfying (6.30) and $\mathbf{z}_i^T P_i \mathbf{z}_i \geq 1$. Refer to Definition 2.4 for more detail.

Using the S-procedure, the condition (6.31) holds if there exists $P_i > 0$ and $\alpha \geq 0$ such that

$$\mathbf{z}_i^T \left(A_i^T P_i + P_i A_i \right) \mathbf{z}_i + 2\mathbf{z}_i^T P_i B_{ri} r_i + \alpha \left(\mathbf{z}_i^T P_i \mathbf{z}_i - r_i^T r_i \right) < 0$$

or

$$\begin{bmatrix} A_i^T P_i + P_i A_i + \alpha P_i & P_i B_{ri} \\ B_{ri}^T P_i & -\alpha \end{bmatrix} \leq 0. \tag{6.32}$$

Equivalently, defining $Q_i = P_i^{-1}$, the matrix inequality (6.32) is equivalent to the following inequality in Q_i and α:

$$\begin{bmatrix} A_i Q_i + Q_i A_i^T + \alpha Q_i & B_{ri} \\ B_{ri}^T & -\alpha \end{bmatrix} \leq 0. \tag{6.33}$$

Refer to Sect. 6.2.2 for more detailed derivations. As done in (6.22), if the matrix A_i is decomposed into $\tilde{A}_i$ and $F\Theta_i$, the ellipsoid $\mathscr{E}(P_i) = \mathscr{E}(Q_i^{-1})$ containing reachable sets can be obtained as follows:

Theorem 6.3 *An ellipsoid* $\mathscr{E}(Q_i^{-1}) = \{\mathbf{z}_i \in \Re^3 | \mathbf{z}_i^T Q_i^{-1} \mathbf{z}_i \leq 1\}$ *contains reachable sets of the augmented error dynamics in* (6.30) *if there exist a diagonal matrix* $Q_i > 0$, $\sigma > 0$, *and* $\alpha \geq 0$ *such that*

$$\begin{bmatrix} \tilde{A}_i Q_i + Q_i \tilde{A}_i^T - \sigma F F^T + \alpha Q_i & B_{ri} \\ B_{ri}^T & -\alpha \end{bmatrix} \leq 0 \tag{6.34}$$

where $i = 1, \ldots, n$ *and the resulting DSC gain matrix is* $\Theta_i = -(\sigma/2) F^T Q_i^{-1}$.

Proof Using (6.22), the Schur complement of (6.33) is

$$\tilde{A}_i Q_i + Q_i \tilde{A}_i^T + FY + Y^T F^T + \alpha Q_i + \frac{1}{\alpha} B_{ri} B_{ri}^T < 0$$

where $Y = \Theta_i Q_i$. As used in the stabilization problem, an equivalent condition using Finsler's lemma is derived as follows:

$$\tilde{A}_i Q_i + Q_i \tilde{A}_i^T - \sigma F F^T + \alpha Q_i + \frac{1}{\alpha} B_{ri} B_{ri}^T < 0$$

with $Y = -(\sigma/2)F^T$, thus $\Theta_i = Y Q_i^{-1} = -(\sigma/2)F^T Q_i^{-1}$. The Schur complement of the above inequality condition is (6.34). Next, if Q is a diagonal and positive definite matrix, all off-diagonal elements of $F\Theta_i = -(\sigma/2)FF^T Q_i^{-1}$ become zero. □

Using the result of Theorem 6.3, we can compute the smallest ellipsoid containing the reachable sets of the error dynamics by minimizing either a volume or a maximum diameter of the ellipsoid. Since Q is a diagonal matrix and the volume of the ellipsoid depends on

$$\sqrt{\det(Q_i)} = \sqrt{\prod_{i=1}^{3} \lambda_i(Q_i)},$$

the ellipsoid with minimum volume can be obtained by minimizing $\mathrm{Tr}(Q_i) = \sum_{i=1}^{3} \lambda_i(Q_i)$. Similarly the minimum of the largest diameter of $\mathcal{E}(Q_i^{-1})$, which is defined as $2\sqrt{\lambda_{\max}(Q_i)}$, can be computed by minimizing $\lambda_{\max}(Q_i)$. Therefore, the convex optimization problem can be written as follows: for a given α,

$$\begin{aligned} &\text{minimize} \quad \mathrm{Tr}(Q_i) \quad \text{or} \quad \lambda_{\max}(Q_i) \\ &\text{subject to} \quad Q_i > 0, \quad Q_i \text{ diagonal}, \quad \sigma > 0, \quad \text{and} \\ &\qquad\qquad\quad \text{LMI (6.34).} \end{aligned} \tag{6.35}$$

However, since the resulting gain matrix is defined as

$$\Theta_i = -(\sigma/2)F^T Q^{-1},$$

the larger magnitude of Θ_i is calculated as the smaller $\mathrm{Tr}(Q_i)$ or $\lambda_{\max}(Q)$ is obtained. Therefore, we suggest to minimize the magnitude of Θ_i subject to

$$\|\mathbf{y}_i\|_2 = \|C_i \mathbf{z}_i\|_2 \leq c_0 \tag{6.36}$$

where $c_0 > 0$ is determined a priori and related with a performance constraint. For example, suppose the control objective is $x_{1i} \to x_{di}$, thus $S_{1i} \to 0$. Then, in general there is a performance constraint such that $|S_{1i}| \leq c_0$ for a tracking problem. The main reason to compute an ellipsoid $\mathcal{E}(Q_i^{-1})$ by solving COP (6.35) is to check whether $\mathcal{E}(Q_i^{-1})$ is contained in

$$\mathcal{C} = \{\mathbf{z}_i \in \Re^3 \,|\, \|C_i \mathbf{z}_i\|_2 \leq c_0 \text{ for } c_0 > 0\}.$$

That is,

$$\mathcal{R}_i \subset \mathcal{E}(Q_i^{-1}) \subset \mathcal{C}.$$

Since the inequality condition for $\mathscr{E}(Q_i^{-1}) \subset \mathscr{C}$ is written as (refer to [12])

$$\begin{bmatrix} Q_i & Q_i C_i \\ C_i^T Q_i & c_0^2 I \end{bmatrix} \geq 0, \tag{6.37}$$

the smallest magnitude of Θ_i can be obtained by both minimizing σ and maximizing $\lambda_{\min}(Q)$. The corresponding convex optimization problem is given thus: for the given $\delta \in [0, 1]$ and $\alpha > 0$,

$$\begin{aligned} &\text{minimize} \quad \delta\sigma - (1-\delta)\lambda_{\min}(Q_i) \\ &\text{subject to} \quad Q_i > 0, \quad Q_i \text{ diagonal}, \quad \sigma > 0, \quad \text{and} \\ &\qquad\qquad \text{LMI (6.34)} \quad \text{and} \quad (6.37). \end{aligned} \tag{6.38}$$

It is important to remark that δ is chosen to make $\lambda_{\min}(Q_i) = c_0^2$ to minimize the magnitude of Θ_i. Furthermore, if $\lambda_{\max}(Q_i)$ is minimized to be $\lambda_{\max}(Q_i) = c_0^2$, the resulting Q_i is $c_0^2 I$ for the given δ. Otherwise, a larger magnitude of Θ_i is generated.

Alternatively, if measurement noise is considered, combining (6.16) with (6.11), the augmented error dynamics is

$$\dot{\mathbf{z}}_i = A_i \mathbf{z}_i + [B_i \; I] \begin{bmatrix} \ddot{x}_{di} \\ \mathbf{w}_i \end{bmatrix} = A_i \mathbf{z}_i + B_{wi} \tilde{\mathbf{w}}_i. \tag{6.39}$$

Then, LMI (6.26) can be modified with B_{wi} instead of I as follows:

$$\begin{bmatrix} \tilde{A}_i Q_i + Q_i \tilde{A}_i^T - \sigma F F^T + B_{wi} B_{wi}^T & Q_i C_i^T \\ C_i Q_i & -\kappa^2 I \end{bmatrix} \leq 0, \quad i = 1, \ldots, n. \tag{6.40}$$

Finally, Θ_i can be chosen for the given K_{1i} and κ^* by solving the following COP:

$$\begin{aligned} &\text{minimize} \quad \delta\sigma - (1-\delta)\lambda_{\min}(Q_i) \\ &\text{subject to} \quad Q_i > 0, \quad Q_i \text{ diagonal}, \quad \sigma > 0, \\ &\qquad\qquad 0 \leq \kappa \leq \kappa^*, \quad \text{and} \quad \text{LMI (6.40)}, \end{aligned} \tag{6.41}$$

for a given $\delta \in [0, 1]$.

Algorithm 6.2 The synthesis procedure of DSC for a tracking problem is summarized as follows:

Step 1. Determine K_{1i}^* using Step 1 of Algorithm 6.1.

Step 2a. For a given $\delta^* \in [0, 1]$ and $c_0 > 0$, solve COP (6.38) iteratively with logarithmic spacing of $\alpha[j] \in [10^{-n}, 10^n]$ for $j = 1, \ldots, N$. If there exist a solution of COP (6.38) such that $\lambda_{\min}(Q_i) = c_0^2$, the resulting gain matrix $\Theta_i = -(\sigma/2)F^T Q_i^{-1}$, thus $K_{2i}^* = \Theta_i(1, 2)$ and $\tau_{2i}^* = 1/\Theta_i(2, 3)$. Otherwise, repeat Step 2a with a different δ.

Step 2b. For the given error dynamics (6.39), solve COP (6.41) with given K_{1i}^* and κ^*. Then, the resulting gain matrix $\Theta_i = -(\sigma/2)F^T Q_i^{-1}$.

Example 6.2 Consider the system in Example 6.1 with different control objectives. Suppose the control objective of the first inverted pendulum is $x_{11} \to x_{d1} := \sin t$ while one of the second one is $x_{12} \to x_{d2} := -\sin t$. In addition, the performance

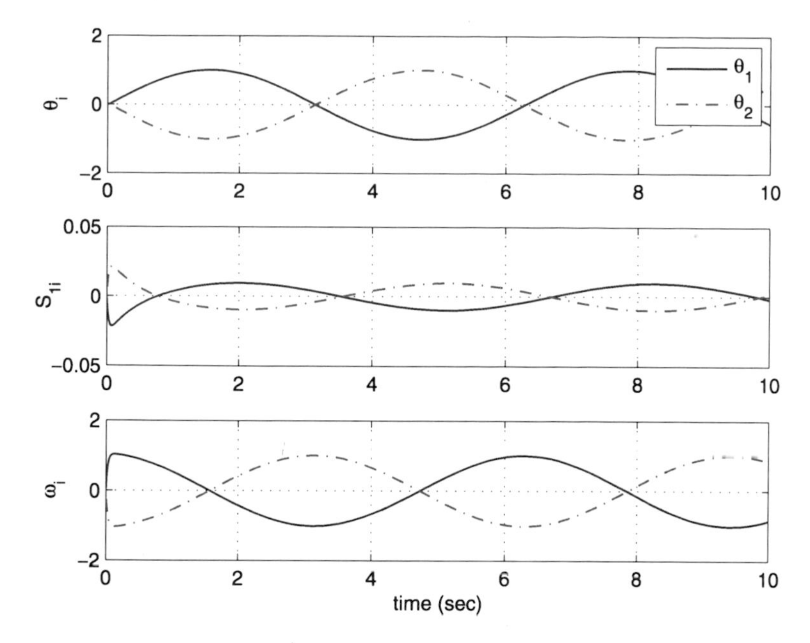

Fig. 6.6 Time responses of θ_1 and $\dot{\theta}_i$

constraint is given as $|x_{1i} - x_{di}| \leq c_0 := 0.1$. Then the augmented error dynamics using (6.11) is

$$\begin{cases} \dot{\mathbf{z}}_i = A_i \mathbf{z}_i + B_i \ddot{x}_{di}, & |\ddot{x}_{di}| \leq 1, \\ y_i = C_i \mathbf{z}_i \end{cases} \quad \text{for } i = 1, 2 \tag{6.42}$$

where A_i is given in (6.11), $B_i = [0\ 0\ -1]^T$, $C_i = [1\ 0\ 0]$, $\ddot{x}_{di} = (-1)^i \sin t$, and $|\ddot{x}_{di}| \leq 1$.

Step 1. Let $K_{1i} = 2.3357$ which is calculated in Example 6.1.
Step 2a. For $\delta = 0.02$ and $c_0 = 0.1$, COP (6.38) is solved iteratively with logarithmic spacing of $\alpha[j] \in [0.1, 10]$ for $j = 1, \ldots, 20$. The solution of COP (6.38) is

$$\sigma = 0.8178 \quad \text{and} \quad Q = c_0^2 I.$$

Then the resulting matrix of Θ_i is

$$\Theta_i = -(\sigma/2) F^T Q^{-1} = -\left(\sigma/2c_0^2\right) F^T = -\frac{0.8178}{0.02} F^T,$$

thus, $K_{21} = 40.8905$ and $\tau_{21} = 1/40.8905 = 0.0245$.

For the given set of controller gains

$$\{K_{1i}, K_{2i}, \tau_{2i}\} = \{2.3357,\ 40.8904,\ 0.0245\}.$$

Figure 6.6 shows the time responses of $\mathbf{x}_i$ satisfying $|S_{1i}| = |x_{1i} - x_{di}| \leq c_0$.

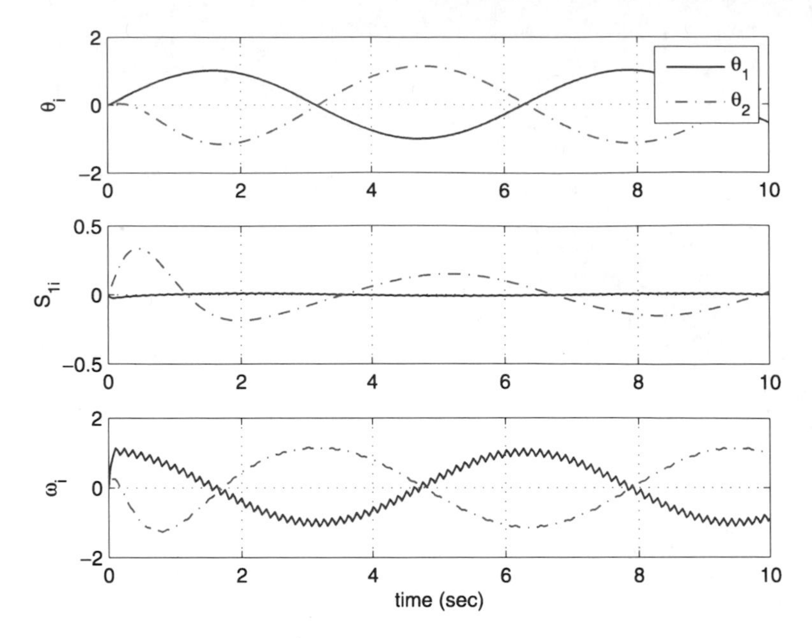

Fig. 6.7 Time responses of θ_1 and $\dot{\theta}_i$ in presence of measurement noise

Alternatively, suppose measurement noise is considered, i.e., $\upsilon_{1i} = \upsilon_{2i} = 0.05\sin(50t)$ for simulation. Based on (6.39), the augmented error dynamics in (6.42) can be rewritten as

$$\begin{cases} \dot{\mathbf{z}}_i = A_i\mathbf{z}_i + [B_i \;\; I]\begin{bmatrix} \ddot{x}_{di} \\ \mathbf{w}_i, \end{bmatrix} = A_i\mathbf{z}_i + B_{wi}\tilde{\mathbf{w}}_i, \\ y_i = C_i\mathbf{z}_i, \end{cases} \quad \text{for } i = 1, 2.$$

Step 2b. As used in Example 6.1, for the given $K^*_{1i} = 2.3357$, $\kappa^* = 0.5$, and $\delta = 0.5$, the calculated Θ_i by solving COP (6.41) is

$$\Theta_i = \begin{bmatrix} 0 & -16.1384 & 0 \\ 0 & 0 & -3.0029 \end{bmatrix}.$$

Thus, $K^*_{2i} = 16.1384$ and $\tau^*_{2i} = 1/3.0029 = 0.333$.

If the set of controller gains calculated via Step 2b is applied only to the second inverted pendulum for comparison, the time responses of $\mathbf{x}_i$ are shown in Fig. 6.7. Due to a larger magnitude of τ_{2i} for the second pendulum, the time response of ω_2 is smoother than ω_1. However, S_{12} is larger than S_{11} due to a smaller magnitude of K_{22}. If a smaller κ^* is given, a larger magnitude of K_{22} may be calculated and the corresponding tracking performance will be improved with a degradation of noise attenuation.

6.2.4 Avoiding Cancelations

Suppose f_i in (6.2) is decoupled into linear and nonlinear terms such as

$$f_i = a_i^T x_i + \tilde{f}_i \tag{6.43}$$

where $a_i = [a_1\ a_2]^T \in \Re^2$, the linear term may not be canceled by redefining u in (6.8) by

$$u_i = \tilde{f}_i + \dot{x}_{2i,d} - K_{2i} S_{2i}. \tag{6.44}$$

Depending on the value of a_i, the decision whether the linear term is canceled by u can be made by a stability analysis of the closed-loop error dynamics.

If u_i defined in (6.44) is applied to the system (6.5), the closed-loop system in (6.9) is written as

$$\begin{cases} \dot{x}_{1i} = S_{2i} + \xi_{2i} + \dot{x}_{di} - K_{1i} S_{1i}, \\ \dot{x}_{2i} = a_i^T \mathbf{x}_i + \dot{x}_{2i,d} - K_{2i} S_{2i}, \\ \dot{\xi}_{2i} = -\frac{1}{\tau_{2i}} \xi_{2i} - \ddot{x}_{di} + K_{1i} \dot{S}_{1i}. \end{cases} \tag{6.45}$$

Furthermore, $\mathbf{x}_i$ can be written as a function of $\mathbf{z}_i$ as follows:

$$\begin{aligned} x_{1i} &= (x_{1i} - x_{di}) + x_{di} = S_{1i} + x_{di}, \\ x_{2i} &= (x_{2i} - x_{2d,i}) + (x_{2d,i} - \bar{x}_{2i}) + \bar{x}_{2i} = S_{2i} + \xi_{2i} + \dot{x}_{di} - K_{1i} S_{1i} \\ \Longrightarrow \quad & \mathbf{x}_i = C_{zi} \mathbf{z}_i + \begin{bmatrix} x_{di} \\ \dot{x}_{di} \end{bmatrix} \end{aligned} \tag{6.46}$$

where

$$C_{zi} = \begin{bmatrix} 1 & 0 & 0 \\ -K_{1i} & 1 & 1 \end{bmatrix}.$$

Using (6.46), the augmented system in (6.45) can be written in the error dynamics form as follows:

$$\begin{aligned} \dot{\mathbf{z}}_i &= \left(A_i + e_2 a_i^T C_{zi}\right)\mathbf{z}_i + B_{ri} r_i \\ &= A_i' \mathbf{z}_i + B_{ri} r_i \end{aligned} \tag{6.47}$$

where $e_2 = [0\ 1\ 0]^T \in \Re^{3\times 1}$, $r_i = [x_{di}\ \dot{x}_{di}\ \dot{x}_{di}]^T \in \Re^3$, and

$$B_{ri} = \begin{bmatrix} 0 & 0 & 0 \\ a_1 & a_2 & 0 \\ 0 & 0 & -1 \end{bmatrix} \in \Re^{3\times 3}.$$

Therefore, using the result in Sect. 2.5.1, the eigenvalues of A_i' can be calculated as

$$\det(\lambda I - A_i') = 0 \quad \text{or} \quad \det(\lambda T_i - \bar{A}_i') = 0.$$

Since

$$\bar{A}_i' = \bar{A}_i + e_2 a_i^T C_{zi} = \begin{bmatrix} -K_{1i} & 1 & 1 \\ a_{1i} - a_{2i} K_{1i} & -K_{2i} + a_{2i} & a_{2i} \\ 0 & 0 & -1/\tau_{2i} \end{bmatrix},$$

the characteristic equation is

$$\det\left(\begin{bmatrix} \lambda + K_{1i} & -1 & -1 \\ -a_{1i} + a_{2i}K_{1i} & \lambda + K_{2i} - a_{2i} & -a_{2i} \\ -K_{1i}\lambda & 0 & \lambda + 1/\tau_{2i} \end{bmatrix}\right) = 0.$$

Equivalently,

$$\lambda^3 + \left(K_{2i} + \frac{1}{\tau_{2i}} - a_{2i}\right)\lambda^2 + \left(\frac{K_{1i} + K_{2i} - a_{2i}}{\tau_{2i}} - a_{1i}\right)\lambda + \frac{1}{\tau_{2i}}(K_{1i}K_{2i} - a_{1i}) = 0.$$

If the Routh stability criterion is used, the conditions for A_i' to be Hurwitz are

$$K_{2i} + \frac{1}{\tau_{2i}} - a_{2i} > 0, \tag{6.48a}$$

$$K_{1i}K_{2i} - a_{1i} > 0, \tag{6.48b}$$

$$\left(K_{2i} + \frac{1}{\tau_{2i}} - a_{2i}\right)\left(\frac{K_{1i} + K_{2i} - a_{2i}}{\tau_{2i}} - a_{1i}\right) > \frac{1}{\tau_{2i}}(K_{1i}K_{2i} - a_{1i}). \tag{6.48c}$$

Remark 6.3 The matrix A_i' in (6.47) can be also decomposed into two matrices:

$$A_i' = \left(\tilde{A}_i + e_2 a_i^T C_{zi}\right) + F\Theta_i.$$

Therefore, two sets of controller gains $\{K_{1i}, K_{2i}, 1/\tau_{2i}\}$ with respect to two different versions of error dynamics in (6.11) and (6.47) can be calculated, respectively, through Algorithm 6.1 or 6.2. Therefore, the decision whether the linear term of f_i in (6.43) is canceled by u_i or not can be made by the optimal value of Algorithm 6.1 or 6.2. That is, if the smaller magnitude of controller gains is calculated with avoiding the cancelation of $a_i^T \mathbf{x}_i$ in f_i, it is preferable to design u_i in (6.44) to avoid cancelation of the linear term.

6.3 Extension to Rigid Body Dynamics

The design methodology of DSC developed in Sect. 6.2 will be extended to the rigid body dynamics described in (6.3). First, define the first error vector as $S_1 := x_1 - x_{1d} \in \Re^n$ where $x_{1d} \in \Re^n$ is the reference or control objective. After differentiating the error vector S_1, the synthetic input, $\bar{x}_2$, is defined as

$$\dot{S}_1 = \dot{x}_1 - \dot{x}_{1d} = x_2 - \dot{x}_{1d},$$

$$\bar{x}_2 = \dot{x}_{1d} - \Lambda_1 S_1$$

where Λ_1 is a 5×5 diagonal gain matrix, $\Lambda_1 = \text{diag}(K_{11}, \dots, K_{51}) > 0$. As done in (6.7) for the ith subsystem, a set of first-order lowpass filters is introduced to calculate the desired value which will be used to define the second error vector as follows:

$$T_2\dot{x}_{2d} + x_{2d} = \bar{x}_2, \quad x_{2d}(0) := \bar{x}_2(0) \tag{6.49}$$

where T_2 is a 5×5 diagonal matrix containing time constants as

$$T_2 = \mathrm{diag}(\tau_{12}, \ldots, \tau_{52}).$$

Next, the second surface error vector is defined as $S_2 := x_2 - x_{2d} \in \Re^n$. After differentiating the error vector, the control input is designed as follows:

$$\begin{aligned}\dot{S}_2 &= \dot{x}_2 - \dot{x}_{2d} = D^{-1}(u - Hx_2 - g) - \dot{x}_{2d},\\ u &= Hx_2 + g + D(\dot{x}_{2d} - \Lambda_2 S_2)\end{aligned} \tag{6.50}$$

where $\Lambda_2 := \mathrm{diag}(K_{12}, \ldots, K_{52})$ is a 5×5 diagonal gain matrix. Furthermore, using (6.49), the control input u in (6.50) can be rewritten as

$$u = Hx_2 + g + D\{T_2^{-1}(\bar{x}_2 - x_{2d}) - \Lambda_2 S_2\}. \tag{6.51}$$

As done in (6.9), similarly the closed-loop system can be written as follows:

$$\begin{cases}\dot{x}_1 = (x_2 - x_{2d}) + (x_{2d} - \bar{x}_2) + \bar{x}_2 = S_2 + \xi_2 + \dot{x}_{1d} - \Lambda_1 S_1,\\ \dot{x}_2 = \dot{x}_{2d} - \Lambda_2 S_2,\\ \dot{\xi}_2 = \dot{x}_{2d} - \dot{\bar{x}}_2 = -T_2^{-1}\xi_2 - \ddot{x}_{1d} + \Lambda_1 \dot{S}_1.\end{cases} \tag{6.52}$$

The corresponding error dynamics can be written in matrix form as

$$\begin{bmatrix} I & 0 & 0\\ 0 & I & 0\\ -\Lambda_1 & 0 & I\end{bmatrix}\begin{bmatrix}\dot{S}_1\\ \dot{S}_2\\ \dot{\xi}_2\end{bmatrix} = \begin{bmatrix} -\Lambda_1 & I & I\\ 0 & -\Lambda_2 & 0\\ 0 & 0 & -T_2^{-1}\end{bmatrix}\begin{bmatrix}S_1\\ S_2\\ \xi_2\end{bmatrix} + \begin{bmatrix}0\\ 0\\ -I\end{bmatrix}\ddot{x}_{1d}$$

or equivalently,

$$\begin{bmatrix}\dot{S}_1\\ \dot{S}_2\\ \dot{\xi}_2\end{bmatrix} = \begin{bmatrix} -\Lambda_1 & I & I\\ 0 & -\Lambda_2 & 0\\ -\Lambda_1^2 & \Lambda_1 & \Lambda_1 - T_2^{-1}\end{bmatrix}\begin{bmatrix}S_1\\ S_2\\ \xi_2\end{bmatrix} + \begin{bmatrix}0\\ 0\\ -I\end{bmatrix}\ddot{x}_{1d}.$$

Therefore, the augmented error dynamics can be written as

$$\dot{z} = Az + B\ddot{x}_{1d} \tag{6.53}$$

where $z = [S_1^T \ S_2^T \ \xi_2^T]^T \in \Re^{3n}$. Furthermore, the matrix A can be decomposed as

$$\dot{z} = (\tilde{A} + F\Theta)z + B\ddot{x}_{1d} \tag{6.54}$$

where

$$\tilde{A} = \begin{bmatrix} -\Lambda_1 & 1 & 1\\ 0 & 0 & 0\\ -\Lambda_1^2 & \Lambda_1 & \Lambda_1\end{bmatrix}, \quad F = \begin{bmatrix}0 & 0\\ I & 0\\ 0 & I\end{bmatrix}, \quad \Theta = \begin{bmatrix}0 & -\Lambda_2 & 0\\ 0 & 0 & -T_2^{-1}\end{bmatrix}.$$

Therefore, a set of controller gain matrices $\{\Lambda_1, \Lambda_2, T_2^{-1}\}$ can be designed via Algorithm 6.1 for a stabilization problem or Algorithm 6.2 for a tracking problem. DSC is applied to the biped walking of a 5-link rigid body system in Chap. 9.

6.4 Decentralized Dynamic Surface Control

The interconnected system of N *strict-feedback* nonlinear subsystems is considered as follows:

$$\dot{\mathbf{x}}_i = U_i \mathbf{x}_i + B_i u_i + \mathbf{f}_i(\mathbf{x}_i) + \sum_{j=1, j\neq i}^{N} \mathbf{g}_{ij}(\mathbf{x}_j) \tag{6.55}$$

where $i = 1, 2, \dots, N$, $\mathbf{x}_i \in \Re^n$ and $u_i \in \Re$ are the state and control input of the ith nonlinear subsystem, respectively. $\mathbf{x}_j \in \Re^n$ is the state of the jth subsystem. Since the system is in the strict-feedback form, the matrix $U_i = \text{diag}([1, \dots, 1], 1) \in \Re^{n\times n}$, $B_i = [0 \ \cdots \ 0 \ 1]^T \in \Re^n$, and the nonlinearities f_{ki}, the k-th element of $\mathscr{C}^2$ nonlinear function vector $\mathbf{f}_i$, are Lipshitz satisfying $\|\mathbf{f}_i\| \leq \gamma_i$. The nonlinear function vector $\mathbf{g}_{ij}$ represents the interconnection among subsystems satisfying

$$\mathbf{h}_i^T \mathbf{h}_i \leq c_i^2 \tag{6.56}$$

where $\mathbf{h}_i = \sum_{j=1, j\neq i}^{N} \mathbf{g}_{ij}(\mathbf{x}_j) \in \Re^n$ and c_i is a known positive constant. It is noted that $\mathbf{h}_i$ is considered as an unit-peak function without loss of generality, i.e., $c = 1$ after (6.55) is multiplied by the known constant $1/c$ [12]. It is noted that a class of large-scale nonlinear systems can be transformed to (6.55) via a global diffeomorphism [55].

The analysis and design methodology of decentralized dynamic surface control (DDSC) proposed in [81] is discussed in this section. After the preliminary design is described, the augmented error dynamics of the N controlled nonlinear systems are derived, thus enabling us to provide a systematic method for designing a set of controller gains for the closed-loop system in the framework of LMI.

6.4.1 *Preliminary Design of DDSC*

The equation in (6.55) can be written componentwise as follows: for $1 \leq k \leq n-1$,

$$\dot{x}_{ki} = x_{(k+1)i} + f_{ki}(x_{1i}, \dots, x_{ki}) + h_{ki}(\mathbf{x}_1, \dots, \mathbf{x}_{i-1}, \mathbf{x}_{i+1}, \dots, \mathbf{x}_N) \tag{6.57a}$$

for $k = n$,

$$\dot{x}_{ki} = u_i + f_{ki}(x_{1i}, \dots, x_{ni}) + h_{ki}(\mathbf{x}_1, \dots, \mathbf{x}_{i-1}, \mathbf{x}_{i+1}, \dots, \mathbf{x}_N) \tag{6.57b}$$

where h_{ki} is the kth function of $\mathbf{h}_i = \sum_{j=1, j\neq i}^{N} \mathbf{g}_{ij}$.

Suppose the objective is $x_{1i} \to x_{1i,d} = 0$ for stabilization. As introduced in Chap. 2, first define the kth error surface of the ith subsystem as $S_{ki} = x_{ki} - x_{ki,d}$ for $1 \leq k \leq n-1$. After differentiating the error surface and using (6.57a) and (6.57b), we get

$$\dot{S}_{ki} = \dot{x}_{ki} - \dot{x}_{ki,d} = x_{(k+1)i} + f_{ki} + h_{ki} - \dot{x}_{ki,d}$$

where $\dot{x}_{1i,d} = 0$. If $x_{(k+1)i}$ is considered the synthetic input, then $S_{ki}\dot{S}_{ki} < 0$ is satisfied if $x_{(k+1)i} = \bar{x}_{(k+1)i}$ where

$$\bar{x}_{(k+1)i} := -f_{ki} - h_{ki} + \dot{x}_{ki,d} - K_{ki}S_{ki}$$

where $K_{ki} > 0$ is the controller gain. However, it is assumed that state information of other interconnected subsystems is not available, thus the interconnection function is assumed to be unknown. Therefore, the synthetic input is calculated as

$$\bar{x}_{(k+1)i} := -f_{ki} + \dot{x}_{ki,d} - K_{ki}S_{ki}. \tag{6.58}$$

In order to force $x_{(k+1)i} \to \bar{x}_{(k+1)i}$, define $S_{(k+1)i} := x_{(k+1)i} - x_{(k+1)i,d}$ where $x_{(k+1)i,d}$ is calculated from $\bar{x}_{(k+1)i}$ passed through a first order low-pass filter, i.e.,

$$\tau_{(k+1)i}\dot{x}_{(k+1)i,d} + x_{(k+1)i,d} = \bar{x}_{(k+1)i} \tag{6.59}$$

where $x_{(k+1)i,d}(0) := \bar{x}_{(k+1)i}(0)$.

After continuing this procedure up to $k = n - 1$, define $S_{ni} = x_{ni} - x_{ni,d}$. Finally, the desired control input is chosen as

$$u_i = -f_{ni} + \dot{x}_{ni,d} - K_{ni}S_{ni}. \tag{6.60}$$

It is noted that $\dot{x}_{ni,d} = (\bar{x}_{ni} - x_{ni,d})/\tau_{ni}$ from (6.59) for $k = n - 1$.

6.4.2 *Augmented Error Dynamics*

Similar with the DSC design methodology developed in Sect. 2.4, a set of augmented error dynamics can be derived for the closed-loop interconnected systems. After adding and subtracting $\hat{x}_{i+1}$, $x_{(i+1)d}$, and $\bar{x}_{i+1}$ in (6.57a), (6.57b) i.e.,

$$\dot{x}_{ki} = [x_{(k+1)i} - x_{(k+1)i,d}] + [x_{(k+1)i,d} - \bar{x}_{(k+1)i}] + \bar{x}_{(k+1)i} + f_{ki} + h_{ki}, \tag{6.61}$$

and using $\bar{x}_{i+1}$ from (6.58) and the definitions of S_i, (6.61) is rewritten as follows,

$$\dot{x}_{ki} = S_{(k+1)i} + \xi_{(k+1)i} + \dot{x}_{ki,d} - K_{ki}S_{ki} + h_{ki}$$

where the filter error is defined as $\xi_{(k+1)i} := x_{(k+1)i,d} - \bar{x}_{(k+1)i}$. After iterating up to $k = n$, it is rewritten as for $1 \le k \le n - 1$,

$$\dot{S}_{ki} = S_{(k+1)i} + \xi_{(k+1)i} - K_{ki}S_{ki} + h_{ki} \tag{6.62a}$$

for $k = n$,

$$\dot{S}_{ki} = -K_{ki}S_{ki} + h_{ki}. \tag{6.62b}$$

Next, since a set of low-pass filters are added for DDSC, the filter error dynamics should also be included. Differentiating the filter error defined by $\xi_i \in \Re^{n-1}$ and using (6.58), the filter error dynamics is for $1 \le k \le n - 1$,

$$\dot{\xi}_{(k+1)i} = \frac{d}{dt}(x_{(k+1)i,d} - \bar{x}_{(k+1)i}) = -\frac{1}{\tau_{(k+1)i}}\xi_{(k+1)i} + \dot{f}_{ki} - \ddot{x}_{ki,d} + K_{ki}\dot{S}_{ki}. \tag{6.63}$$

With $x_{i1,d} = \dot{x}_{i1,d} = \ddot{x}_{i1,d} = 0$ for stabilization, (6.63) can be rewritten as follows; for $k = 1$,

$$-K_{ki}\dot{S}_{ki} + \dot{\xi}_{(k+1)i} = -\frac{1}{\tau_{(k+1)i}}\xi_{(k+1)i} + \dot{f}_{ki} \tag{6.64a}$$

for $2 \le k \le n-1$,

$$-K_{ki}\dot{S}_{ki} + \dot{\xi}_{(k+1)i} - \frac{1}{\tau_{ki}}\dot{\xi}_{ki} = -\frac{1}{\tau_{(k+1)i}}\xi_{(k+1)i} + \dot{f}_{ki}. \tag{6.64b}$$

Finally, combining (6.64a), (6.64b) with (6.62a), (6.62b) we have the augmented error dynamics in the form of block matrices as follows:

$$\begin{bmatrix} \mathbf{I}_n & \mathbf{0} \\ -\mathbf{T}_{S_i} & \mathbf{T}_{\xi_i} \end{bmatrix} \begin{bmatrix} \dot{\mathbf{S}}_i \\ \dot{\xi}_i \end{bmatrix} = \begin{bmatrix} \mathbf{A}_{1i} & \mathbf{I}_{n-1} \\ \mathbf{0} & \mathbf{A}_{2i} \end{bmatrix} \begin{bmatrix} \mathbf{S}_i \\ \xi_i \end{bmatrix} + \begin{bmatrix} \mathbf{0} \\ \mathbf{I}_{n-1} \end{bmatrix} \mathbf{p}_i + \begin{bmatrix} \mathbf{I}_n \\ \mathbf{0} \end{bmatrix} \mathbf{h}_i \tag{6.65}$$

where

$$\mathbf{S}_i := [S_{1i}, \ldots, S_{ni}]^T \in \Re^n, \qquad \xi_i := [\xi_{2i}, \ldots, \xi_{ni}]^T \in \Re^{n-1},$$
$$\mathbf{p}_i := [\dot{f}_1, \ldots, \dot{f}_{n-1}]^T \in \Re^{n-1}$$

and the submatrices are the following:

$$\mathbf{A}_{1i} = \mathrm{diag}([1, \ldots, 1], 1) - \mathrm{diag}(K_{1i}, \ldots, K_{ni}) = U_i - \mathrm{diag}(K_{1i}, \ldots, K_{ni}),$$
$$\mathbf{A}_{2i} := -\mathrm{diag}(1/\tau_{2i}, \ldots, 1/\tau_{ni}),$$
$$\mathbf{T}_{S_i} := \left[\mathrm{diag}(K_{1i}, \ldots, K_{(n-1)i})\ 0_{n-1}\right] \in \Re^{(n-1)\times n},$$
$$\mathbf{T}_{\xi_i} := \mathbf{I}_{n-1} + \Gamma_i, \quad \Gamma_i = \mathrm{diag}\left(\left[-\frac{1}{\tau_{2i}}, \ldots, -\frac{1}{\tau_{(n-1)i}}\right], -1\right).$$

Since the matrix on the left hand side of (6.65) is invertible with an inverse given by

$$\begin{bmatrix} \mathbf{I}_n & \mathbf{0} \\ -\mathbf{T}_{S_i} & \mathbf{T}_{\xi_i} \end{bmatrix}^{-1} = \begin{bmatrix} \mathbf{I}_n & \mathbf{0} \\ \mathbf{T}_{\xi_i}^{-1}\mathbf{T}_{S_i} & \mathbf{T}_{\xi_i}^{-1} \end{bmatrix},$$

the augmented closed-loop error dynamics of N nonlinear subsystems with DDSC can be reformulated as

$$\dot{\mathbf{z}}_i = \mathbf{A}_i \mathbf{z}_i + \mathbf{B}_{pi}\mathbf{p}_i + \mathbf{B}_{hi}\mathbf{h}_i \tag{6.66}$$

where $\mathbf{z}_i := [\mathbf{S}_i^T\ \xi_i^T]^T \in \Re^{2n-1}$ and the matrices are

$$\mathbf{A}_i = \begin{bmatrix} \mathbf{A}_{i1} & \mathbf{I} \\ \mathbf{T}_{\xi_i}^{-1}\mathbf{T}_{S_i}\mathbf{A}_{i1} & \mathbf{T}_{\xi_i}^{-1}(\mathbf{T}_{S_i}\mathbf{I} - \mathbf{A}_{i2}) \end{bmatrix},$$
$$\mathbf{B}_{pi} = \begin{bmatrix} \mathbf{0} \\ \mathbf{T}_{\xi}^{-1} \end{bmatrix}, \qquad \mathbf{B}_{hi} = \begin{bmatrix} \mathbf{I}_n \\ \mathbf{T}_{\xi_i}^{-1}\mathbf{T}_{S_i} \end{bmatrix}.$$

Remark 6.4 While the error dynamics of centralized DSC for stabilization is derived in the form of $\dot{\mathbf{z}} = \mathbf{A}\mathbf{z} + \mathbf{B}_p\mathbf{p}$ in Chap. 2, the closed-loop error dynamics in (6.66) is extended with an additional term, $\mathbf{h}_i$. Therefore, with the assumption (6.56), it can be regarded as a diagonal norm-bounded Linear Differential Inclusion (LDI) if there exist a matrix C_{z_i} such that $|\mathbf{p}_i| \le |C_{z_i}\mathbf{z}_i|$. It is interesting to note that the augmented error dynamics with DDSC in (6.66) for stabilization is similar with the versions with centralized DSC for a tracking problem introduced in Sect. 2.6. This fact leads us to estimate the performance of DDSC in terms of quadratic tracking, which will be addressed next.

6.4.3 Decentralized Stabilization

The overall augmented error dynamics of N nonlinear subsystems with decentralized DSC is written as follows:

$$\begin{aligned}
\begin{bmatrix} \dot{\mathbf{z}}_1 \\ \vdots \\ \dot{\mathbf{z}}_N \end{bmatrix} &= \begin{bmatrix} \mathbf{A}_1 & \cdots & \mathbf{0} \\ \vdots & \ddots & \vdots \\ \mathbf{0} & \cdots & \mathbf{A}_N \end{bmatrix} \begin{bmatrix} z_1 \\ \vdots \\ z_N \end{bmatrix} + \begin{bmatrix} \mathbf{B}_{p1} & \cdots & \mathbf{0} \\ \vdots & \ddots & \vdots \\ \mathbf{0} & \cdots & \mathbf{B}_{pN} \end{bmatrix} \begin{bmatrix} \mathbf{p}_1 \\ \vdots \\ \mathbf{p}_N \end{bmatrix} \\
&\quad + \begin{bmatrix} \mathbf{B}_{h1} & \cdots & \mathbf{0} \\ \vdots & \ddots & \vdots \\ \mathbf{0} & \cdots & \mathbf{B}_{hN} \end{bmatrix} \begin{bmatrix} \mathbf{h}_1 \\ \vdots \\ \mathbf{h}_N \end{bmatrix} \\
\Longrightarrow \quad & \dot{\mathbf{z}} = \mathbf{A}\mathbf{z} + \mathbf{B}_p\mathbf{p} + \mathbf{B}_h\mathbf{h}.
\end{aligned}$$

If the centralized DSC proposed in Chap. 2 is used, the augmented error dynamics is derived in the form of

$$\dot{\mathbf{z}} = \mathbf{A}\mathbf{z} + \mathbf{B}_p\mathbf{p}$$

because the interconnection term $\mathbf{h}$ is known and canceled by either a synthetic input or control input. Then, the calculation of a positive definite matrix $P \in \Re^{M\times M}$ where $M = n \times N$ is necessary to guarantee quadratic stabilization. This may be a heavier burden for appropriate assignment of controller gains and filter time constants compared with the decentralized approach. This is a main motivation to develop DDSC.

The augmented error dynamics shown in (6.66) is decomposed into a linear low dimensional part and two potentially nonlinear and high dimensional parts. As done in Chap. 3, the first nonlinear part $\mathbf{p}_i$ can be decoupled into a vanishing and a nonvanishing perturbation. That is,

$$\begin{aligned}
\dot{f}_{1i} &= \frac{\partial f_{1i}}{\partial x_{1i}}\dot{x}_{1i} = \frac{\partial f_{1i}}{\partial x_{1i}}(S_{2i} + \xi_{2i} - K_{1i}S_{1i} + h_{1i}) = \frac{\partial f_{1i}}{\partial x_{1i}}c_{z1}\mathbf{z}_i + \frac{\partial f_{1i}}{\partial x_{1i}}h_{1i} \\
&= \mathbf{J}_{1i}C_{zi}\mathbf{z}_i + \mathbf{J}_{1i}h_{1i}, \\
\dot{f}_{ki} &= \sum_{j=1}^{k}\frac{\partial f_{ji}}{\partial x_{ji}}\dot{x}_{ji} = \sum_{j=1}^{k}\frac{\partial f_{ji}}{\partial x_{ji}}(S_{(j+1)i} + \xi_{(j+1)i} - \xi_{ji}/\tau_{ji} - K_{ji}S_{ji} + h_{ji}) \\
&= \sum_{j=1}^{k}\left(\frac{\partial f_{ji}}{\partial x_{ji}}c_{zj}\mathbf{z}_i + \frac{\partial f_{ji}}{\partial x_{ji}}h_{ji}\right) = \mathbf{J}_{ki}C_{zi}\mathbf{z}_i + \mathbf{J}_{ki}\mathbf{h}_i
\end{aligned}$$

where $k = 2, \ldots, n-1$, $\mathbf{J}_{ki}$ is the kth row of $\mathbf{J}_i = \partial \mathbf{f}_i / \partial \mathbf{x}_i \in \Re^{(n-1)\times(n-1)}$, and

$$c_{z1} = [-K_{1i}\ 1\ 0\ \cdots\ 0\ 1\ 0\ \cdots\ 0] \in \Re^{1\times(2n-1)},$$

$$c_{zj} = \left[\cdots\ -K_{ji}\ 1\ 0\ \cdots\ 0\ -\frac{1}{\tau_j}\ 1\ 0\ \cdots\ 0\right] \in \Re^{1\times(2n-1)} \quad \text{for } j = 2, \ldots, n-1,$$

$$C_{zi} = \begin{bmatrix} c_{z1} \\ \vdots \\ c_{z(n-1)} \end{bmatrix} \in \Re^{(n-1)\times(2n-1)}.$$

Therefore, the nonlinear function $\mathbf{p}_i$ can be written as

$$\mathbf{p}_i = \mathbf{J}_i C_z \mathbf{z}_i + \mathbf{J}_i \mathbf{h}_i := \mathbf{w}_i + \mathbf{J}_i \mathbf{h}_i. \tag{6.67}$$

Combining (6.66) with (6.67), the error dynamics can be written as follows:

$$\dot{\mathbf{z}}_i = \mathbf{A}_i \mathbf{z}_i + \mathbf{B}_{wi} \mathbf{w}_i + [\mathbf{B}_{hi}\ \mathbf{B}_{wi}] \begin{bmatrix} \mathbf{h}_i \\ \mathbf{J}_i \mathbf{h}_i \end{bmatrix} \tag{6.68}$$

where $\mathbf{B}_{wi} = \mathbf{B}_{pi}$. Furthermore, using assumptions of the given system in (6.55), the following additional constraints can be obtained:

$$\|\mathbf{w}_i\| = \|\mathbf{J}_i C_z \mathbf{z}_i\| \le \gamma_i \|C_z \mathbf{z}_i\|,$$

$$\begin{bmatrix} \mathbf{h}_i^T & (\mathbf{J}_i \mathbf{h}_i)^T \end{bmatrix} \begin{bmatrix} \mathbf{h}_i \\ \mathbf{J}_i \mathbf{h}_i \end{bmatrix} = \|\mathbf{h}_i\|^2 + \|\mathbf{J}_i \mathbf{h}_i\|^2 \le c_i^2 + \gamma_i^2 c_i^2 = \mu_i^2.$$

Finally, the error dynamics is written as follows:

$$\begin{aligned} \dot{\mathbf{z}}_i &= \mathbf{A}_i \mathbf{z}_i + \mathbf{B}_{wi} \mathbf{w}_i + \mathbf{B}_{ri} \mathbf{r}_i, \\ \|\mathbf{w}_i\| &\le \gamma_i \|C_z \mathbf{z}_i\|, \qquad \|\mathbf{r}_i\| \le 1 \end{aligned} \tag{6.69}$$

where

$$\mathbf{r}_i = \frac{1}{\mu_i} \begin{bmatrix} \mathbf{h}_i \\ \mathbf{J}_i \mathbf{h}_i \end{bmatrix}, \qquad \mathbf{B}_{ri} = \mu_i [\mathbf{B}_{hi}\ \mathbf{B}_{wi}].$$

Therefore, the closed-loop error dynamics in (6.69) can be regarded as a set of norm-bounded LDIs. Then, the ellipsoidal bound of the reachable sets is obtained as follows:

Theorem 6.4 *Suppose the matrices* $\mathbf{A}_i$, $\mathbf{B}_{wi}$ *and* $\mathbf{B}_{ri}$ *are given. The ellipsoid,* $\mathcal{E}_i = \{z_i \in \Re^n | z_i^T P z_i \le 1\}$, *contains the reachable sets of the augmented error dynamics in* (6.69) *if there exist* $P > 0$, $\sigma \ge 0$, *and* $\alpha \ge 0$ *such that*

$$\begin{bmatrix} \mathbf{A}_i^T P + P\mathbf{A}_i + \alpha P + \sigma C_{zi}^T C_{zi} & P\mathbf{B}_{wi} & P\mathbf{B}_{ri} \\ \mathbf{B}_{wi}^T P & -\sigma I & 0 \\ \mathbf{B}_{ri}^T P & 0 & -\alpha I \end{bmatrix} \le 0. \tag{6.70}$$

It is noted that the proof of the theorem is omitted due to similarity with one of Theorem 2.4. As introduced in Sect. 2.6, by minimizing a size of the ellipsoid (refer to Algorithm 2.2), the tracking performance can be estimated for the given set of controller gains.

Remark 6.5 If a holonomic system of N particles is considered in (6.2), the augmented error dynamics (6.66) is simplified to

$$\begin{cases} \dot{\mathbf{z}}_i = \mathbf{A}_i \mathbf{z}_i + \mathbf{B}_{hi} \mathbf{h}_i, \\ \mathbf{y}_i = C_i \mathbf{z}_i. \end{cases} \tag{6.71}$$

With the assumption in (6.56), (6.71) are the same as (6.30). Therefore, a set of controller gains can be designed via Algorithm 6.2 in Sect. 6.2.3.

Example 6.3 (Decentralized stabilization of coupled inverted double pendulums) Consider the coupled inverted double pendulums in Example 6.1 and the performance constraint is given as $|x_{1i}(t)| \leq 0.05$ after $t \geq T$. The model can be written in the form of (6.55) as follows:

$$\begin{aligned} \dot{\mathbf{x}}_i &= \begin{bmatrix} 0 & 1 \\ 0 & 0 \end{bmatrix} \mathbf{x}_i + \begin{bmatrix} 0 \\ 1 \end{bmatrix} u_i + \begin{bmatrix} 0 \\ \frac{g}{l} \sin x_{i1} - \frac{ka^2}{m_i l^2} \sin x_{i1} \cos x_{i1} \end{bmatrix} \\ &\quad + \begin{bmatrix} 0 \\ \frac{ka^2}{m_i l^2} \sin x_{1j} \cos x_{1j} \end{bmatrix}_{j \neq i} \\ &= U_i \mathbf{x}_i + B_i u_i + \mathbf{f}_i + \sum_{j=1, j\neq i}^{2} \mathbf{g}_{ij}(\mathbf{x}_j) \end{aligned}$$

where $\mathbf{x}_i = [\theta_i \ \dot{\theta}_i]^T$ and the assumption in (6.55) is satisfied as follows:

$$h_i^T h_i = \left(\frac{ka^2}{m_i l^2}\right)^2 \sin^2 x_{j1} \cos^2 x_{j1} = \left(\frac{b_i}{2} \sin 2x_{j1}\right)^2 \leq \left(\frac{b_i}{2}\right)^2 := c_i^2$$

where $b_i = \frac{ka^2}{m_i l^2}$ and the value for simulations is given in Example 6.1.

Augmented Error Dynamics Define the first error surface as $S_{1i} := x_{1i}$. After differentiating it, the synthetic input $\bar{x}_{2i}$ and the first low-pass filter are derived as follows:

$$\begin{aligned} &\bar{x}_{2i} = -K_{1i} S_{1i}, \\ &\tau_{2i} \dot{x}_{2i,d} + x_{2i,d} = \bar{x}_{2i}, \quad x_{2i,d}(0) := \bar{x}_{2i}(0). \end{aligned}$$

Next, defining the second surface as $S_{2i} := x_{2i} - x_{2i,d}$ and differentiating it, we get the control input as

$$\begin{aligned} \dot{S}_{2i} &= u_i + \frac{g}{l} \sin x_{1i} - \frac{ka^2}{m_i l^2} \sin x_{1i} \cos x_{1i} - \dot{x}_{2i,d} + \frac{ka^2}{m_i l^2} \sin x_{1j} \cos x_{1j} \\ \Longrightarrow \quad u_i &= -\frac{g}{l} \sin x_{1i} + \frac{ka^2}{m_i l^2} \sin x_{1i} \cos x_{1i} + \dot{x}_{2i,d} - K_{2i} S_{2i}. \end{aligned}$$

It is noted that the interconnection term between subsystems, $\frac{ka^2}{m_i l^2} \sin x_{1j} \cos x_{1j}$, is not used in u_i while it is in u_i in Example 6.1. Therefore, using the result of (6.71), we can derive the augmented closed-loop error dynamics as follows:

$$\dot{\mathbf{z}}_i = A_i \mathbf{z}_i + B_{hi} \tilde{h}_i, \quad |\tilde{h}_i| \leq 1 \tag{6.72}$$

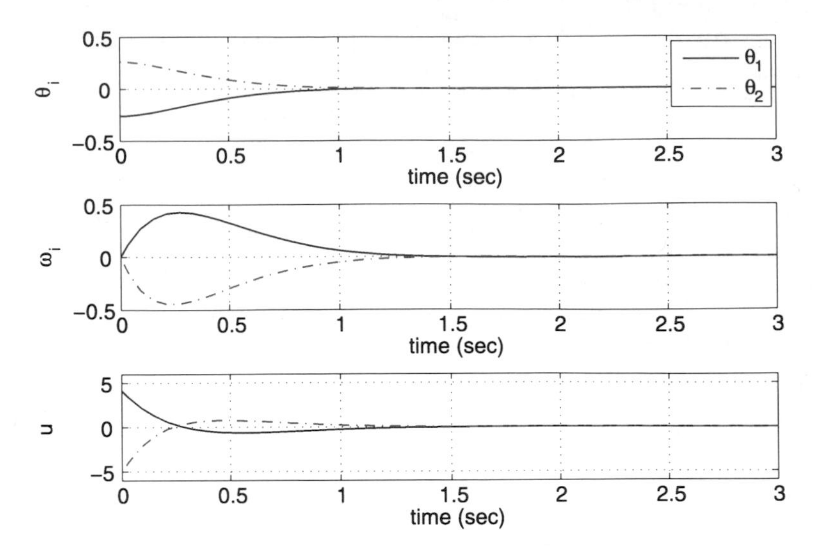

Fig. 6.8 Time responses of $\mathbf{x}_i$ in the presence of a parametric uncertainty $b_1(t) = 0.25[1 + \sin(50t)]$ for the given $\{\theta_1(0), \theta_2(0)\} = \{-\pi/12, \pi/12\}$

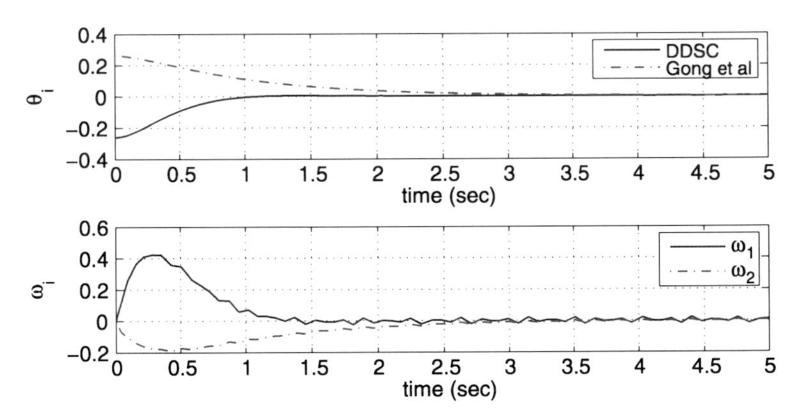

Fig. 6.9 Time responses of $\mathbf{x}_i$ in presence of measurement noise $\mathbf{y}_i = \mathbf{x}_i + [0.05\ 0.05]^T \sin(50t)$

where $\mathbf{z}_i$ and A_i are given in (6.11), $\tilde{h}_i = \sin 2x_{1j}$, and $B_{hi} = [0\ b_i/2\ 0]^T$.

Design of Controller Gains

Step 1. Use $K_{1i} = 2.3357$ which is calculated in Example 6.1.

Step 2a. With $\delta = 0.02$ and $c_0 = 0.05$, COP (6.38) is solved iteratively with logarithmic spacing of $\alpha[j] \in [0.1, 10]$ for $j = 1, \ldots, 20$. The solutions of COP (6.38) are, respectively,

$$\sigma_1 = 0.032, \qquad \sigma_2 = 0.0401, \quad \text{and} \quad Q = c_0^2 I.$$

Then, the resulting matrix of Θ_i is

$$\Theta_i = -(\sigma_i/2)F^T Q^{-1} = -\left(\sigma_i/2c_0^2\right)F^T.$$

Thus, a set of controller gains are calculated as follows:

$$\{K_{1i}, K_{2i}, \tau_{2i}\}_{i=1} = \{2.3357, 6.4033, 0.1561\},$$
$$\{K_{1i}, K_{2i}, \tau_{2i}\}_{i=2} = \{2.3357, 8.0269, 0.1246\}.$$

When the initial condition is given by $\{\theta_1(0), \theta_2(0)\} = \{-\pi/12, \pi/12\}$ and $\dot{\theta}_i = 0$, and the parametric uncertainty is considered as $b_1(t) = 0.25[1 + \sin(50t)]$ only for the first inverted pendulum for comparison, Fig. 6.8 shows the time responses of $\mathbf{x}_i$. It is validated that θ_i stay in a boundary defined by $|x_{1i}| \leq 0.05$ via DDSC.

Furthermore, in the presence of measurement noise as considered in Example 6.1, the results of DDSC are compared with those of the nonlinear control, $u_i = -(1.25 + |2.92560_i + 2.8251\dot{\theta}_i|)(2.92560_i + 2.8251\dot{\theta}_i)$, proposed by Gong et al. [31]. It is shown in Fig. 6.9 that performance of both controllers is robust and similar in terms of trajectories of $\mathbf{x}_i$. It is noted that the response of θ_i with DDSC is faster.

Part II
Applications

Chapter 7
Automated Vehicle Control

Reliable and robust control has received more attention due to a growing demand for more safety and reliability of large-scale dynamic systems such as aircraft, chemical processing plants, and automated highway systems (AHS). A major objective has been to synthesize a control structure so that the system performs satisfactorily under anomalistic conditions as well as in the presence of uncertainties and disturbances. In this chapter, we consider the problem of designing reliable control systems and work to develop a system that enhances the reliability and safety for an AHS. The idea of AHS has been realized due to the rapid advances in microprocessors and electronics including communication and sensor technologies, for instance, "platooning" of transit buses demonstrated by the California PATH Program in 2003 (refer to Fig. 7.1). However, there are still several open areas in need of further research for deployment of an AHS. One critical study is to consider and accommodate any malfunction or failure in both sensors and actuators, which may endanger the passengers or cause a catastrophe on the highway.

For large-scale nonlinear systems, the design of a reliable control system is generally very complicated and requires a systematic procedure to realize its ultimate advantages. First, we need to develop a systematic analysis and control design methodology to achieve control objectives as well as stability under the assumption of no malfunction or failure. As discussed in Chap. 3, this problem is an interesting question itself in the area of robust nonlinear control. Next, a hierarchical hybrid architecture for the reliable control system will be developed not only to build from a simple to a complex system hierarchically, but also to integrate its components *modularly* in the sense of their individual integration and extension. Finally, specific design of components needs to be investigated in the framework of the hierarchical hybrid structure to accommodate faults occurring in the system.

In the area of fault tolerant control (FTC), a fault is regarded as any kind of malfunction in a system, and may lead to system instability or result in unacceptable performance degradation. Such a fault can occur in any component of the system such as sensors, actuators, and system components. Fault tolerant control aims to make the system insensitive to model uncertainties as well as to accommodate these potential faults in the design of the control system. Many methodologies of

B. Song, J.K. Hedrick, *Dynamic Surface Control of Uncertain Nonlinear Systems*, Communications and Control Engineering,
DOI 10.1007/978-0-85729-632-0_7,

Fig. 7.1 Automated transit buses for DEMO'03 (California PATH at U.C. Berkeley)

FTC design can be found in the literature, however most of these techniques depend on a mathematical model of both the physical system and the effects of faults upon that system. Model-based FTC is typically characterized as either being passive or active [66]. Passive approaches consider faults as a type of bounded uncertainty, and rely on robust control theory to design a controller which is robust to both model uncertainties and faults [65, 103]. Hence, the passive approach does not require additional controllers to compensate for the considered faults. However, these passive FTC designs tend to rely on critical assumptions, such that the faulty system model is known completely or a specific class of faults are considered [92].

Active approaches typically divide the FTC into three parts which are designed independently: a Fault Detection and Diagnosis (FDD) System, a Fault Management System (FMS) [113], and a set of reconfigurable controllers. The goal of the FDD system is to detect and identify faults in sensors, actuators, and components of a system, and subsequently supplies the FMS with information about the current status of the controlled system. The FMS then decides on the appropriate controller to achieve the control objectives of the current mode based on this information, the objectives, and the control reconfiguration strategy. Finally, the set of controllers contains all of the possible control laws available to the FTC for both normal and faulty scenarios. Although a great deal of effort has been devoted to the problem of FDD using model-based techniques as shown in survey papers [25, 26, 30, 43, 106] and references therein, few studies have been dedicated to the other two parts of the FTC or to the integrated design of the FTC [113].

Figure 7.2 presents a hybrid hierarchical structure of FTC incorporating the three tasks. The goal of the FDD system is to detect and identify faults in sensors, actua-

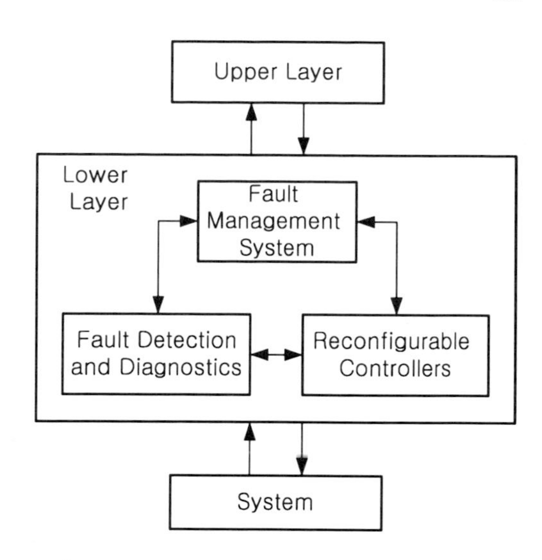

Fig. 7.2 Schematic framework of Fault Tolerant Control

tors, and components of a system, and subsequently supplies the fault management system (FMS) with information about the current status of the controlled system. The FMS then evaluates performance and decides on the appropriate controller to achieve the control objectives of the current mode based upon this information, the objectives, and the control reconfiguration strategy. Finally, the set of controllers contains all of the possible control laws available to the FTC for both normal and faulty scenarios. The proposed FTC approach shares ideas of both the passive and active FTC approaches to accommodate faults in a large-scale system. In other words, the initial controller for the non-faulty system will be designed to provide robust performance with respect to model uncertainties as well as pre-defined faults consistent with the passive approach. However, if the performance degradation due to faults is not acceptable through the performance evaluation task, the initial controller will be switched to one of a set of reconfigurable controllers via controller reconfiguration. In this chapter, it will be shown how a fault classification scheme leads quite naturally to the performance evaluation task, which provides a decision logic between the passive and active approach.

This chapter is divided as follows: In Sect. 7.1, DSC is applied to automated longitudinal control of a passenger vehicle and a simulation study is performed for stability and performance analysis. Section 7.2 presents a statement of the non-faulty system as well as the faulty system under consideration, outlines the development of the DSC law for the non-faulty system, and shows the derivation of the closed-loop error dynamics for the faulty system. Furthermore, Sect. 7.2 will give specific details on explaining FDD and the fault classification for the proposed FTC. Finally, the fault classification is applied to the problem in Sect. 7.1 for both single and multiple fault classification.

Table 7.1 System variables and descriptions

Variable	Description	Variable	Description
v	velocity of the vehicle (m/s)	T_e	net engine torque (N m)
T_b	braking torque (N m)	P_m	manifold pressure (KPa)
P_w	brake pressure (KPa)	α	throttle angle (rad)
P_{mc}	master cylinder pressure (KPa)	TC	empirical throttle characteristic

7.1 Application to Longitudinal Vehicle Control

In this section, we will show the application of all the theoretical results developed in Chap. 2 and 3 to automated longitudinal control of a passenger vehicle. A longitudinal vehicle model for controller design and the coordinated throttle and brake control using DSC has been developed and demonstrated by the California PATH Program [28, 37, 59]. In this section, the choice of DSC gains and stability verification via convex optimization will be investigated.

7.1.1 Engine and Brake Control via DSC

A three state longitudinal vehicle model is used for the purpose of developing a controller for the case of no faults [19, 28, 37]. The model has one state based on the vehicle kinematics and another two states which model the engine and brake system dynamics as follows:

$$\dot{v} = \frac{T_e - R_g(T_b + M_r + h \cdot C_a v^2)}{J_{eq}} + \Delta f_1(v), \tag{7.1a}$$

$$\dot{P}_m = \frac{R_{air} T_m}{V_m}\left[\mathrm{MAX} \cdot \mathrm{TC}(\alpha) \cdot \mathrm{PRI}(P_m) - \dot{m}_{ao}\right] + \Delta f_{2e}(P_m), \tag{7.1b}$$

$$\dot{P}_w = \frac{\partial P_w}{\partial V} \sigma C_q \sqrt{|P_{mc} - P_w|} + \Delta f_{2b}(P_w) \tag{7.1c}$$

with the hypotheses that T_e and T_b can be approximated in a certain range of velocity as

$$T_e = C_e P_m \quad \text{and} \quad T_b = \begin{cases} K_b(P_w - P_{po}), & \text{if } P_w \geq P_{po}, \\ 0, & \text{otherwise} \end{cases} \tag{7.2}$$

where Δf_1, Δf_{21}, and Δf_{22} are the model uncertainties and

$$\mathrm{PRI} = 1 - \exp\left[9\left(\frac{P_m}{P_{atm}} - 1\right)\right], \qquad \sigma = \mathrm{sgn}(P_{mc} - P_w).$$

Furthermore, the system variables and parameters are listed in Tables 7.1 and 7.2, respectively, and this model can be extended to include a more detailed component dynamics [19, 28].

Table 7.2 System parameters and values for simulation

Parameter	Description	Value
h	effective wheel radius	0.323 (m)
M_r	rolling resistance moment	80.75 (N m)
C_a	aerodynamic drag coefficient	0.3693 (Kg/m)
J_{eq}	equivalent inertia	179.5821 (Kg m)
R_{air}	gas constant for air	0.287 (KJ/Kg/K)
V_m	manifold volume	0.00447 (m^3)
C_q	brake pressure coefficient	1.49
C_e	engine torque coefficient	2.4562(N m/KPa)
K_b	brake torque coefficient	0.9 (N m/KPa)
MAX	maximum flow rate	0.684 (Kg/s)
T_m	manifold temperature	301 (K)
P_{po}	push-out pressure	115 (KPa)

Here discussion is limited to a velocity tracking problem of the three state system for the longitudinal vehicle control where the desired trajectory v_{des} and its first and second derivatives $\dot{v}_{des}, \ddot{v}_{des}$ are bounded. Then, the control laws for determining the commanded engine torque and brake pressure will be designed using a dynamic surface controller [37]. Moreover, this controller was implemented on the California PATH vehicles in DEMO'97, an automated highway technology demonstration that occurred in San Diego, California in August of 1997 [37]. Using the design procedure introduced in Sect. 2.3, the first error surface representing the velocity tracking error is defined as $S_{1e} = v - v_{des}$ where the subscript e represents engine control. After differentiating it and using (7.1a) and (7.2),

$$\dot{S}_{1e} = \frac{C_e P_m - R_g(T_b + M_r + h \cdot C_a \cdot v^2)}{J_{eq}} + \Delta f_1 - \dot{v}_{des}.$$

The sliding condition for S_{1e} is satisfied if $S_{1e}\dot{S}_{1e} < 0$, however there is no direct control over the surface dynamics. But if P_m is considered as the forcing term for the surface dynamics, then the sliding condition on some boundary layer is satisfied if $P_m = \bar{P}_m$ where

$$\bar{P}_m = \frac{R_g}{C_e}\left(T_b + M_r + C_a h v^2\right) + \frac{J_{eq}}{C_e}(\dot{v}_{des} - K_{1e}S_{1e}) \tag{7.3}$$

where $\bar{P}_m$ is chosen to drive $S_{11} \to 0$. However, since P_m is not controlled directly, a second surface is required to ensure that P_m tracks $\bar{P}_m$. Next, define the second surface as $S_{2e} = P_m - P_{mdes}$ where P_{mdes} equals $\bar{P}_m$ passed through a first order low-pass filter, i.e.,

$$\tau_{2e}\dot{P}_{mdes} + P_{mdes} = \bar{P}_m, \quad P_{mdes}(0) := \bar{P}_m(0).$$

The subsequent desired error dynamics is

$$\dot{S}_{2e} = \dot{P}_m - \dot{P}_{mdes} = -K_{2e}S_{2e}.$$

Using (7.1b), the desired throttle angle is given by

$$\alpha_d = \mathrm{TC}^{-1}\left(\frac{\dot{m}_{ao} + V_m(\dot{P}_{mdes} - K_{2e}S_{2e})/(R_{air}T_m)}{\mathrm{MAX}\cdot\mathrm{PRI}}\right) \tag{7.4}$$

where TC^{-1} is the inverse function of the empirical throttle characteristic.

A control law for the brake system can be derived similarly by defining $S_{1b} = S_{1e}$ where the subscript b represents brake control. The resulting equation for the desired brake pressure is

$$\bar{P}_w = \frac{1}{K_b R_g}\left[T_{ect} - R_g\left(M_r + C_a h v^2\right) - J_{eq}(\dot{v}_{des} - K_{1b}S_{1b})\right] + P_{po} \tag{7.5}$$

where T_{ect} is the minimum or closed throttle torque. For the second surface, define $S_{2b} = P_w - P_{wdes}$ where

$$\tau_{2b}\dot{P}_{wdes} + P_{wdes} = \bar{P}_w, \quad P_{wdes}(0) := \bar{P}_w(0).$$

After differentiating S_{2b} and using (7.1c), the master cylinder pressure commanded is

$$P_{mc} = P_w + \sigma\left(\frac{\dot{P}_{wdes} - K_{2b}S_{2b}}{\frac{\partial P_w}{\partial V}C_q}\right)^2 \tag{7.6}$$

where $\sigma = \mathrm{sgn}(\dot{P}_{wdes} - K_{2b}S_{2b})$. The specific control mode of the vehicle is determined by switching condition based on the desired and residual acceleration computed by the engine control law. The residual acceleration is defined as

$$a_{resid} = \frac{1}{J_{eq}}\left[T_{ect} - R_g\left(M_r + C_a h v^2\right)\right]$$

and represents the acceleration of the vehicle as a result of the closed throttle torque, rolling resistance, and aerodynamic drag. For example, if the engine controller computes $a_{syn} \geq a_{resid}$ where $a_{syn} := \dot{v}_{des} - K_{1e}S_{1e}$, then engine control is used. Otherwise, brake control is activated [28, 37].

7.1.2 Switched Closed Loop Error Dynamics

In order to analyze and design the DSC system, consider the closed-loop dynamics for engine control as follows:

$$\begin{cases} \dot{v} = \frac{\bar{P}_m + (P_m - P_{mdes}) + (P_{mdes} - \bar{P}_m)}{c_1} - \frac{R_g M_r}{J_{eq}} - f_1(v) + \Delta f_1, \\ \dot{P}_m = \frac{R_{air}T_m}{V_m}[\mathrm{MAX}\cdot\mathrm{TC}(\alpha_d)\cdot\mathrm{PRI}(P_m) - \dot{m}_{ao}] + \Delta f_{2e}, \\ \dot{\xi}_{2e} = \dot{P}_{mdes} - \dot{\bar{P}}_m \end{cases} \tag{7.7}$$

where ξ_{2e} is the low-pass filter error,

$$a_e = \frac{J_{eq}}{C_e}, \quad \text{and} \quad f_1(v) = \frac{R_g C_a h}{J_{eq}}v^2.$$

By use of (7.3) and (7.4), the error dynamics for engine control in (7.7) is written in terms of the surface and filter error

$$\begin{cases} \dot{S}_{1e} = -K_{1e}S_{1e} + (S_{2e} + \xi_{2e})/a_e + \Delta f_1, \\ \dot{S}_{2e} = -K_{2e}S_{2e} + \Delta f_{2e}, \\ \dot{\xi}_{2e} = -\xi_{2e}/\tau_{2e} + a_e[K_{1e}\ \dot{S}_{1e} - \ddot{v}_{des} - \dot{f}_1]. \end{cases} \tag{7.8}$$

Similarly, the longitudinal equation for brake control ($P_w \geq P_{po}$) is presented as follows:

$$\dot{v} = \frac{T_{ect} - R_g K_b[\bar{P}_w - P_{po} + (P_w - P_{wdes}) + (P_{wdes} - \bar{P}_w)]}{J_{eq}} - \frac{R_g M_r}{J_{eq}} - f_1(v) + \Delta f_1.$$

Then, using (7.5) and (7.6), the closed-loop error dynamics for braking control is given by

$$\begin{cases} \dot{S}_{1b} = -K_{1b}S_{1b} + (S_{2b} + \xi_{2b})/a_b + \Delta f_1, \\ \dot{S}_{2b} = -K_{2b}S_{2b} + \Delta f_{2b}, \\ \dot{\xi}_{2b} = -\xi_{2b}/\tau_{2b} + a_b[K_{1b}\dot{S}_{1b} - \ddot{v}_{des} - \dot{f}_1] \end{cases} \tag{7.9}$$

where $a_b = -\frac{J_{eq}}{R_g K_b}$.

Finally, combining (7.8) with (7.9), we obtain the following switched error dynamics:

$$T_i\dot{z}_i = \bar{A}_i z_i + \bar{B}_{di}\dot{f}_1 + [e_1\ e_2\ \bar{B}_{di}]\begin{bmatrix} \Delta f_1 \\ \Delta f_{2i} \\ \ddot{v}_{des} \end{bmatrix}$$

$$\Longrightarrow \quad \dot{z}_i = A_i z_i + B_{di}\dot{f}_1 + B_{ri}r_{ui} \quad \text{for } i = e, b \tag{7.10}$$

where $z_i = [S_{1i}\ S_{2i}\ \xi_{2i}]^T$, $r_{ui} = [\Delta f_1\ \Delta f_{2i}\ \ddot{v}_{des}]^T$, and the system matrices are

$$T_i = \begin{bmatrix} 1 & 0 & 0 \\ 0 & 1 & 0 \\ -a_i K_{1i} & 0 & 1 \end{bmatrix}, \qquad \bar{A}_i = \begin{bmatrix} -K_{1i} & 1/a_i & 1/a_i \\ 0 & -K_{2i} & 0 \\ 0 & 0 & -1/\tau_{2i} \end{bmatrix},$$

$$A_i = T_i^{-1}\bar{A}_i, \qquad B_{ri} = T_i^{-1}[e_1\ e_2\ \bar{B}_{di}], \quad B_{di} = T_i^{-1}\bar{B}_{di} = \bar{B}_{di} = [0\,0\ -a_i]^T,$$

$$e_1 = [1\,0\,0]^T, \qquad e_2 = [0\,1\,0]^T.$$

It is interesting to note that the third-order longitudinal model above is not strictly in a class of nonlinear systems with strict-feedback form in (2.3) but in a more general class of nonlinear systems in (2.13) discussed in Chap. 2. It is shown that the augmented error dynamics in form of a linear system subject to perturbation terms is still available as shown in (7.10). Moreover, using the result of Lemma 3.1,

(7.8) and (7.9), $\dot{f}_1$ can be written as follows:

$$\begin{aligned}\dot{f}_1 &= \frac{\partial f_1}{\partial v}\dot{v} = \frac{\partial f_1}{\partial v}(\dot{S}_{1i} + \dot{v}_{des}) \\ &= \frac{2R_g C_a h v}{J_{eq}}\{-K_{1i}S_{1i} + (S_{2i} + \xi_{2i})/a_i + \Delta f_1 + \dot{v}_{des}\} \\ &= J_{11}c_{zi}z_i + J_{11}\dot{v}_{des} + J_{11}\Delta f_1 \end{aligned} \tag{7.11}$$

where

$$J_{11} = \frac{\partial f_1}{\partial v} = \frac{2R_g C_a h v}{J_{eq}} \quad \text{and} \quad c_{zi} = \left[-K_{1i}\frac{1}{a_i}\,\frac{1}{a_i}\right].$$

7.1.3 Simultaneous Quadratic Boundedness

Although the error dynamics for engine and brake control is a switched system, the stability and tracking performance of this system can be posed as a convex optimization problem by extending the methodology proposed in Chaps. 2 and 3. Quadratic tracking for the given error dynamics will be investigated in this section. Without considering model uncertainty, the error dynamics in (7.10) can be simplified as

$$\begin{aligned}\dot{z}_i &= A_i z_i + B_{di}\dot{f}_1 + B_{di}\ddot{v}_{des} = A_i z_i + B_{wi}w_i + [B_{di}\ B_{di}]\begin{bmatrix} J_{11}\dot{v}_{des} \\ \ddot{v}_{des}\end{bmatrix} \\ &:= A_i z_i + B_{wi}w_i + B_{ri}r \end{aligned} \tag{7.12}$$

where $w_i = J_{11}c_{zi}z_i$ and $B_{wi} = B_{di}$. Suppose a convex and compact set $\mathscr{D}_i$ is defined as

$$\mathscr{D}_i = \left\{\mathbf{x}_i \in \Re^2 \,|\, |v| \le v_{\max},\ \left\|\left[v\ \dot{v}\ \ddot{v}^2\right]^T\right\| \le c,\ \text{for } v_{\max}, c > 0\right\} \tag{7.13}$$

where $\mathbf{x}_e = [v\ P_m]^T \in \Re^2$ for engine control and $\mathbf{x}_b = [v\ P_b]^T \in \Re^2$ for brake control. If $v_{des}(t)$ is a feasible output trajectory in $\mathscr{D}_i$ (refer to Definition 2.2), there exists a positive constant γ such that

$$|J_{11}| = \frac{\partial f_1}{\partial v} = \left|\frac{2R_g C_a h v}{J_{eq}}\right| \le \left|\frac{2R_g C_a h v_{\max}}{J_{eq}}\right| := \gamma, \quad \forall v \in \mathscr{D}_i \tag{7.14}$$

where C_a and J_{eq} are given as in Table 7.2, and R_g is determined depending on the current gear engagement status. Moreover, r in (7.12) is norm-bounded as

$$\|r\|_2^2 = (J_{11}\dot{v}_{des})^2 + (\ddot{v}_{des})^2 \le (\gamma^2 + 1)c^2 := r_0^2. \tag{7.15}$$

After normalizing r as $\tilde{r} = r/r_0$ and $\tilde{B}_{ri} = r_0 B_{ri}$, and defining $\tilde{c}_{zi} = \gamma c_{zi}$, the error dynamics in (7.12) is summarized in the DNLDI form as follows:

$$\begin{cases} \dot{z}_i = A_i z_i + B_{wi}w_i + \tilde{B}_{ri}\tilde{r}, \\ |w_i| \le |\tilde{c}_{zi}z_i|, \\ \|\tilde{r}\| \le 1. \end{cases} \quad \text{for } i = e, b, \tag{7.16}$$

While each error dynamics in (7.16) for $i = e, b$ is equivalent to (2.57) in Sect. 2.6, they are in a class of the switched DNLDI. Therefore, the result of Theorem 2.4 and corresponding COP (2.62) can be extended to the switched error dynamics in (7.16).

That is, simultaneous quadratic boundedness of the switched error dynamics can be defined as follows:

Definition 7.1 Suppose a set of controller gains, Θ, is given, $\|r\| \neq 0$, and v_{des} in r is the feasible output trajectory in $\mathscr{D}_i$. The switched error dynamics in (7.16) is *simultaneously quadratically bounded with Lyapunov matrix P* if there exists $P > 0$ such that

$$z_i^T P z_i > 1 \quad \text{implies} \; \left(A_i z + B_{wi} w_i + \tilde{B}_{ri}\tilde{r}\right)^T P z + z^T P \left(A_i z + B_{wi} w_i + \tilde{B}_{ri}\tilde{r}\right) < 0$$

for all nonzero $z_i \in \mathscr{E}_P = \{z_i \in \Re^{n_z} | z_i^T P z_i \leq 1\}$.

Then, the smallest ellipsoid containing reachable sets of the error dynamics (7.16) can be calculated as follows: for a fixed $\alpha \in [a,\ b]$,

$$\begin{array}{ll} \text{maximize} & \lambda_{\min}(P) \\ \text{subject to} & P > 0, \qquad \sigma \geq 0, \\ & \begin{bmatrix} A_i^T P + P A_i + \alpha P + \sigma \tilde{c}_{zi}^T \tilde{c}_{zi} & P B_{wi} & P \tilde{B}_{ri} \\ B_{wi}^T P & -\sigma & 0 \\ \tilde{B}_{ri}^T P & 0 & -\alpha I \end{bmatrix} < 0, \quad i = e, b. \end{array} \tag{7.17}$$

Example 7.1 (Performance estimation of longitudinal controller) Suppose the velocity trajectory is given as in Fig. 7.3. Then, a convex set $\mathscr{D}_i$ in (7.13) can be defined and let $v_{\max} = 26$ and $c^2 = 2$ for $\mathscr{D}_i$. Furthermore, γ in (7.14) and r_0 in (7.15) are calculated as follows:

$$\gamma = \frac{2 R_g C_a h v_{\max}}{J_{eq}} = 0.0106, \quad r_0 = \sqrt{(\gamma^2 + 1)c^2} = 1.4143$$

where the system parameters are given in Table 7.2. The convex optimization problem in (7.17) can be solved numerically for the given set of controller gains using MATLAB and CVX (refer to Sect. 2.6). Two different sets of controller gains are tested to investigate how they affect a size of ellipsoid. That is, the first set is $\Theta_1 = \{K_{1i}, K_{2e}, K_{2b}, \tau_{2i}\} = \{1, 2.5, 5, 0.02\}$ and the other one is $\Theta_2 = \{1, 5, 10, 0.02\}$. It is noted that only different magnitudes of K_{2i} are used.

To compute the ultimate bound of the closed-loop error dynamics for the given set of controller gain Θ_1, COP (7.17) is solved iteratively for a fixed α. That is, after 20 logarithmically equally spaced points between 10^{-1} and 10^1 are generated for the α, the minimum of the maximum diameter, which is defined as $d_{\max} = 2/\sqrt{\lambda_{\min}(P)}$, is obtained when $\alpha = 1.4384$ (see in the left plot of Fig. 7.4). Then the 20 linearly equally spaced points between 1.1288 and 1.833 are generated and the iterative computation of COP (7.17) is performed for each α. Finally, for $\alpha = 1.6477$, the maximum diameter of the ellipsoid, $d_{\max}$, is 208.5897 which is the semi-axis in the ξ_2 axis. Furthermore, the corresponding ellipsoid is shown in Fig. 7.6.

Similarly, the maximum diameter of the ellipsoid for Θ_2 is 210.4577 when $\alpha = 1.6746$ (see in the right plot of Fig. 7.5). The corresponding ellipsoid is drawn in Fig. 7.6. By comparing two ellipsoids in Fig. 7.6, it is shown that the size of the

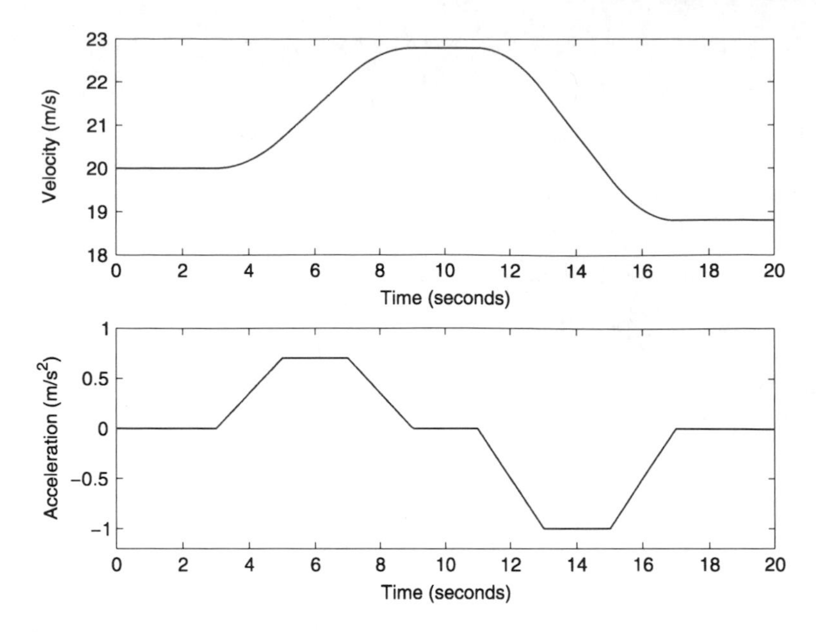

Fig. 7.3 Desired velocity and acceleration trajectory

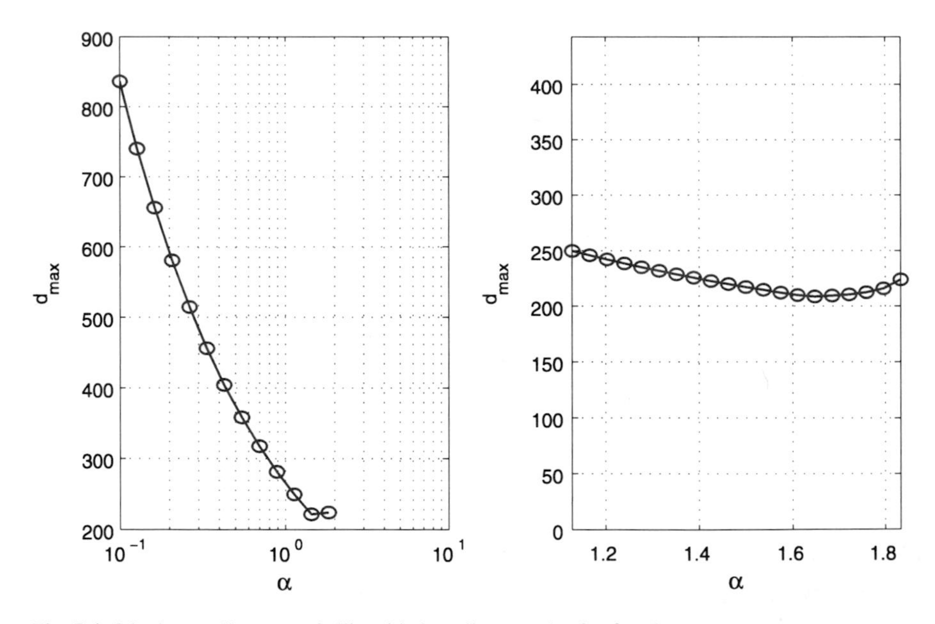

Fig. 7.4 Maximum diameter of ellipsoid along line search of α for Θ_1

reachable set can be decreased by increasing magnitude of K_{2i}. This result makes intuitive sense, because the error dynamics converges more quickly and the system is more robust to model uncertainty when higher surface gains are used.

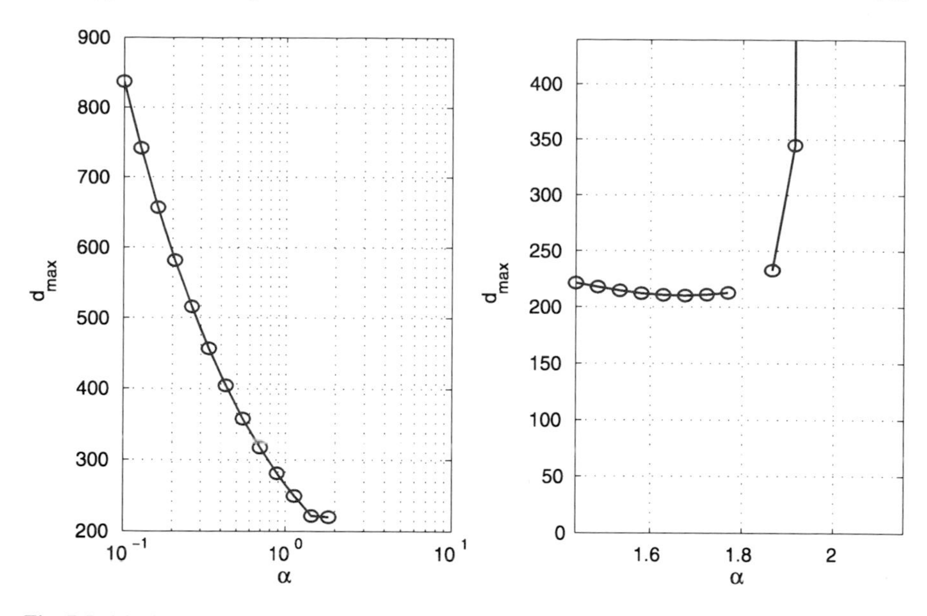

Fig. 7.5 Maximum radius of ellipsoid along line search of α for Θ_2

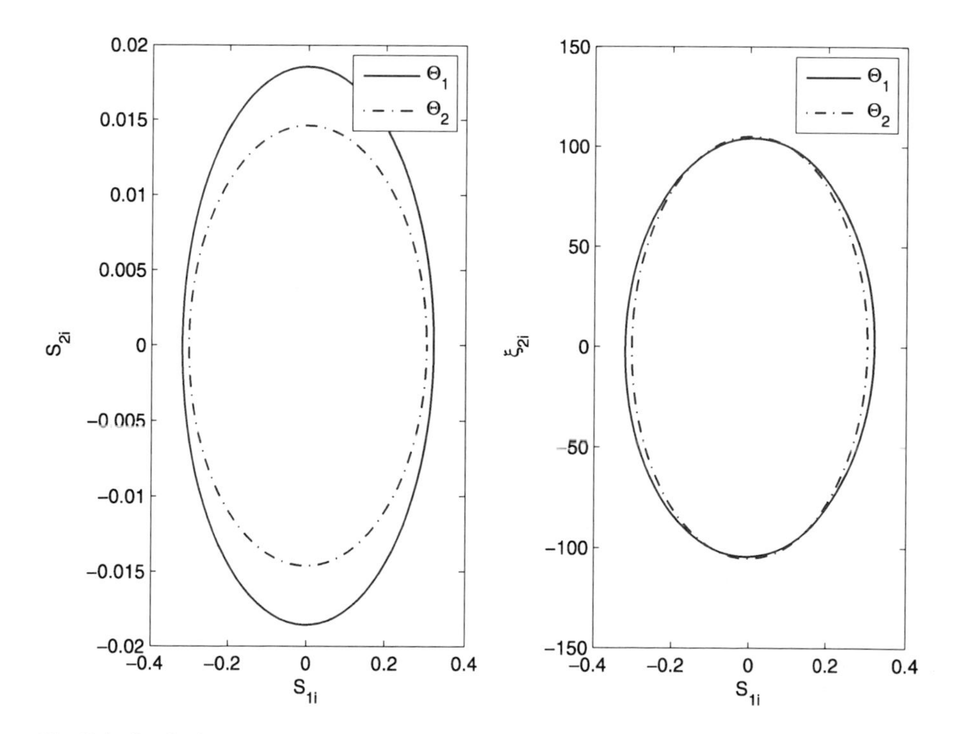

Fig. 7.6 Quadratic bound in S_1–S_2 and S_1–ξ_2 planes

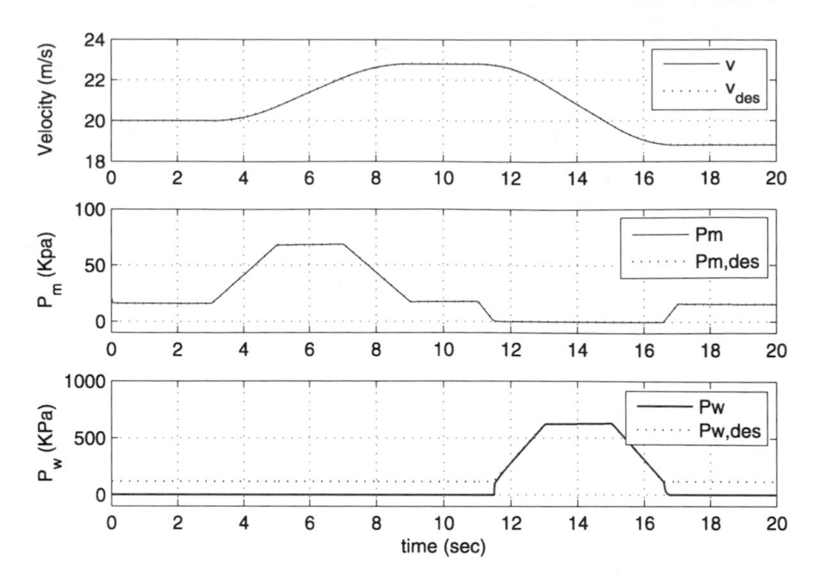

Fig. 7.7 Time responses of $\mathbf{x}_i$

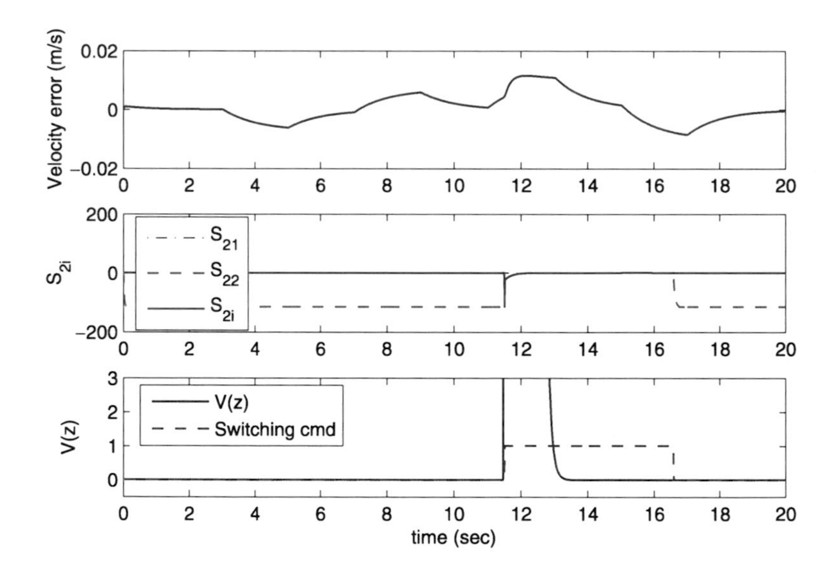

Fig. 7.8 Time responses of S_{1i}, S_{2i} and Lyapunov-like function level $V(z)$

For the given Θ_1, the time responses of $\mathbf{x}_i$ are shown in Fig. 7.7 and corresponding errors are shown in Fig. 7.8. When the engine control is switched to brake control about 11.5 s (see switching mode plot in Fig. 7.8), there is a discrepancy between P_w and P_{wdes} because $\bar{P}_w$ in (7.5) includes a constant push-out pressure P_{po} while P_w is released to zero during engine control. This results in a large magnitude of S_{2i}, thus a Lyapunov-like function level $V(z) = z^T P z$ is greater than 1. However, once $V(z)$ is less than one after a certain time, $z(t)$ stays within the ellipsoid.

7.1.4 Input–Output Stability

As done in Sect. 3.3, disturbances and uncertainties are considered for analysis of stability and performance. If Δf_1 and Δf_{2i} are considered as exogenous inputs and (7.11) is used, the error dynamics in (7.10) is written as

$$\begin{aligned}
\dot{z}_i &= A_i z_i + B_{wi} \dot{f}_1 + B_{ri} r_{ui} \\
&= A_i z_i + B_{wi} w_i + [B_u \; e_2 \; B_{wi} \; B_{wi} \; B_{wi}] \begin{bmatrix} \Delta f_1 \\ \Delta f_{2i} \\ \ddot{v}_{des} \\ J_{11} \Delta f_1 \\ J_{11} \dot{v}_{des} \end{bmatrix} \\
&= A_i z_i + B_{wi} w_i + [B_{ri} \; B_{wi} \; B_{wi}] \begin{bmatrix} r_{ui} \\ J_{11} \Delta f_1 \\ J_{11} \dot{v}_{des} \end{bmatrix} \\
&= A_i z_i + B_{wi} w_i + B_{li} r_{li}
\end{aligned} \tag{7.18}$$

where $B_{ui} = T_i^{-1} e_1$. Furthermore, the error dynamics in (7.18) is rewritten as

$$\begin{cases} \dot{z}_i = A_i z_i + B_{wi} w_i + B_{li} r_{li}, \\ |w_i| \le |\tilde{c}_{zi} z_i|, \\ y_i = C_y z_i, \end{cases} \quad \text{for } i = e, b, \tag{7.19}$$

where $C_y = [1 \; 0 \; 0]$. Then, COP (7.20) to calculate the induced $\mathscr{L}_2$ gain in Definition 3.1 can be extended to the switched system in (7.19) as follows: for $i = e, b$,

$$\begin{aligned}
&\text{minimize} \quad \kappa^2 \\
&\text{subject to} \quad P_i > 0, \quad \sigma_i > 0, \quad \kappa > 0, \quad \text{and} \\
&\begin{bmatrix} A_i^T P_i + P_i A_i + C_y^T C_y + \sigma_i \tilde{c}_{zi}^T \tilde{c}_{zi} & P B_{wi} & P B_{li} \\ B_{wi}^T P & -\sigma_i & 0 \\ B_{li}^T P & 0 & -\kappa^2 I \end{bmatrix} < 0.
\end{aligned} \tag{7.20}$$

Remark 7.1 If either Δf_1 or Δf_{2i} is bounded and classified as a vanishing perturbation, the error dynamics in (7.18) can be rearranged. For instance, if there exists a matrix c_{f1} such that

$$|\Delta f_1| \le |c_{f1} z_i|,$$

the error dynamics in (7.18) can be rewritten as

$$\begin{aligned}\dot{z}_i &= A_i z_i + [B_{ui}\ B_{wi}\ B_{wi}]\begin{bmatrix}\Delta f_1\\ J_{11}\Delta f_1\\ w_i\end{bmatrix} + [e_2\ B_{wi}\ B_{wi}]\begin{bmatrix}\Delta f_{2i}\\ \ddot{v}_{des}\\ J_{11}\dot{v}_{des}\end{bmatrix}\\ &= A_i z_i + \tilde{B}_{wi}\tilde{w}_i + B_{li} r_{li}.\end{aligned}$$

Since there exist componentwise upper bounds of $\tilde{w}_i$ such as

$$|\tilde{w}_i| = \left|\begin{bmatrix}\Delta f_1\\ J_{11}\Delta f_1\\ w_i\end{bmatrix}\right| \le \left|\begin{bmatrix}c_{f1}z_i\\ \gamma c_{f1}z_i\\ c_{zi}z_i\end{bmatrix}\right| := |C_{zi}z_i|,$$

the error dynamics in (7.19) can be rewritten as

$$\begin{cases}\dot{z}_i = A_i z_i + \tilde{B}_{wi}\tilde{w}_i + B_{li} r_{li},\\ |\tilde{w}_i| \le |C_{zi}z_i|, & \text{for } i = e, b\\ y_i = C_y z_i,\end{cases} \tag{7.21}$$

where $C_z \in \Re^{3\times 3}$. Then, COP (7.20) can be solved with the inequality constraint of $\tilde{w}$ to calculate more accurate $\mathscr{L}_2$ gain.

Example 7.2 (Input–output stability of longitudinal controller) Suppose the model uncertainties in (7.1a), (7.1b) and (7.1c) are given, respectively, as follows:

$$\begin{aligned}\Delta f_1(v) &= \frac{C_\Delta h R_g}{J_{eq}} v^2,\\ \Delta f_{2e}(P_m) &= d_1 P_m \sin(\omega_1 \cdot P_m),\\ \Delta f_{2b}(P_w) &= d_2 P_w \sin(\omega_2 \cdot P_w),\end{aligned}$$

where C_Δ represents a parametric uncertainty of C_a, d_i and ω_i are positive numbers. Let $d_1 = d_2 = 0.02$, $\omega_1 = \omega_2 = 5$, and C_Δ is the normally distributed random number between $\pm 20\%$ of the nominal C_a for simulation.

For the given sets of controller gains, Θ_1 and Θ_2 in Example 7.1, COP (7.20) can be solved numerically as shown in MATLAB Program 7-1. The corresponding $\mathscr{L}_2$ gain is 1.0008 for both sets of gains. When a new set of controller gains are given as

$$\Theta_3 = \{K_{1i}, K_{2e}, K_{2b}, \tau_{2i}\} = \{2.1, 2.5, 5, 0.02\},$$

A_i is Hurwitz and its eigenvalues are

$$\lambda(A_e) = \{-2.1965, -2.5, -47.8035\}, \qquad \lambda(A_b) = \{-2.1965, -5.0, -47.8035\}.$$

COP (7.20) can be solved numerically based on MATLAB Program 7-1, and the calculated $\mathscr{L}_2$ gain is 0.4766. Figure 7.9 shows the tracking error of DSC with respect to two different gain sets Θ_1 and Θ_3 in the presence of uncertainties. As expected by the magnitude of the calculated $\mathscr{L}_2$ gain, it is shown that DSC with Θ_3 gives better performance of disturbance rejection. As mentioned in Example 7.1, large errors about 11.5 s (i.e., when engine control is switched to brake control) result from a discrepancy between P_w and P_{wdes} initially.

MATLAB Program 7-1

```
%***** Define the dimension of matrices *****
n = size(A1, 2);
m = size(Bd1, 2);
l = size(Bl1, 2);

%***** cvx version *****
cvx_begin sdp
   variable P1(n,n) symmetric;
   variable P2(n,n) symmetric;
   variables kappa sigma1 sigma2;
   minimize(kappa);
   kappa >= 0;    sigma1 >= 0;    sigma2 >= 0;
   P1 == semidefinite(n);    P2 == semidefinite(n);
   -[A1'*P1 + P1*A1 + Cy'*Cy - sigma1*Cz1'*Cz1 P1*Bd1' P1*Bl1;
   Bd1'*P1 -sigma1*eye(m) zeros(m,l);
   Bl1'*P1 zeros(l,m) -kappa*eye(l)] = semidefinite(n+m+l);
   -[A2'*P2 + P2*A2 + Cy'*Cy - sigma1*Cz2'*Cz2 P2*Bd2' P2*Bl2;
   Bd2'*P2 -sigma2*eye(m) zeros(m,l);
   Bl2'*P2 zeros(l,m) -kappa*eye(l)] = semidefinite(n+m+l);
cvx_end

%***** Calculation of L2 gain *****
L2 = sqrt(kappa);
```

7.2 Passive Fault Tolerant Control

As mentioned in the introduction of the chapter, the proposed FTC approach shares ideas of both the passive and active FTC approaches to accommodate faults in a large-scale system. That is, a default controller for a non-faulty system will be designed to provide robust performance with respect to model uncertainties as well as pre-defined faults consistent with the passive approach. However, if the performance degradation due to faults is not acceptable through the performance evaluation task, the default controller will be switched to one of a set of reconfigurable controllers via controller reconfiguration which will be discussed in the next chapter. In this section, it will be shown how a fault classification scheme leads quite naturally to the performance evaluation task, which provides a decision logic between the passive and active approach. Moreover, the proposed fault classification technique will inherit robustness.

Two main ideas will be presented and their applications to longitudinal vehicle control will be shown through simulations; one is to evaluate if DSC is robust enough to compensate for a specific class of faults as well as model uncertainties. The other is to use a fault classification approach to explain the combined effect of multiple faults.

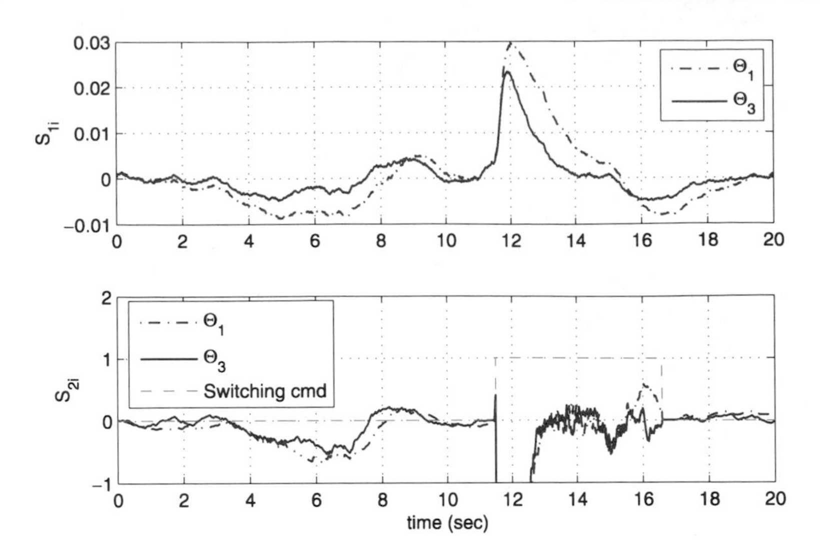

Fig. 7.9 Velocity tracking error S_{1i} and S_{2i} for Θ_1 and Θ_3

7.2.1 Problem Statement

Consider the following nonlinear system in a *parametric strict-feedback* form with actuator saturation:

$$\begin{cases} \dot{x}_i = x_{i+1} + a_i f_i(x_1, \ldots, x_i) + \Delta f_{im}(x_1, \ldots, x_i) \quad \text{for } 1 \le i \le n-1, \\ \dot{x}_n = u + a_n f_n(x), \\ y = x, \end{cases} \tag{7.22}$$

where the state $x \in \Re^n$, a_i is the system parameter, Δf_{im} is the model uncertainty, $y \in \Re^n$ stands for output measurement and is assumed to measure the full state, and

$$u = \begin{cases} u, & \text{if } |u| \le u_0, \\ u_0 \operatorname{sign}(u), & \text{otherwise.} \end{cases}$$

The "parametric strict-feedback" comes from the reason that the nonlinearity (f_i) and model uncertainty (Δf_{im}) are only functions of $x_1, \ldots, x_i$ and the system parameter (a_i) is linearly parameterized [55, 111].

Three different types of faults can be considered here: parametric (f_P), sensor (f_S), and actuator faults (f_A). In this chapter, we will focus on faults which enter as a multiplicative term for parametric (or component) faults, and/or for sensor and actuator faults. If all these faults are considered with respect to the non-faulty system in (7.22), then the faulty system is obtained as follows:

$$\begin{cases} \dot{x}_i = x_{i+1} + (1-\beta_i)a_i f_i + \Delta f_{im} = x_{i+1} + a_i f_i + \Delta f_{im} + \Delta f_{ip}, \\ \dot{x}_n = (1-\mu)u + (1-\beta_n)a_n f_n = (1-\mu)u + a_n f_n + \Delta f_{np}, \\ y = [I_n + F_s(\nu)]x, \end{cases} \tag{7.23}$$

where Δf_{ip} is an additive term due to the ith multiplicative parametric fault of a_i, and $\mu(t), \beta_i(t) \in [0, 1]$ are multiplicative terms which represent reduction in control input u and parameter a_i, respectively. Furthermore, $F_s(\nu) = \mathrm{diag}[\nu_1(t), \dots, \nu_n(t)] \in \Re^{n \times n}$ is also a multiplicative term for representing the faulty measurement of the state x and $\nu_i(t) \in [0, 1]$.

The challenging problems considered in this section for designing FTC can be summarized as follows:

- A mathematical model of the non-faulty system has nonlinearities and/or model uncertainties.
- An exact model of the faulty system is not known by FDD. In other words, the isolated fault from FDD also has some degree of uncertainty and some multiple faults could be detected but not all of them isolated by FDD.
- Any single or multiple faults among sensor, actuator, and parametric faults can happen in the components. Furthermore, they could be abrupt or incipient, and partial or total failure.
- All components in FTC structure are integrated *modularly*, in the sense that a number of FTC components including different FDD and reconfigurable controllers can be integrated and extended individually.

This section provides how to deal with the above problems in Fig. 7.10. Although a complete FTC design cannot be shown here in detail, most of results can be extended to integrate all components for the FTC system.

7.2.2 Error Dynamics for a Faulty Nonlinear System

In this section, the default controller for the non-faulty system will be designed and investigated. Then, performance degradation due to certain faults will be analyzed, based on the closed-loop error dynamics of DSC. While the standard design procedure for DSC is described in Sect. 2.3, the only difference is that the forcing term ($\bar{x}$) is calculated based on the measurement y, not the state x. That is, with no ith sensor and parametric fault assumption ($\beta_i = 0$, $\nu_i = 0$),

$$\bar{x}_{i+1} = -a_i f_i(y) + \dot{x}_{id} - K_i \tilde{S}_i, \quad i = 1, \dots, n-2 \tag{7.24}$$

where $\tilde{S}_i := y_i - x_{id}$ is the measured surface error. Similarly, the desired control input u_d is chosen as

$$u_d = -a_n f_n(y) + \dot{x}_{nd} - K_n \tilde{S}_n = -a_n f_n(y) + \frac{\bar{x}_n - x_{nd}}{\tau_n} - K_n \tilde{S}_n \tag{7.25}$$

where $\tilde{S}_n = y_n - x_{nd}$. As done in Chap. 5, similarly the control input can be defined as

$$u = \gamma \cdot u_d \tag{7.26}$$

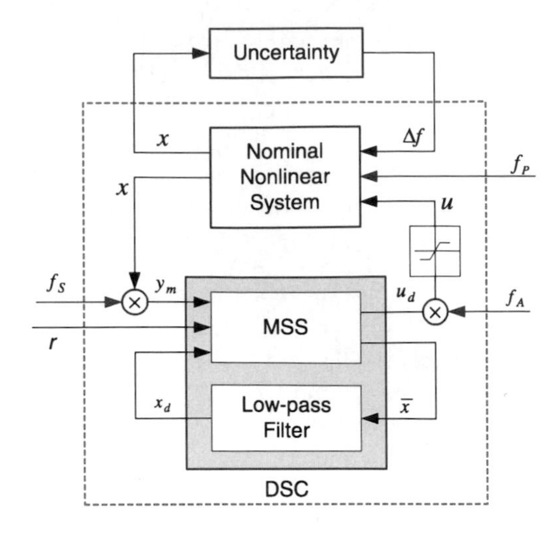

Fig. 7.10 Schematic block diagram of DSC in the presence of faults

where the variable $|\gamma| \in (0, 1]$ is

$$\gamma = \begin{cases} 1, & \text{if } |u_d| \le u_0, \\ \frac{u_0}{|u_d|}\,\mathrm{sign}(u_d), & \text{otherwise.} \end{cases}$$

By adding and subtracting $\bar{x}_{i+1}$ and $x_{(i+1)d}$ in (7.23), and substituting the control input u with (7.25) and (7.26), the closed-loop faulty system can be derived as

$$\begin{cases} \dot{x}_i = \bar{x}_{i+1} + (x_{i+1} - x_{(i+1)d}) + (x_{(i+1)d} - \bar{x}_{i+1}) + a_i f_i + \Delta f_{im} + \Delta f_{ip}, \\ \dot{x}_n = \gamma(1-\mu)\{-a_n f_n(y) + \dot{x}_{nd} - K_n \tilde{S}_n\} + a_n f_n + \Delta f_{np}. \end{cases} \tag{7.27}$$

Then, by the definitions of S, ξ, and $\bar{x}$, (7.27) can be rewritten as an augmented error dynamics (Ω_f) such that

$$\begin{aligned} \dot{S}_i &= -K_i \tilde{S}_i + S_{i+1} + \xi_{i+1} + a_i\{f_i(x) - f_i(y)\} + \Delta f_{im} + \Delta f_{ip} \\ &= -K_i S_i + S_{i+1} + \xi_{i+1} + \Delta f_{im} + \Delta f_{ip} + \Delta f_{is}, \\ \dot{S}_n &= \gamma(1-\mu)(\dot{x}_{nd} - K_n \tilde{S}_n) - \dot{x}_{nd} + a_n f_n(x) - \gamma(1-\mu) a_n f_n(y) + \Delta f_{np} \\ &= -\gamma K_n (1-\mu) S_n + \frac{1-\gamma(1-\mu)}{\tau_n}\xi_n + \Delta f_{np} + \Delta f_{ns} \end{aligned} \tag{7.28}$$

where $\tilde{S}_i = S_i + \nu_i x_i$ and

$$\begin{aligned} \Delta f_{is} &= a_i\{f_i(x) - f_i(y)\} - \nu_i K_i x_i, \\ \Delta f_{ns} &= a_n\{f_n(x) - \gamma(1-\mu) f_n(y)\} - \gamma(1-\mu)\nu_i K_n x_n. \end{aligned}$$

Since $(n-1)$th order filter dynamics is incorporated in DSC design, we can consider the filter error y as an augmented state in the error dynamics. Hence, the filter error dynamics is as follows:

$$\dot{\xi}_{i+1} = \dot{x}_{(i+1)d} - \dot{\bar{x}}_{i+1} = -\frac{\xi_{i+1}}{\tau_{i+1}} - \left[-a_i \dot{f}_i(y) + \ddot{x}_{id} - K_i \dot{\tilde{S}}_i\right] \quad \text{for } i = 1, \ldots, n-1.$$

This can be rewritten as

$$\begin{aligned}
\dot{\xi}_2 - K_1\dot{S}_1 &= -\frac{\xi_2}{\tau_2} + \Delta\dot{f}_1 - \ddot{x}_{1d},\\
\dot{\xi}_3 - \frac{\dot{\xi}_2}{\tau_2} - K_2\dot{S}_2 &= -\frac{\xi_3}{\tau_3} + \Delta\dot{f}_2,\\
&\vdots\\
\dot{\xi}_n - \frac{\dot{\xi}_{n-1}}{\tau_{n-1}} - K_{n-1}\dot{S}_{n-1} &= -\frac{\xi_n}{\tau_n} + \Delta\dot{f}_{n-1}
\end{aligned} \tag{7.29}$$

where

$$\Delta\dot{f}_i = a_i f_i + K_i(\nu_i\dot{x}_i + \dot{\nu}_i x_i) \quad \text{and} \quad \dot{f}_i := \sum_{j=1}^{j=i}\frac{\partial f_i}{y_j}\dot{y}_j.$$

Combining (7.28) with (7.29), the augmented error dynamics can be written in matrix form as follows:

$$\begin{aligned}
&T\begin{bmatrix}\dot{S}\\ \dot{\xi}\end{bmatrix} = \begin{bmatrix}\tilde{A}_{11} & \tilde{A}_{12}\\ \mathbf{0} & A_{22}\end{bmatrix}\begin{bmatrix}S\\ \xi\end{bmatrix} + \begin{bmatrix}\mathbf{I} & \mathbf{0}\\ \mathbf{0} & \mathbf{I}\end{bmatrix}\begin{bmatrix}\Delta f_e\\ \Delta\dot{f}_e\end{bmatrix} + \bar{B}_e\ddot{x}_{1d}\\
&\Longrightarrow \quad T\dot{z} = \bar{A}_f z + \bar{B}_w w_f + \bar{B}_e\ddot{x}_{1d}
\end{aligned} \tag{7.30}$$

where the matrices T and $\bar{B}_e$ are defined in Lemma 2.1, the submatrices are

$$\tilde{A}_{11} = \begin{bmatrix} -K_1 & 1 & \cdots & 0\\ 0 & \ddots & \ddots & \vdots\\ \vdots & \cdots & -K_{n-1} & 1\\ 0 & \cdots & 0 & -\gamma K_n(1-\mu)\end{bmatrix},$$

$$\tilde{A}_{12} = \begin{bmatrix} 1 & 0 & \cdots & 0\\ 0 & \ddots & \ddots & \vdots\\ \vdots & \cdots & 1 & 1\\ 0 & \cdots & 0 & \frac{1-\gamma(1-\mu)}{\tau_n}\end{bmatrix},$$

and the perturbation terms are

$$\begin{aligned}
\Delta f_e &= \Delta f_m + \Delta f_p + \Delta f_s \in \Re^n,\\
\Delta\dot{f}_e &= \Theta\dot{f} + K(F_s\dot{x} + \dot{F}_s x) \in \Re^{n-1}, \quad \Theta = \mathrm{diag}(a_1,\ldots,a_{n-1}).
\end{aligned}$$

After T^{-1} is multiplied on both sides of (7.30) , we can obtain the closed-loop error dynamics for the faulty system with DSC as

$$\dot{z} = A_f z + B_w w_f + B_e\ddot{x}_{1d} \tag{7.31}$$

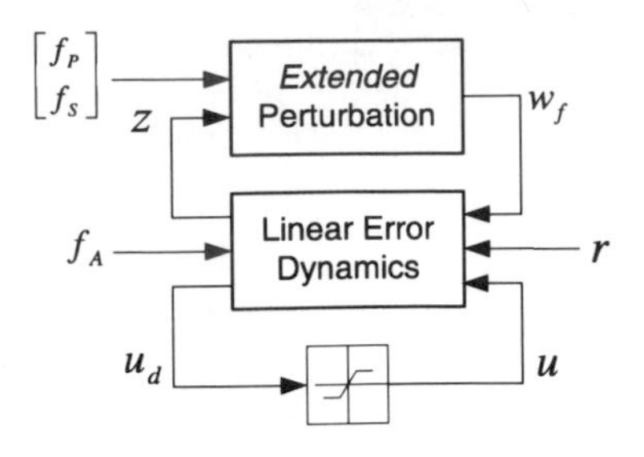

Fig. 7.11 Error dynamics of a closed-loop faulty system via DSC

where

$$A_f = T^{-1}\bar{A}_f = \begin{bmatrix} \tilde{A}_{11} & \tilde{A}_{12} \\ T_\xi^{-1}K\tilde{A}_{11} & T_\xi^{-1}(K\tilde{A}_{12} + A_{22}) \end{bmatrix},$$

$$B_w = T^{-1}\bar{B}_w = \begin{bmatrix} \mathbf{I} & \mathbf{0} \\ T_\xi^{-1}K & T_\xi^{-1} \end{bmatrix},$$

and B_e is defined in Lemma 2.1.

Remark 7.2 As shown in Fig. 7.11, the actuator fault directly affects the system error matrix A_f, while both parametric and sensor faults can be described as an external input to a perturbation term w_f. It there is no actuator fault, i.e., $\mu = 0$ and $\gamma = 1$, the system error matrix A_f becomes A_{cl} which is defined in Lemma 2.1. Furthermore, if both parametric and sensor fault does not occur, i.e., $\beta_i = 0$, and $F_s = \mathbf{0}$, the perturbation term w_f becomes $w = [\Delta f \ \Theta \dot{f}]^T$. Then, the error dynamics in (7.31) is written as

$$\dot{z} = A_{cl}z + B_w w + B_e \ddot{x}_{1d} \tag{7.32}$$

which is equivalent to (3.3). If w is decoupled into vanishing and nonvanishing perturbation terms, the error dynamics in Lemma 3.1 is derived.

If the uncertainties as well as faults are neglected, the error dynamics in (7.31) is written as

$$\dot{z} = A_{cl}z + \begin{bmatrix} \mathbf{0} \\ T_\xi^{-1} \end{bmatrix} \Theta \dot{f} + B_e \ddot{x}_{1d}.$$

Again, if $\dot{f}$ is decoupled into $J(x)C_z z + J_1 \dot{x}_{1d}$, the error dynamics in Lemma 2.1 is derived.

In consequence, all degradation due to multiplicative faults can be expressed as the faulty system matrix A_f and *extended* perturbation term w_f, in the sense that more nonlinear functions such as Δf_p and Δf_s are included in w_f, compared with the perturbation term w of the non-faulty system. This interpretation motivates us to classify a fault based on performance degradation, which will be discussed next.

7.2.3 *Fault Classification*

The role of FMS in Fig. 7.2 is to check the condition of the components as well as performance, and to choose an appropriate control law, which ensures stability and some prescribed level of performance, i.e., a control law reconfiguration mechanism. Therefore, it is apparent that knowledge of the effects of each fault upon the system performance and the control redundancies are critical. Before designing controller reconfiguration, the variety of faults should be classified first.

7.2.3.1 Fault Detection and Diagnosis

In the literature, they are usually classified on the basis of characteristics or isolability in FDD. Based on the time characteristic of faults, there are typically abrupt and incipient faults [25]. The abrupt faults occur suddenly, i.e., step-like changes. The incipient faults are slowing developing faults such as bias or drift and are typically small.

Regarding the isolability of the fault, model-based FDD is one of the well-known approaches. Most model-based FDD are dependent on the idea of *redundancy*, and there are generally two kinds of redundancies: *physical* and *analytical redundancy* [30]. Physical redundancy uses multiple physical sensors for each measurement, coupled with simple majority voting logic. However, this type of redundancy can be quite expensive and adds complexity to the hardware.

Alternatively, analytical redundancy uses a mathematical model of the system to relate different physical sensors via *residuals*. These residuals are quantitative differences between one sensor's measurements and estimates based on a mathematical model and the other available sensor measurements. Furthermore, the residuals should be sensitive to the occurrence of faults as well as robust to model uncertainties and disturbances to allow for correct detection and identification. Many techniques for designing these residuals exist in the literature such as state estimation techniques including the observer [26], detection filter, parity space approaches and parameter identification techniques [43].

Detection and identification of faults in the system are conducted by processing this set of residuals. Typically, this involves determining when the signal properties of the residuals have changed via change detection and pattern recognition techniques [7, 10, 99]. Unfortunately, it may not be possible to uniquely identify the cause of a fault depending of the design of the residual vector. From the viewpoint of FDD, faults are typically classified into two categories [89],

- *Isolatable* fault (IF)—if the faulty component can be uniquely determined by the FDD. That is, specific location and magnitude of the fault can be identified by the FDD.
- *Detectable* fault (DF)—if the faulty component cannot be uniquely determined by the FDD. At least, occurrence of the fault is detected while detailed information such as the location and magnitude cannot be provided.

7.2.3.2 Fault Classification for Switched System

While the faults can be classified into the isolable and detectable fault with respect to FDD, an additional criterion for the classification is defined from a controller point of view as follows [89]:

Definition 7.2 If the faulty system in (7.23) with certain faults is quadratically trackable (or stabilizable) via a given DSC law for the non-faulty system, then the faults are called *tolerable* for the current controller. Otherwise, they are said to be *intolerable.*

The proposed fault classification is basically a binary classification: the tolerable and intolerable fault. If the error trajectory for the faulty system (7.23) is quadratically bounded with Lyapunov matrix P, which is obtained for the non-faulty system with DSC, i.e., the tracking error stays within the ellipsoidal bound $\mathcal{E}_P = \{z \in \Re^{2n-1} | z^T P z \le 1\}$, then the isolated and/or detected faults are regarded as the tolerable faults. For instance, if there is either parametric or sensor fault or both, the error dynamics in (7.31) becomes

$$\dot{z} = A_{cl} z + B_w w_f + B_e \ddot{x}_{1d} \tag{7.33}$$

where it is assumed that there is no actuator fault. Then, if the extended perturbation w_f can be decomposed into vanishing and nonvanishing perturbation terms

$$w_f = p + q, \quad |p_i| \le \|C_{zi} z\|$$

where C_z is chosen at the initial controller design stage, (7.33) is rewritten in diagonal norm-bounded LDI form as

$$\begin{cases} \dot{z} = A_{cl} z + B_w p + [B_w \ B_e] \begin{bmatrix} q \\ \ddot{x}_{1d} \end{bmatrix} = A_{cl} z + B_w p + B_u r_u, \\ |p_i| \le \|C_{zi} z\|. \end{cases}$$

Then, with an assumption of $\|r_u\| \le r_0$, stability and performance analysis can be performed in the sense of quadratic tracking as discussed in Chap. 3. In the case of actuator fault, if the degraded tracking performance still satisfies the robust tracking performance that is achieved by the non-fault system with DSC, the fault is classified to be tolerable.

As sliding mode control is robust with respect to model uncertainties as well as a class of parametric faults [92], DSC also inherits the similar robustness and is a passive approach in the sense that no additional controller is required to compensate for specific class of faults [89]. In other words, the tracking error will stay in the ellipsoidal invariant set $\mathcal{E}_P$ under model uncertainty and certain faults. Therefore, this ellipsoidal error bound provides a means of classifying the severity of a fault based on the performance degradation. Thus it can be used as a switching criterion between the nominal and degraded mode controllers. This can also be said to be a switching criterion between passive and active fault tolerant control approach.

Four categories of faults based on FDD and tracking performance estimated at the initial controller design stage are shown in Table 7.3. If the occurred fault is

Table 7.3 Fault classification and handling decision table

Classification	Isolatable fault	Detectable fault
Tolerable fault	Specific and gradual warning	Gradual warning
Intolerable fault	Controller reconfiguration	Emergency warning and handling

tolerable, no additional controller reconfiguration is required and a warning signal may be generated to alert the operator. At the same time, if the fault is isolatable, a specific warning can be provided, e.g., the location of the fault. Otherwise, a gradual warning signal can be given, based on a quadratic Lyapunov-like function level, e.g., if $V(z) = z^T Pz$ is greater than one and becomes larger, it indicates that the fault degrades the system performance more severely even though the cause is not known. If the fault is intolerable, then some types of controller reconfiguration will be required, and a required specific action will depend on the isolability of the fault. That is, if the fault is isolatable, then the FMS controller reconfiguration scheme will choose a new control law to deal with the specific fault appropriately. However, if the fault is only detectable or particularly severe, an emergency situation is declared and a supervisory controller or operator is given control of the system [58]. These aspects of the problem will be discussed later in Chap. 8.

One of the advantages of this classification is that unnecessary and frequent switching of controllers can potentially be reduced. Also, the relationship between fault magnitude and its effect upon the system performance can be readily seen. Furthermore, even if there is a *hidden* fault which cannot be possibly isolated by the FDD, it can be coped with when it is classified as a tolerable fault.

If the error dynamics in (7.31) and (7.32) is considered as a switched system discussed in Sect. 7.1, the error dynamics for non-faulty and faulty systems can be written as

$$\begin{cases} \dot{z} = A_n z + B_w w_n + B_e \ddot{x}_{1d}, \\ \dot{z} = A_f z + B_w w_f + B_e \ddot{x}_{1d} \end{cases} \tag{7.34}$$

where $A_n = A_{cl}$, $w_n = w \in \Re^{n_w}$, and the subscripts n and f represent normal and faulty, respectively. If w_n and w_f are decomposed into vanishing and nonvanishing perturbation terms and there exist matrices C_{ni} and C_{fi} such that

$$w_n = p_n + q_n, \quad |p_{in}| \le \|C_{in} z\|,$$
$$w_f = p_f + q_f, \quad |p_{if}| \le \|C_{if} z\|$$

where p_{in} and p_{if} are ith elements of p_n and p_f, respectively, the switched error dynamics in (7.34) is written as

$$\dot{z} = A_k z + B_w p_k + [B_w \ B_e] \begin{bmatrix} q_k \\ \ddot{x}_{1d} \end{bmatrix} = A_k z + B_w p_k + B_r r_k, \\ |p_{ik}| \le \|C_{ik} z\| \tag{7.35}$$

where $k = n, f$. If $r_k \in \Re^{n_r}$ is norm-bounded as $\|r_k\| \leq r_0$, the error dynamics is summarized as

$$\begin{cases} \dot{z} = A_k z + B_w p_k + \tilde{B}_r \tilde{r}_k, \\ |p_{ik}| \leq \|C_{ik} z\|, \qquad \|\tilde{r}_k\| \leq 1 \end{cases} \tag{7.36}$$

where $\tilde{r}_k = r_k / r_0$ and $\tilde{B}_r = r_0 B_r$.

As used in Sect. 7.1, ellipsoidal approximation of the error bounds in Sect. 2.6 can be used to determine when a fault drives the error, as measured by the level sets of the Lyapunov-like function, beyond the level which may be caused by limitations of the tracking performance and modeling uncertainty. As developed in COP (7.17), similarly the ellipsoidal error bound for the hybrid error dynamics in (7.36) can be cast into a convex optimization problem as follows:

$$\begin{aligned} &\text{maximize} \quad \lambda_{\min}(P) \\ &\text{subject to} \quad P > 0, \qquad \Sigma \geq 0, \quad \text{and} \\ &\begin{bmatrix} A_k^T P + P A_k + \alpha P + C_k^T \Sigma_B C_k & P B_w & P \tilde{B}_r \\ B_w^T P & -\Sigma & 0 \\ \tilde{B}_r^T P & 0 & -\alpha I \end{bmatrix} < 0 \end{aligned} \tag{7.37}$$

where

$$C_k = \begin{bmatrix} C_{1k}^T & \cdots & C_{n_r k}^T \end{bmatrix}^T, \qquad \Sigma = \operatorname{diag}(\sigma_1, \ldots, \sigma_{n_w}) \in \Re^{n_w \times n_w},$$

and $\Sigma_B = \operatorname{diag}(\sigma_1 I, \ldots, \sigma_{n_w} I)$ is the diagonal block matrix.

The resulting matrix P can then be used in the quadratic Lyapunov-like function to classify the severity of a fault as follows: if $V(z) = z^T P z$ is less than one, the fault is regarded as a tolerable one. Otherwise, it is considered a severe fault and the corresponding level set shows how much the performance is degraded. It is noted that the above problem gives only a sufficient condition for the stability of the switched nonlinear system. An alternative strategy to approach such a problem is to use a piecewise quadratic Lyapunov function as in [49].

Example 7.3 (Fault classification for longitudinal vehicle control) Consider parametric uncertainty and faults in the system (7.1a), (7.1b), (7.1c). Suppose there is the parametric uncertainty in C_a. Then, Δf_1 is derived as in Example 7.2 and there exists a γ_0 such that

$$\Delta f_1 = \left| \frac{C_\Delta h R_g}{J_{eq}} v^2 \right| \leq \frac{0.2 C_a h R_g v_{\max}}{J_{eq}} |v| := \gamma_0 |v|$$

where it is assumed that $|C_\Delta| \leq 0.2 C_a$ and $|v| \leq v_{\max} = 26$. Using (7.10), the error dynamics for a normal (or nonfaulty) system is

$$\dot{z}_i = A_i z_i + B_{di} \dot{f}_1 + [B_{ui} \ B_{di}] \begin{bmatrix} \Delta f_1 \\ \ddot{v}_{des} \end{bmatrix}, \tag{7.38}$$

where

$$A_i = \begin{bmatrix} -K_{1i} & 1/c_i & 1/c_i \\ 0 & -K_{2i} & 0 \\ -a_i K_{1i}^2 & K_{1i} & K_{1i} - 1/\tau_{2i} \end{bmatrix},$$
$$B_{di} = \begin{bmatrix} 0 \\ 0 \\ -a_i \end{bmatrix}, \qquad B_{ui} = \begin{bmatrix} 1 \\ 0 \\ a_i K_{1i} \end{bmatrix}.$$

If Δf_1 are decomposed into a vanishing and nonvanishing term

$$|\Delta f_1| =\leq \gamma_0 |v| = \gamma_0 |S_1 + v_{des}| \leq \gamma_0 \big|[1\ 0\ 0] z_i\big| + \gamma_0 |v_{des}|,$$

there exist functions p and q such that

$$\Delta f_1 = p + q, \quad |p| \leq \gamma_0 \big|[1\ 0\ 0] z_i\big|, \ |q| \leq \gamma_0 |v_{des}|. \tag{7.39}$$

Using (7.11) and (7.39), $\dot{f}_1$ is written as

$$\dot{f}_1 = J_{11} c_{zi} z_i + J_{11} \dot{v}_{des} + J_{11} \Delta f_1 = J_{11} (c_{zi} z_i + \dot{v}_{des} + p + q). \tag{7.40}$$

The error dynamics versions combining (7.38) with (7.39) and (7.40) are rewritten as

$$\begin{aligned} \dot{z}_i &= A_i z_i + [B_{ui}\ B_{di}\ B_{di}] \begin{bmatrix} p \\ J_{11} p \\ J_{11} c_{zi} z_i \end{bmatrix} + [B_{ui}\ B_{di}\ B_{di}\ B_{di}] \begin{bmatrix} q \\ J_{11} q \\ J_{11} \dot{v}_{des} \\ \ddot{v}_{des} \end{bmatrix} \\ &= A_i z_i + B_{wi} w_i + B_{ri} r, \end{aligned} \tag{7.41}$$

where w_i is bounded componentwise and r is norm-bounded as follows:

$$|w_i| = \left| \begin{bmatrix} p \\ J_{11} p \\ J_{11} c_{zi} z_i \end{bmatrix} \right| \leq \left| \begin{bmatrix} \gamma_0 [1\ 0\ 0] \\ \gamma \gamma_0 [1\ 0\ 0] \\ \gamma c_{zi} \end{bmatrix} z_i \right| := |C_{zi} z_i|,$$

$$\|r\|^2 \leq (1+\gamma^2) q^T q + \gamma^2 \dot{v}_{des}^2 + \ddot{v}_{des}^2 \leq (1+\gamma^2) \gamma_0^2 v_{des}^2 + \gamma^2 \dot{v}_{des}^2 + \ddot{v}_{des}^2 := r_0^2.$$

Suppose two parametric faults such as C_e and K_b in (7.2) are considered. If they are abrupt faults, they can be written as multiplicative terms mathematically

$$C_e(t) = \begin{cases} C_e, & \text{if } t \leq t_f, \\ (1-\beta_e) C_e, & \text{otherwise,} \end{cases} \quad \text{and} \quad K_b(\beta_b) = \begin{cases} K_b, & \text{if } t \leq t_f, \\ (1-\beta_b) K_b, & \text{otherwise,} \end{cases}$$

where $\beta_e, \beta_b \in [0, 1]$ and $t_f > 0$ is the time faults occur. Then, a_i in (7.8) and (7.9) are written as

$$a_i(t) = \begin{cases} a_i, & \text{if } t \leq t_f, \\ \frac{a_i}{1-\beta_i}, & \text{otherwise,} \end{cases} \quad \text{for } i = e, b.$$

Then, the error dynamics for the fault system is

$$\dot{z}_i = A_i^f z_i + B_{wi}^f w_i + B_{ri}^f r, \tag{7.42}$$

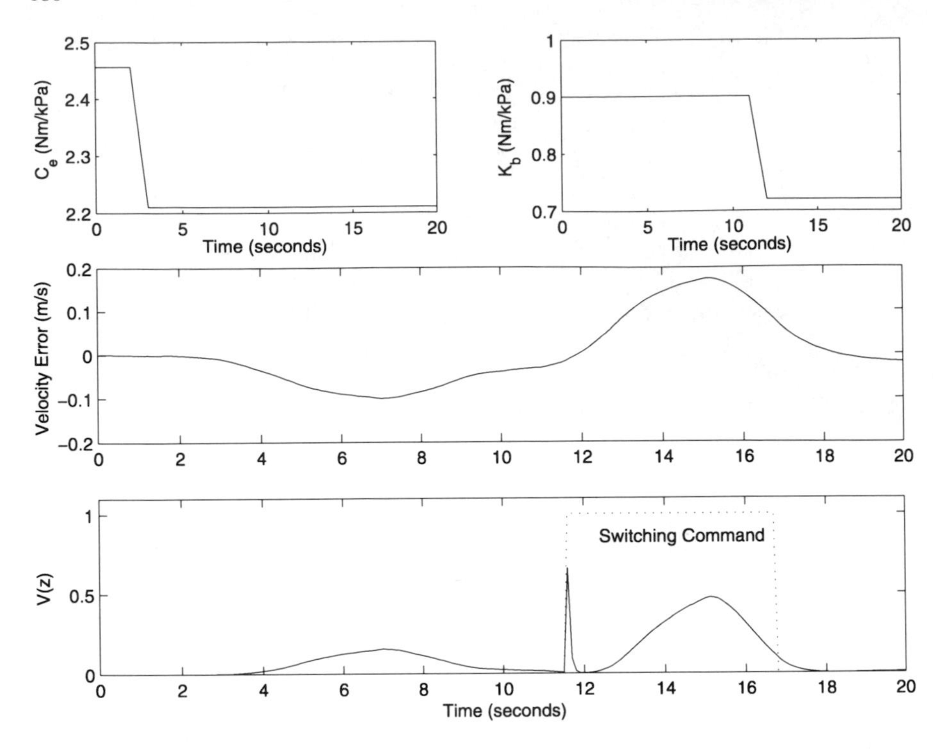

Fig. 7.12 Tracking error and quadratic function L level—parametric fault in C_e and K_b

where the matrices are

$$A_i^f = T_{fi}^{-1} \bar{A}_{fi}, \qquad B_{wi}^f = \begin{bmatrix} E_1 \; B_{di}^f \; B_{di}^f \end{bmatrix}, \qquad B_{ri}^f = \begin{bmatrix} E_1 \; B_{di}^f \; B_{di}^f \; B_{di}^f \end{bmatrix},$$

$$T_{fi} = \begin{bmatrix} 1 & 0 & 0 \\ 0 & 1 & 0 \\ -a_i K_{1i}/(1-\beta_i) & 0 & 1 \end{bmatrix},$$

$$\bar{A}_{fi} = \begin{bmatrix} -K_{1i} & (1-\beta_i)/a_i & (1-\beta_i)/a_i \\ 0 & -K_{2i} & 0 \\ 0 & 0 & -1/\tau_{2i} \end{bmatrix},$$

$$B_{di}^f = \begin{bmatrix} 0 \; 0 \; -a_i/(1-\beta_i) \end{bmatrix}^T.$$

Therefore, combining (7.41) with (7.42), the switched error dynamics is summarized as

$$\begin{cases} \dot{z}_i = A_i z_i + B_{wi} w_i + B_{ri} r, & \text{if } t \le t_f, \\ \dot{z}_i = A_i^f z_i + B_{wi}^f w_i + B_{ri}^f r, & \text{otherwise,} \end{cases} \quad \text{for } i = e, b, \tag{7.43}$$

where t_f is unknown.

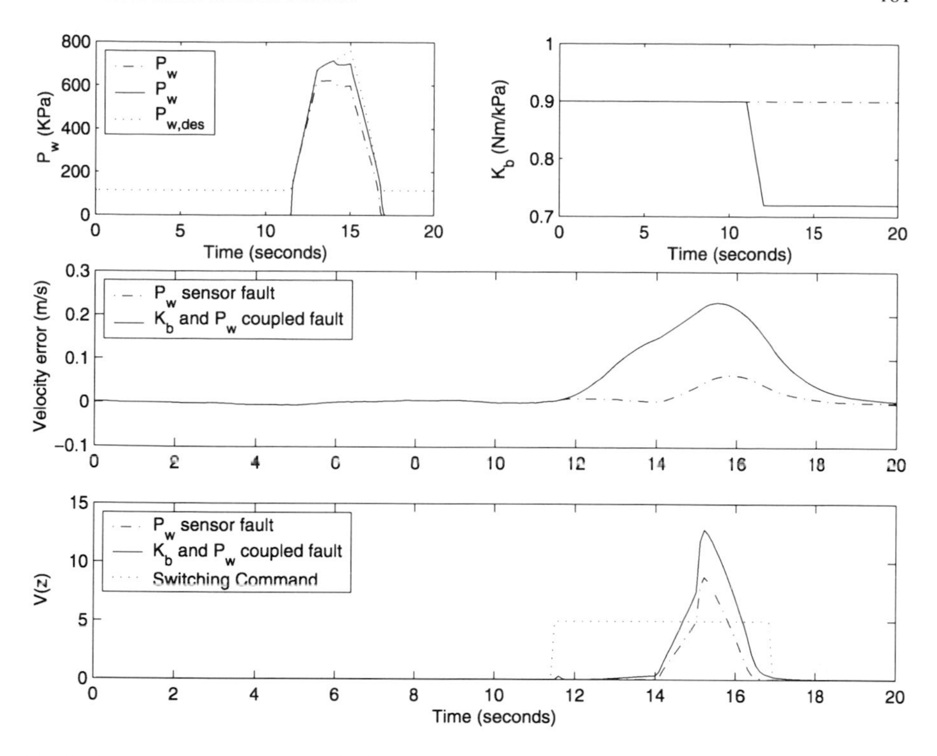

Fig. 7.13 Tracking error and quadratic function level—faults in P_w and K_b

Finally, COP (7.17) to calculate the smallest ellipsoid containing reachable sets of (7.41) can be extended to the following COP for the switched error dynamics in (7.43): For a fixed $\alpha \in [a,\ b]$ and $i = e, b$,

$$
\begin{aligned}
&\text{maximize} \quad \lambda_{\min}(P) \\
&\text{subject to} \quad P > 0, \qquad \Sigma_1 \geq 0, \qquad \Sigma_2 \geq 0 \quad \text{and} \\
&\begin{bmatrix} A_i^T P + PA_i + \alpha P + C_{zi}^T \Sigma_1 C_{zi} & PB_{wi} & P\tilde{B}_{ri} \\ B_{wi}^T P & -\Sigma_1 & 0 \\ \tilde{B}_{ri}^T P & 0 & -\alpha I \end{bmatrix} < 0, \\
&\begin{bmatrix} (A_i^f)^T P + PA_i^f + \alpha P + C_{zi}^T \Sigma_2 C_{zi} & PB_{wi}^f & P\tilde{B}_{ri}^f \\ (B_{wi}^f)^T P & -\Sigma_2 & 0 \\ (\tilde{B}_{ri}^f)^T P & 0 & -\alpha I \end{bmatrix} < 0
\end{aligned}
\tag{7.44}
$$

where $\tilde{B}_{ri} = r_0 B_{ri}$ and $\tilde{B}_{ri}^f = r_0 B_{ri}^f$.

Figure 7.12 presents the velocity tracking error and a Lyapunov-like function level in the presence of parametric uncertainty and faults in C_e and K_b. In this case, there is a 10% reduction in C_e at 3 seconds and 20% reduction in K_b after 12 seconds. That is, $\beta_e = 0.1$ and $\beta_b = 0.2$. When the parametric faults happen, it does not severely degrade the system performance, i.e. the level of Lyapunov-like function does not increase much (see third row in Fig. 7.12). Therefore, the DSC is inherently robust enough to handle this type and degree of fault. It is noted that a peak of the Lyapunov function level in Fig. 7.12 comes from nonzero initial conditions, $z_i(0) \neq 0$, during switching from throttle to brake control and it dies out quickly because the hybrid closed-loop error dynamics is quadratically stable.

Figure 7.13 shows how the velocity errors and Lyapunov-like function level evolve in time under a fault in the P_w sensor measurement, as well as a coupled fault in both P_w and K_b. The figure shows a simulation for a 0.2% proportional bias in P_w after 14 seconds, and a 20% reduction in K_b after 12 seconds. It is interesting to note that the P_w sensor fault is small relative to the K_b parametric fault, however its effect on the Lyapunov-like function level is much more pronounced. As expected from the sensor fault case, the coupled fault is also classified as an intolerable fault. Consequently, although the individual fault is tolerable, respectively, the coupled faults have a large impact upon the system performance and are therefore regarded as intolerable faults.

Chapter 8
Fault Tolerant Control for AHS

While the passive fault tolerant control approach is proposed to compensate for a certain type of fault in Chap. 7, its performance is in general limited to these particular faults. For active fault tolerant control with applications to automated highway systems (AHS), a general hierarchical architecture composed of fault detection and diagnosis (FDD) working in conjunction with a hierarchy of control laws was proposed by Lygeros et al. [58]. More specifically, the FDD for longitudinal vehicle control in AHS was developed by Rajamani et al. [75, 76] and a fault management system (FMS) was first addressed to accommodate a single fault in [109].

The analysis and design of the lowest layer in this hierarchical control architecture, a regulation layer in Fig. 8.1, will be the focus of this chapter. Similar to the general lowest layer architecture for the passive and active FTC approaches as discussed in Chap. 7, the regulation layer for AHS contains three main components: FDD, a set of longitudinal controllers, and FMS. The FDD uses sensor measurements and actuator commands to detect and identify faults in vehicle components [76]. The longitudinal controller uses sensor feedback to provide good velocity and distance tracking performance in the presence of uncertainties. Finally, the FMS decides upon the specific control law based upon the current status of the vehicle components. In general, these systems are designed independently to meet their specific goals.

If the regulation layer is viewed as an integrated system, the overall design requires a compromise between two competing objectives; the FDD should be sensitive to faults in order for the FMS to choose the most appropriate control strategy, while the longitudinal controller should attempt to maintain good performance and minimize the effects of uncertainties coming from disturbances and noise, and even faults. For example, if the longitudinal controller is very robust, it may be hard for FDD to distinguish a fault from uncertainties and disturbances based on the sensor measurements unless the fault is severe. On the other hand, if FDD is too sensitive, it will cause frequent false alarms and result in unnecessary switching among different control laws. Therefore, the design of the fault tolerant control architecture should include some means for a tradeoff between these competing objectives. In Sect. 7.2.3, the fault classification was demonstrated to show a method of analyzing

B. Song, J.K. Hedrick, *Dynamic Surface Control of Uncertain Nonlinear Systems*, Communications and Control Engineering,
DOI 10.1007/978-0-85729-632-0_8,

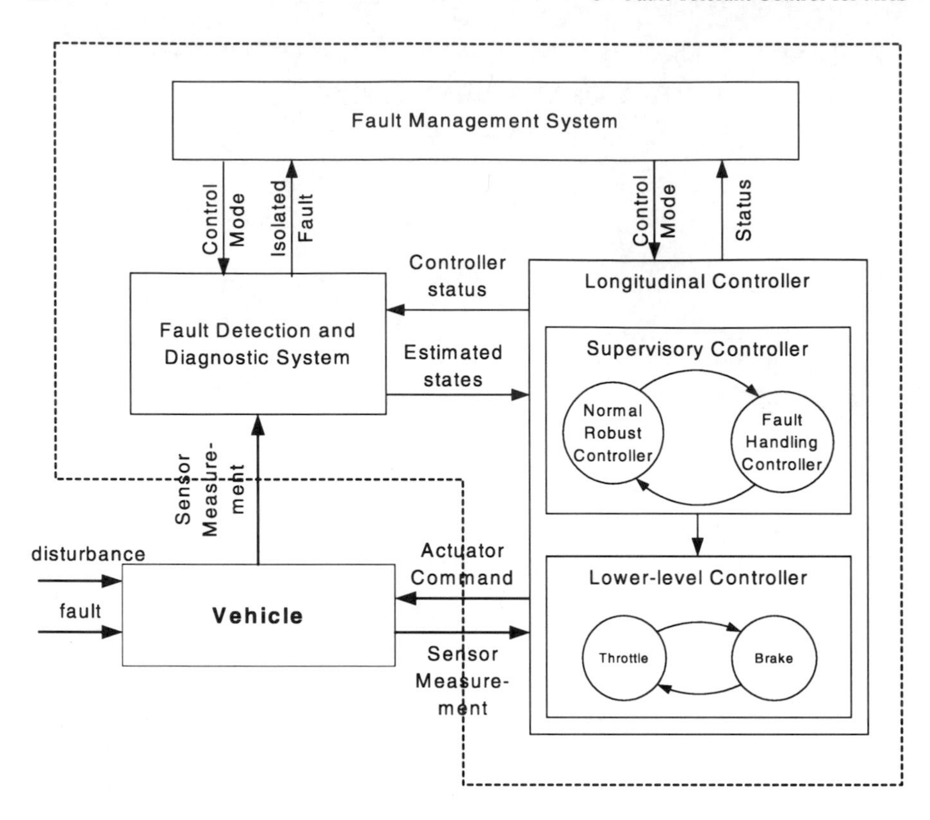

Fig. 8.1 Schematic structure of regulation layer in fault tolerant control architecture

performance degradation caused by a fault after the longitudinal controller is designed. The key idea is to measure the performance degradation by the level sets of a quadratic Lyapunov-like function, which is a quadratic error bound calculated via convex optimization.

If the fault is classified an intolerable fault as discussed in Chap. 7, we may need to choose one of a set of reconfigurable controllers to stabilize the faulty system and achieve either allowable or degraded mode performance. Here, we assume there is no control channel failure to simplify design of reconfigurable controllers, even though multiple controllers for a faulty system can be used to deal with the control channel failure [95]. Then, this assumption enables us to use a decentralized multiple control structure. In Chaps. 4 and 5, the nonlinear compensator and constrained stabilization of the nonlinear system were investigated and the theoretical results can be applied to a synthesis problem of reconfigurable controllers. In this chapter, two reconfigurable controller techniques will be discussed: an observer-based control scheme for sensor faults and a trajectory reconfiguration scheme for actuator related parametric faults. Finally, the proposed fault tolerant control sys-

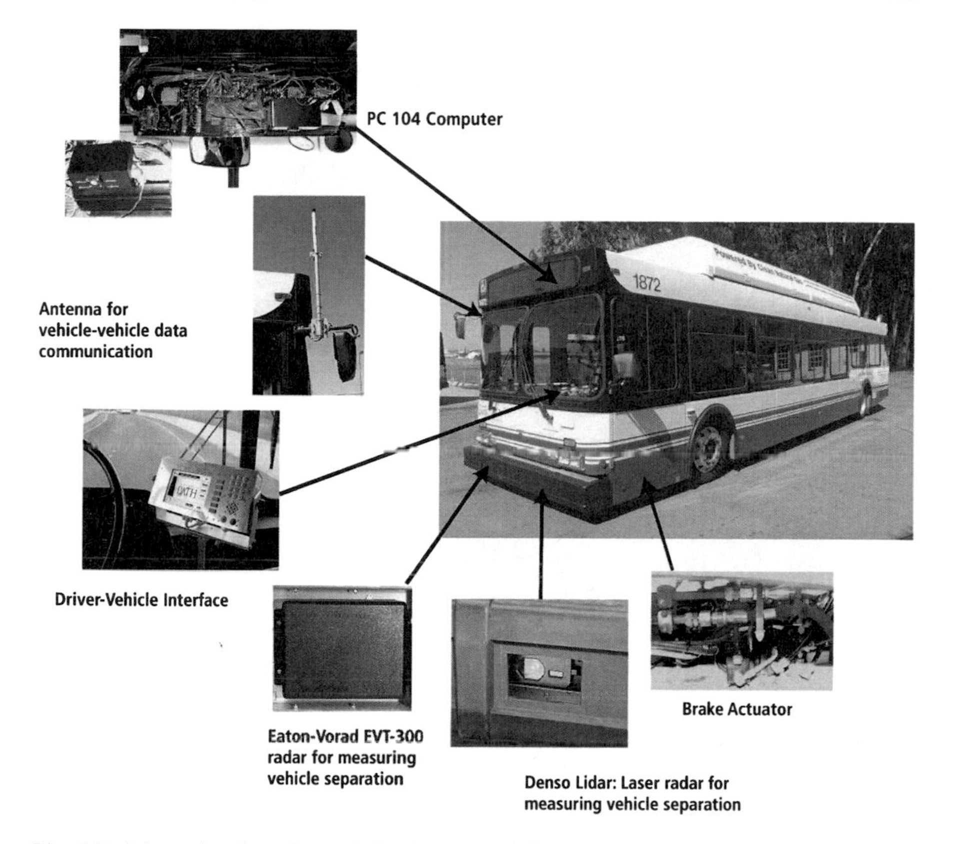

Fig. 8.2 Schematic of an electronic hardware setup for an automated transit bus of California PATH

tem will be applied to longitudinal control of an automated transit bus shown in Fig. 8.2.

8.1 Controller Reconfiguration

A supervisory controller of the longitudinal controller in Fig. 8.1 consists of a set of controllers as shown in Fig. 8.3. One of them is designed as a default robust controller which works for both a non-faulty system and a system with tolerable faults. If the occurring fault is classified an isolatable and intolerable fault in FMS, the default controller is switched to another controller which is able to compensate for this type of the intolerable fault [85]. Otherwise, the fault could cause a large amount of performance degradation, and even potentially drive the system to be unstable. While the multiple controller structure is proposed to accommodate all possible worst cases, only two types of reconfiguration mechanisms will be considered to deal with isolatable faults in either sensors or actuators.

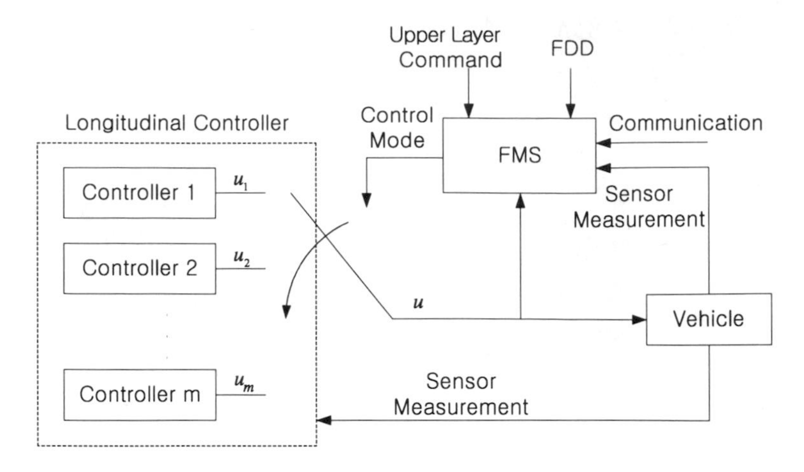

Fig. 8.3 Parallel architecture for multiple controllers in regulation layer

8.1.1 Observer-Based DSC

Consider the system given in (7.22). It can be written in matrix form as follows:

$$\begin{cases} \dot{x} = Ux + B_u u + \Theta f + \Delta f_m, \\ y = x \end{cases} \tag{8.1}$$

where $U = \text{diag}([1, \dots, 1], 1) \in \Re^{n \times n}$, $\Theta = \text{diag}(a_1, \dots, a_n) \in \Re^{n \times n}$, and $B_u = [0 \cdots 0\ 1]^T \in \Re^n$. If there is a fault in the jth measurement, y in (8.1) is written as

$$y = \big[I_n - F_s(\nu)\big]x$$

where $F_s(\nu) = \text{diag}[0, \dots, \upsilon_j, \dots, 0]$ and $\upsilon_j(t) \in [0, 1]$. Furthermore, if it is identified and isolated from FDD, the non-faulty healthy measurements are written as

$$y_h = C_j y$$

where $C_j \in \Re^{(n-1)\times n}$ is the measurement matrix such that its row corresponding to the faulty measurement is eliminated from the identity matrix.

As proposed in Sect. 4.1, let the nonlinear observer be

$$\dot{\hat{x}} = U\hat{x} + B_u u + \Theta f(\hat{x}) + L(y_h - C_j\hat{x}) \tag{8.2}$$

where it is assumed that the pair (U, C_j) is observable and a coordinate transformation matrix T_c is an identity matrix for the sake of clarity. Then, combining (4.44) with (7.31), the augmented error dynamics with possible faults can be given by

$$\dot{z}_e = A_{ef} z_e + B_w \tilde{w}_f + B_e \ddot{x}_{1d} \tag{8.3}$$

where

$$\tilde{w}_f = \begin{bmatrix} \Theta f(x) - \Theta f(\hat{x}) + \Delta f_m + \Delta f_p \\ \Theta \dot{f} \end{bmatrix} := \begin{bmatrix} \phi \\ \psi \end{bmatrix},$$

$$A_{ef} = \left[\begin{array}{cc|c} \tilde{A}_{11} & LC_j & \tilde{A}_{12} \\ \mathbf{0} & U - LC_j & \mathbf{0} \\ \hline T_\xi^{-1} K\tilde{A}_{11} & T_\xi^{-1} KLC_j & T_\xi^{-1}(K\tilde{A}_{12} + A_{33}) \end{array}\right], \qquad B_e = \left[\begin{array}{c} 0 \\ 0 \\ \hline -T_\xi^{-1} b_r \end{array}\right],$$

the submatrices $\tilde{A}_{11}, \tilde{A}_{12}$ are defined in (7.30), and other vectors and matrices in (4.44). Moreover, if there is no parametric fault, i.e., $\Delta f_p = 0$, $\mu = 0$, and $\gamma = 1$, (8.3) becomes equivalent to the augmented error dynamics of a nonlinear compensator in (4.59).

Using the separation principle stated in Chap. 4, first we can design the nonlinear observer independently, i.e., find an observer gain matrix L to guarantee the convergence of estimation errors. In other words, for the given L, we need to show the existence of a positive definite matrix P such that

$$\left[(U - LC_j)e + \phi\right]^T Pe + e^T P\left[(U - LC_j)e + \phi\right] < 0$$

for all nonzero $e = x - \hat{x} \in \Re^n$. As discussed in Sect. 4.1, the observer gain matrix L can be designed if ϕ is Lipschitz and the Lipschitz constant is known. However, $\Delta f_m + \Delta f_p$ may not bounded by a linear function of the estimation error e. Then, the proposed observer can be designed only to guarantee boundedness of the error in terms of the induced $\mathscr{L}_2$ gain, not convergence to the origin (refer to Sect. 4.4 for more details). Finally, once the gain matrix L of the robust observer is obtained, with appropriate choice of controller gains (K_i and τ_i), we can compute a quadratic Lyapunov matrix P which satisfies Theorem 4.5 or 4.6 for quadratic stability via convex optimization.

8.1.2 Trajectory Reconfiguration

If there is a complete failure in the actuator, it is assumed that emergency handling will be carried out by an upper layer (see Fig. 8.3), so only partial failure of the actuator will be considered here. When an intolerable fault occurs in the actuator, the capability of the actuator is degraded and so is the tracking performance. In this case, a two-step trajectory reconfiguration procedure is suggested to provide robust stabilization and reduce the degradation of the tracking performance.

First, the closed-loop system should be stabilized in the current situation using the remaining capability of the actuators. This is called the *stabilization* step, in the sense that the external inputs $\ddot{x}_{1d}$ and $\dot{x}_{1d}$ become zero. Thus, the error dynamics in (7.31) becomes

$$\dot{z} = A_f z + B_w w_f. \tag{8.4}$$

If there is no sensor and parametric fault in the system, the perturbation term w_f is written as

$$w_f = \begin{bmatrix} \Delta f_m \\ \Theta \dot{f} \end{bmatrix} = \begin{bmatrix} \Delta f_m \\ \Theta(JC_z z + J\Delta f_m) \end{bmatrix}$$

where the last equality is derived in Appendix A.2. In Chaps. 2 and 3, three different cases regarding Δf_m are considered. If no model uncertainty is considered as in Chap. 2, the term w_f becomes $w_f = \Theta J C_z z$. If f and $[\partial f / \partial x]$ are continuous, there exists an upper norm-bound of w_f such that

$$\|w_f\| = \|\Theta J C_z z\| \le \gamma \|\Theta C_z z\| := \|\tilde{C}_z z\|$$

where γ is a Lipschitz constant defined in (2.4). Then, the above equation in (8.4) is classified a norm-bounded LDI and the quadratic stability test based on Theorem 2.1 can be performed for the given $\tilde{C}_{zi}$. That is, if there exists a solution matrix $P > 0$ satisfying LMI (2.33), it can be said that the system can be stabilized using the current faulty actuator. It is noted that the system matrix A_f instead of A_{cl} in LMI (2.33) includes the effect of the actuator fault.

If Δf_m is Lipschitz, as summarized in Lemma 3.1, there exists a componentwise norm-bound of w_f such that

$$|w_{if}| \le \|C_{zi} z\|,$$

where w_{if} is the ith component of w_f and C_{zi} is defined in Appendix A.2. Then, the error dynamics in (8.4) becomes a diagonal norm-bounded LDI. As mentioned in Remark 3.1, the quadratic stability can be tested using the result of Theorem 2.3. In a more general case such that Δf_m is locally Lipschitz, nonlinear damping may be used to have an inequality constraint regarding the model uncertainty as discussed in Sect. 3.2. Another approach is to use input–output stability as suggested in Sect. 3.3. If the nonlinear damping is not used, (8.4) can be rewritten as

$$\dot{z} = A_f z + B_w w_f = A_f z + \begin{bmatrix} 0 \\ T_\xi^{-1} \end{bmatrix} w + B_w \begin{bmatrix} \Delta f_m \\ \Theta J \Delta f_m \end{bmatrix} \tag{8.5}$$

where $w = \Theta J C_z z$. After defining the output as $y = C_y z$ and considering the last term in (8.5) as an external input, the induced $\mathscr{L}_2$ gain of the error dynamics can be calculated using the result of Theorem 3.3.

Once the faulty error dynamics in (8.4) is quadratically stable for the given assumptions, the *trajectory reconfiguration* step is performed. It allows us to modify the reference input for the closed-loop system to be quadratically bounded in the presence of faults. For instance, the error dynamics is given in (7.31) with the exogenous input satisfying an inequality constraint $\|r\| = \|[x_{1d}\ \dot{x}_{1d}\ \ddot{x}_{1d}]^T\| \le r_0$. Then, the bounded peak input can be regenerated as $\|r\| \le r_{\max} < r_0$. However, the question is how $r_{\max}$ is chosen to guarantee quadratic boundedness. If the matrix A_f is available via the isolation process of FDD, the value of $r_{\max}$ can be maximized via a line search and convex optimization. However, since those procedures are not usually performed online due to their computational burden, the values of $r_{\max}$ can be predetermined under specific worst case scenarios for a range of intolerable actuator faults. This procedure can also be used for the stabilization step by setting $r_{\max} = 0$.

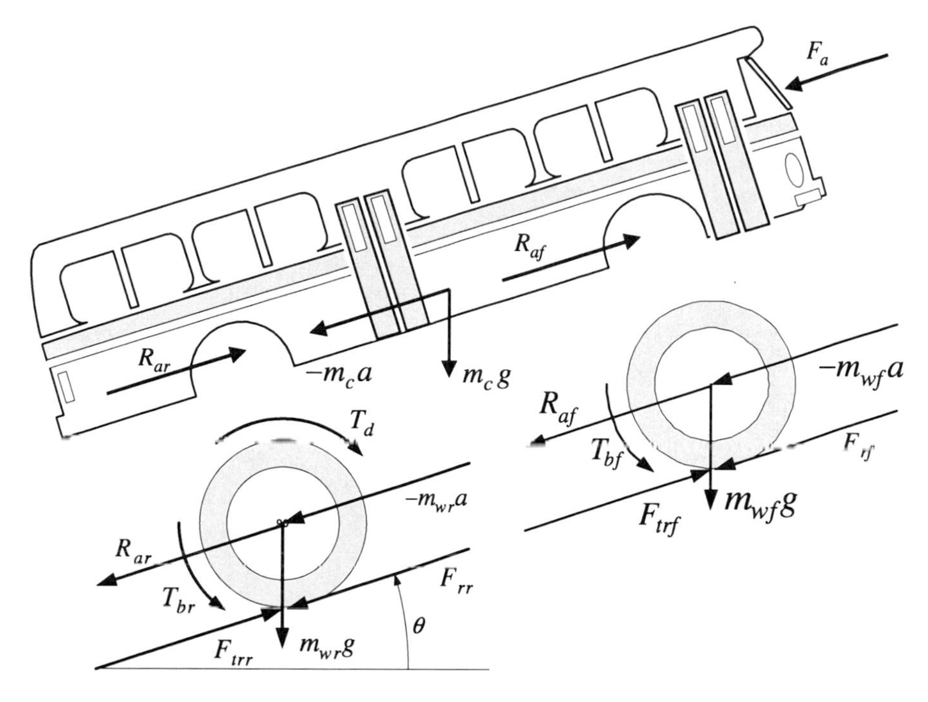

Fig. 8.4 Free body diagram for longitudinal motion of a transit bus

8.2 Longitudinal Control for an Automated Transit Bus

The active fault tolerant control technique with controller reconfiguration proposed in Sect. 8.1 will be applied to automated transit buses shown in Figs. 7.1 and 8.2. Furthermore, due to similarity of kinetics between a passenger vehicle and a transit bus, the proposed technique can also be applied to the passenger vehicle introduced in Chap. 7.

8.2.1 Longitudinal Control via DSC

Under the assumptions that no slip occurs at the wheels and the torque converter is locked ($v = R_g h w_e$ and $T_d = T_e/R_g$), the longitudinal equation of motion can be derived by balancing the forces in the longitudinal direction (see in Fig. 8.4). The resulting single state dynamics is [82]

$$\begin{aligned} \dot{v} &= \frac{T_e - T_{acc} - R_g\{T_b + hF_r + hF_a + mgh\sin(\theta)\}}{J_{eq}} + \Delta f_{1m}(v) \\ &= \frac{T_e - R_g T_b}{J_{eq}} - f_1 + \Delta f_{1m}(v) \end{aligned} \tag{8.6}$$

Table 8.1 System parameters and variables

Parameter	Description	Parameter	Description
v	velocity of the vehicle [m/s]	ω_e	engine speed [rad/s]
T_e	net engine torque [N m]	T_b	braking torque [N m]
T_{acc}	accessory engine torque [N m]	P_b	brake line pressure [kPa]
F_r	rolling resistance force [N]	F_a	aerodynamic drag force [N]
C_r	rolling resistance coefficient	C_a	aerodynamic drag coefficient [kg/m]
h	effective wheel radius [m]	R_g	effective gear ratio
θ	slope angle [rad]	m	total weight of the vehicle [kg]
J_e	inertia of the engine [kg m^2]	J_w	inertia of the axle [kg m^2]

where J_{eq} and f_1 are

$$J_{eq} = \frac{J_e + R_g^2 \cdot (J_\omega + m \cdot h^2)}{R_g h}, \qquad f_1 = \frac{T_{acc}}{J_{eq}} + \frac{R_g h}{J_{eq}} \{C_a v^2 + F_r + mg\sin(\theta)\}$$

and $\Delta f_{1m}(v)$ is the model uncertainty of the vehicle. Furthermore, a list of parameters in the model is summarized in Table 8.1. It is noted that (8.6) is quite similar to (7.1a) since both equations are only based on kinetics in the longitudinal direction with same assumptions. However, the accessory engine torque due to an air conditioner (A/C), a fan, and thermal conditions and road grade are added since they have large impact on the longitudinal vehicle dynamics [85]. In addition, it is assumed that the accessory engine power is constant, e.g.,

$$T_{acc} \times \omega_e = \begin{cases} C_1, & \text{if A/C is on,} \\ C_2, & \text{if A/C is off.} \end{cases}$$

Furthermore, when the throttle is minimum or closed, additional engine torque loss due to the Jake brake and/or a transmission retarder may be added in T_{acc}.

Next, the dynamics of the brake and engine can be represented by the following first order lag systems (see [57, 85, 92]):

$$\dot{T}_e = \frac{1}{\tau_e}(-T_e + u_e) + \Delta f_{2m}, \tag{8.7}$$

$$\dot{P}_b = \begin{cases} \frac{1}{\tau_{bf}}[-P_b + u_b(t - \Delta t_b)] + \Delta f_{3m}, & \text{for filling,} \\ \frac{1}{\tau_{be}}(-P_b + u_b) + \Delta f_{3m}, & \text{for emptying} \end{cases} \tag{8.8}$$

where u_e and u_b are the torque and brake pressure command which may be available via an in-vehicle network [85]. If they are not available, they can be controlled by throttle and brake pedal positions (u_α and u_β), respectively, such as

$$u_e = T_{map}(u_\alpha, \omega_e) \quad \text{and} \quad u_b = P_b(u_\beta)$$

where T_{map} is the empirical engine map. Furthermore, Δf_{2m} and Δf_{3m} represent the modeling uncertainties, $\tau_e > 0$ is the overall time constant for the engine and

Table 8.2 System parameters for modeling

Parameter	Value	Parameter	Value
m	13,381 (kg)	J_w	12.825 (kg m^2)
R_g	0.1873	J_e	1.315 (kg m^2)
h	0.447 (m)	C_a	0.65 (kg/m)
P_o	34.48 (kPa)	K_b	10 (N m/Kpa)
τ_e	0.1 (s)	F_r	1313 (N)
τ_{bf}	0.25 (s)	τ_{be}	0.03 (s)
T_{ect}	0 (N m)	θ	0 (rad)
$T_{e,limit}$	1016.85 (N m)	$P_{b,limit}$	689.48 (kPa)
C_1	26,845 (hp)	C_2	22,968 (hp)

throttle actuator delay, $\tau_{bf}, \tau_{be} > 0$ are the overall time constants of the brake actuator during both the filling and emptying procedure, respectively. It is noted that τ_{be} is much smaller than τ_{bf} due to the quick release value (refer to [57]). Moreover, the brake torque T_b has a proportional relation with the brake pressure P_b in the diaphragm chamber such that [57]

$$T_b = \begin{cases} K_b(P_b - P_o), & \text{if } P_b \geq P_o, \\ 0, & \text{otherwise} \end{cases} \tag{8.9}$$

where P_o is the push-out pressure and K_b is a brake coefficient [57]. Finally, the proposed model without uncertainty models, Δf_m, is validated by comparing simulation results with the experimental test data in [85]. The accuracy of the model with system parameters listed in Table 8.2 is 5% in terms of velocity and engine speed deviation for a certain driving scenario.

Since all states as well as additional information for control can be obtained from sensors such as a pressure sensor, a torque sensor, and a range-rate sensor, the measurement output y is defined as

$$y = [v \ v - v_{des} \ \omega_e \ T_e \ P_b \ \theta]^T \in \Re^6.$$

It is interesting in practice to note that $v - v_{des}$ can be measured by a range-rate sensor if v_{des} is the velocity of a leading vehicle, v, T_e, P_b, and ω_e are available via the standard in-vehicle network [77], and road grade θ can be estimated [5]. Here, discussion is limited to a velocity tracking problem of the three state system for longitudinal vehicle control where the desired trajectory v_{des} is a feasible output trajectory and its first and second derivatives $\dot{v}_{des}$, $\ddot{v}_{des}$ are bounded. However, this can be extended for spacing control as shown in [37, 82].

The objective of the longitudinal controller is to follow a given desired velocity profile via the available two control inputs, namely the engine torque and brake pressure. Realistically, the engine and brakes should not be used simultaneously in order to reduce wear and tear on the vehicle. The remainder of this section will provide derivations for the nominal control laws for engine and brake control. The control laws for determining the commanded engine torque and brake pressure will be designed using a dynamic surface controller [89]. As in Sect. 7.1, after defining the first error surface $S_{1e} = v - v_{des}$, its derivative is given by

$$\dot{S}_{1e} = \frac{T_e - R_g T_b}{J_{eq}} - f_1 + \Delta f_{1m} - \dot{v}_{des}$$

and a synthetic input of T_e is derived as

$$\bar{T}_e = R_g T_b + J_{eq}(f_1 + \dot{v}_{des} - K_{1e} S_{1e}).$$

Moreover, since the information of v and S_{1e} comes from the measurement output and $T_b = 0$ for engine control, the calculated $\bar{T}_e$ is

$$\bar{T}_e = J_{eq}\{f_1(y_1, y_3, y_6) + \dot{v}_{des} - K_{1e} y_2\} \tag{8.10}$$

where y_i is the ith measurement of y and the corresponding first order low-pass filter is

$$\tau_{2e}\dot{T}_{edes} + T_{edes} = \bar{T}_e, \quad T_{edes}(0) := \bar{T}_e(0).$$

Next, define the second surface as $S_{2e} = T_e - T_{edes}$ and its derivative is

$$\dot{S}_{2e} = \dot{T}_e - \dot{T}_{edes} = \frac{1}{\tau_e}(-T_e + u_e) + \Delta f_{2m} - \dot{T}_{edes}.$$

Then, the commanded engine torque is

$$u_e = y_4 + \tau_e(\dot{T}_{edes} - K_{2e}\tilde{S}_{2e}) \tag{8.11}$$

where $\tilde{S}_{2e} := y_4 - T_{edes}$.

A control law for the brake system can be derived similarly by defining $S_{1b} := S_{1e}$. The resulting equation for the desired brake pressure is

$$\bar{P}_b = -\frac{J_{eq}}{K_b R_g}(f_1 + \dot{v}_{des} - K_{1b} y_2) + P_o \tag{8.12}$$

where T_{ect} is the minimum or closed throttle torque. For the second surface, define $S_{2b} := P_b - P_{bdes}$ where

$$\tau_{2b}\dot{P}_{bdes} + P_{bdes} = \bar{P}_b, \quad P_{bdes}(0) := \bar{P}_b(0).$$

After differentiating S_{2b} and using (8.8), the commanded brake pressure is

$$u_b = y_5 + \tau_b(\dot{P}_{bdes} - K_{2b}\tilde{S}_{2b}) \tag{8.13}$$

where

$$\tilde{S}_{2b} = y_5 - P_{bdes} \quad \text{and} \quad \tau_b = \begin{cases} \tau_{bf}, & \text{for filling,} \\ \tau_{be}, & \text{for emptying.} \end{cases}$$

It is remarked that the switching condition between engine and brake control is introduced in Sect. 7.1.

8.2.2 Quadratic Tracking and Validation

If the longitudinal controller derived in Sect. 8.2.1 is applied to the equation of motion for the transit bus, the closed-loop error dynamics is derived as in (7.10) with different definitions of a_i and f_1. For the quadratic boundedness of the closed-loop error equation (7.10), either Theorem 2.4 or Corollary 2.2 provides a sufficient condition for the existence of a quadratic Lyapunov-like function $V(z) = z^T Pz$ that gives quadratic tracking for the switched nonlinear system. Once these controller gains have been chosen to meet some desired performance, a quadratic Lyapunov-like function can be computed which guarantees that the error stays within the ellipsoid of the form

$$\mathcal{E}_P = \left\{z|z^T Pz < 1,\ P > 0\right\}$$

under the modeling uncertainties for the vehicle. It is noted that the size of the ellipsoidal bound depends on the magnitude of controller gain and characteristics of the model uncertainty Δf and reference input r.

Remark 8.1 If a coordinate transformation is used to scale or normalize the component of z_i in (7.10) such as $\bar{z}_i = Dz_i$, the error dynamics in (7.10) is rewritten as

$$\begin{aligned} \dot{\bar{z}}_i &= A_i \bar{z}_i + DB_{wi} w_i + DB_{di} \ddot{v}_{des}, \\ |w_i| &\le \left|C_{zi} D^{-1} \bar{z}_i\right| \end{aligned} \tag{8.14}$$

where D is a diagonal matrix and $A_i = DA_i D^{-1}$. The corresponding ellipsoidal error bound is defined as

$$\mathcal{E}_P = \left\{\bar{z}|\bar{z}^T P\bar{z} \le 1,\ P > 0\right\}$$

where $\bar{P}$ is a solution of LMI (2.61) in Corollary 2.2 for a given α and (8.14). This transformation may help finding a solution P if the error dynamics is ill-conditioned numerically.

Example 8.1 (Quadratic boundedness of longitudinal control for a transit bus) Suppose the system parameters for the bus model are given in Table 8.2. In addition, it is assumed that a direct control of T_e is not available. That is, u_e in (8.7) is controlled by throttle pedal position u_α such that

$$u_e = T_{map}(u_\alpha, \omega_e)$$

where T_{map} is an empirical engine map of the engine speed and throttle pedal position. For instance, Fig. 8.5 shows the engine map of a compressed natural gas engine with 280 (hp) in a transit bus for simulation [89].

While the parametric uncertainty of C_a in (8.6) is considered in Example 7.3, two additional parametric uncertainties in (8.7) and (8.8) are considered as follows:

$$\tau_e \quad \longrightarrow \quad \tau_e + \Delta\tau_e, \qquad \tau_b \quad \Longrightarrow \quad \tau_b + \Delta\tau_b$$

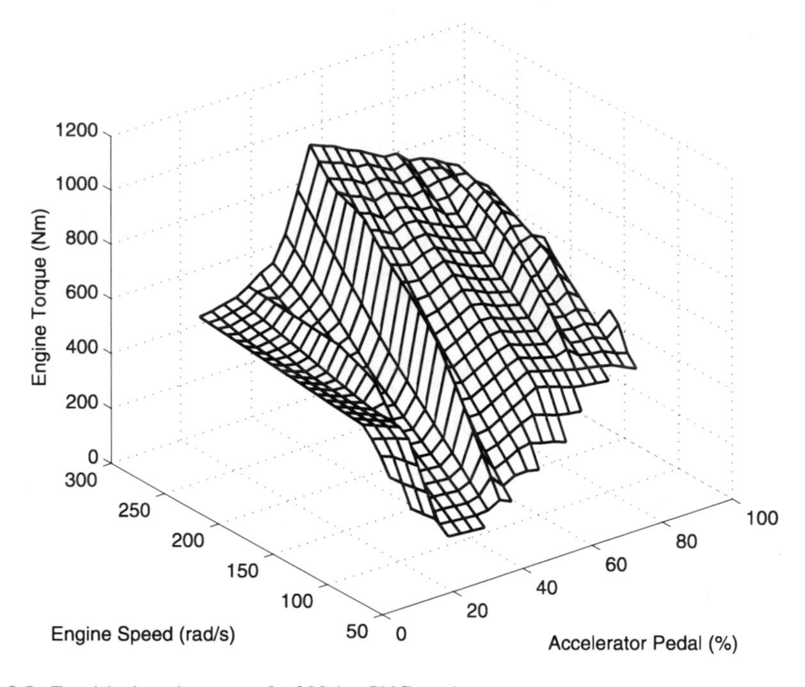

Fig. 8.5 Empirical engine map of a 280-hp CNG engine

where $\Delta\tau_e$ and $\Delta\tau_b$ are the normally distributed random numbers between $\pm 5\%$ of the nominal τ_e and τ_b, respectively. Then, (8.7) and (8.8) are simplified for controller design as

$$\dot{T}_e = \frac{1}{\tau_e + \Delta\tau_e}(-T_e + u_e), \tag{8.15}$$

$$\dot{P}_b = \frac{1}{\tau_b + \Delta\tau_b}(-P_b + u_b) \tag{8.16}$$

where only the filling behavior model in (8.8) is considered with the assumption of no time delay. If (8.11) is applied to (8.15),

$$\begin{aligned} \dot{T}_e &= \frac{\tau_e}{\tau_e + \Delta\tau_e}(\dot{T}_{edes} - K_{2e}S_{2e}), \\ \tau_e(\dot{T}_e - \dot{T}_{edes}) &= -\tau_e K_{2e}S_{2e} - \Delta\tau_e \dot{T}_e, \\ \dot{S}_{2e} &= -K_{2e}S_{2e} + \Delta f_{em} \end{aligned}$$

where

$$\Delta f_{em} = -\frac{\Delta\tau_e}{\tau_e + \Delta\tau_e}\left(-\frac{\xi_{2e}}{\tau_{2e}} - K_{2e}S_{2e}\right).$$

Similarly, the error dynamics of S_{2b} for the brake system in (8.16) is

$$\dot{S}_{2b} = -K_{2b}S_{2b} + \Delta f_{bm}$$

where

$$\Delta f_{bm} = -\frac{\Delta\tau_b}{\tau_b + \Delta\tau_b}\left(-\frac{\xi_{2b}}{\tau_{2b}} - K_{2b}S_{2b}\right).$$

As done in Example 7.3, Δf_1 is written as

$$\Delta f_{1m}(v) = \frac{\Delta C_a h}{J_{eq}} v^2$$

where ΔC_a stands for the parametric uncertainty of C_a.

If DSC is applied as derived above, the error dynamics derived in Example 7.3 is expanded due to include additional parametric uncertainties as follows:

$$\dot{z}_i = A_i z_i + [B_{ui}\ e_2\ B_{di}]\begin{bmatrix}\Delta f_{1m}\\ \Delta f_{im}\\ \dot{f}_1\end{bmatrix} + B_{di}\ddot{v}_{des} \tag{8.17}$$

where $e_2 = [0\ 1\ 0]^T \in \Re^3$, the system matrix A_i is

$$A_i = \begin{bmatrix} -K_{1i} & 1/a_i & 1/a_i \\ 0 & -K_{2i} & 0 \\ -a_i K_{1i}^2 & K_{1i} & K_{1i} - 1/\tau_{2i} \end{bmatrix}, \qquad a_i = \begin{cases} J_{eq}, & \text{if } i = e, \\ -J_{eq}/(R_g K_b), & \text{if } i = b, \end{cases}$$

and all other definitions of vectors and matrices are given in (7.41). As discussed in Example 7.3, Δf_1 is decomposed into vanishing and nonvanishing perturbation functions p and q satisfying

$$|p| \le \gamma_0 \big|[1\ 0\ 0]z_i\big|, \qquad |q| \le \gamma_0 |v_{des}|$$

where $\gamma_0 = \frac{0.2 C_a h R_g v_{\max}}{J_{eq}}$ for the given $v_{\max} = 25$ as defined in Example 7.3 and Δf_{im} is bounded by a vanishing perturbation function such that

$$\begin{aligned} |\Delta f_{im}| &= \left|-\frac{\Delta\tau_i}{\tau_i + \Delta\tau_i}\left(-\frac{\xi_{2i}}{\tau_{2i}} - K_{2i}S_{2i}\right)\right| = \left|\frac{\Delta\tau_i}{\tau_i + \Delta\tau_i}\right| \big|[0\ K_{2i}\ 1/\tau_{2i}]z_i\big| \\ &\le \gamma_1 \big|[0\ K_{2i}\ 1/\tau_{2i}]z_i\big|, \end{aligned}$$

where

$$\gamma_1 = \left|\frac{0.05\tau_i}{\tau_i - 0.05\tau_i}\right| = \frac{1}{19}.$$

Using (7.40), $\dot{f}_1$ is written as

$$\dot{f}_1 = J_{11}c_{zi}z_i + J_{11}\dot{v}_{des} + J_{11}\Delta f_1 = J_{11}(c_{zi}z_i + \dot{v}_{des} + p + q).$$

Therefore, after decomposing $[\Delta f_{1m}\ \Delta f_{im}\ \dot{f}_1]^T \in \Re^3$ into the vanishing and non-vanishing terms, (8.17) is rewritten as

$$\begin{aligned} \dot{z}_i &= A_i z_i + [B_{ui}\ e_2\ B_{di}\ B_{di}]\begin{bmatrix} p \\ \Delta f_{im} \\ J_{11}p \\ J_{11}c_{zi}z_i \end{bmatrix} + [B_{ui}\ B_{di}\ B_{di}\ B_{di}]\begin{bmatrix} q \\ J_{11}q \\ J_{11}\dot{v}_{des} \\ \ddot{v}_{des} \end{bmatrix} \\ &= A_i z_i + B_{wi}w_i + B_{ri}r \quad \text{for } i = e, b. \end{aligned} \tag{8.18}$$

Furthermore, the componentwise upper bounds of w_i are

$$|w_i| = \left| \begin{bmatrix} p \\ \Delta f_{im} \\ J_{11} p \\ J_{11} c_{zi} z_i \end{bmatrix} \right| \le \left| \begin{bmatrix} \gamma_0 [1\ 0\ 0] \\ \gamma_1 [0\ K_{2i}\ 1/\tau_{2i}] z_i \\ \gamma \gamma_0 [1\ 0\ 0] \\ \gamma [-K_{1i}\ 1/a_i\ 1/a_i] \end{bmatrix} z_i \right| := |C_{zi} z_i|$$

where

$$|J_{11}| = \left| \frac{\partial f_1}{\partial v} \right| = \left| \frac{1}{J_{eq}} \frac{\partial T_{acc}}{\partial v} + \frac{2 R_g h C_a}{J_{eq}} v \right| = \left| \frac{1}{J_{eq}} \frac{\partial}{\partial v} \frac{C_j R_g h}{v} + \frac{2 R_g h C_a}{J_{eq}} v \right|$$
$$= \frac{R_g h}{J_{eq}} \left| -\frac{C_j}{v^2} + 2 C_a v \right| \le \frac{R_g h}{J_{eq}} \left| -\frac{C_j}{v_{\max}^2} + 2 C_a v_{\max} \right| := \gamma$$

and r is norm-bounded as

$$\|r\|^2 \le (1+\gamma^2) q^T q + \gamma^2 \dot{v}_{des}^2 + \ddot{v}_{des}^2 \le (1+\gamma^2) \gamma_0^2 v_{des}^2 + \gamma^2 \dot{v}_{des}^2 + \ddot{v}_{des}^2 := r_0^2.$$

Therefore, the closed-loop error dynamics in (8.18) can be summarized in DNLDI form as follows:

$$\dot{z}_i = A_i z_i + B_{wi} w_i + \tilde{B}_{ri} \tilde{r}, \quad |w_i| \le |C_{zi} z_i| \tag{8.19}$$

where $\tilde{r} = r/r_0$ is the unit-peak input satisfying $\tilde{r}^T \tilde{r} \le 1$ and $\tilde{B}_{ri} = r_0 B_{ri}$. If $\bar{z}_i = D z_i$ where $D = \mathrm{diag}(1, 0.01, 0.01)$, the error dynamics is written as in (8.14)

$$\dot{\bar{z}}_i = A_i z_i + D B_{wi} w_i + D \tilde{B}_{ri} \tilde{r}, \quad |w_i| \le \left| C_{zi} D^{-1} \bar{z}_i \right|.$$

Consider two sets of controller gains, such that

$$\{K_{1e}, K_{2e}, \tau_{2e}\} = \{1, 20, 0.02\} \quad \text{and} \quad \{K_{1b}, K_{2b}, \tau_{2b}\} = \{1, 10, 0.02\}.$$

To compute the ultimate bound of the closed-loop error dynamics for the given set of controller gains, the result of Corollary 2.2 is used and COP (2.62) is solved iteratively for a fixed α. That is, after 20 logarithmically equally spaced points between 10^{-1} and 10^1 are generated for the given α, the minimum of the maximum diameter, which is defined as $d_{\max} = 2/\sqrt{\lambda_{\min}(P)}$, is obtained when $\alpha = 0.6952$ (see in Fig. 8.6(a)). Then the 20 linearly equally spaced points between 0.5456 and 0.8859 are generated and the iterative computation of COP (2.62) is performed for each α. Finally, for $\alpha = 0.6709$, the maximum diameter of the ellipsoid, $d_{\max}$, is 107.1893 which is the semi-axis in the ξ_2 axis. Furthermore, projections of the corresponding ellipsoid are shown in Figs. 8.6(c) and (d). As discussed in Example 7.1, the size of the reachable set may be decreased by increasing the magnitude of controller gains, K_{1i} and K_{2i}. However, the larger magnitude of the controller gains may produce a larger error due to measurement noise and more frequent chattering around dynamic surfaces due to switching between engine and brake control.

For the given controller, the velocity tracking performance in the presence of parametric model uncertainties is shown in Figs. 8.7(a) and (b). The corresponding Lyapunov-like function level, $V(z)$ calculated above, is shown in Fig. 8.7(c). It implies that $z_i(t)$ stays within the ellipsoid once it enters the ellipsoidal error bound.

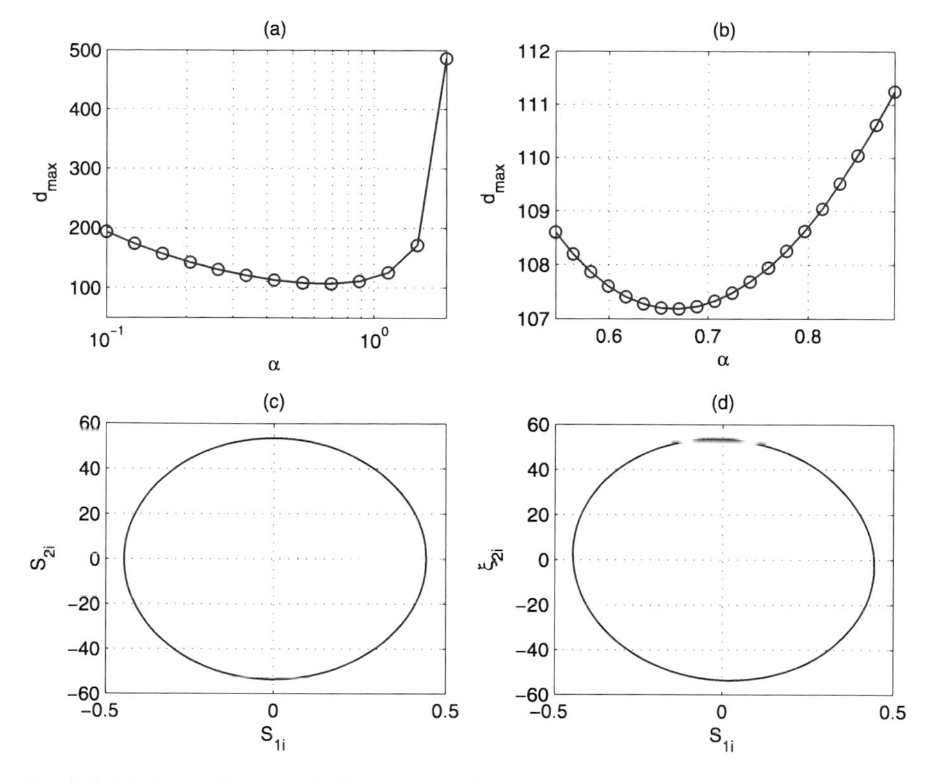

Fig. 8.6 Maximum diameter of ellipsoid along line search of α (*upper*) and quadratic bound in S_1–S_2 and S_1–ξ_2 planes (*bottom*)

The controller proposed in this section was implemented successfully for both 40-foot CNG and 60-foot articulated transit buses. The demonstration of automated public transportation technology was conducted on I-15 in San Diego, California in August of 2003 in cooperation with demonstration team members from the California PATH program (see Figs. 7.1 and 8.2). Figure 8.8 shows time responses of the 40-foot bus with respect to velocity and accelerator pedal position and demonstrates how well the longitudinal control algorithm track a desired speed profile. It also shows that the longitudinal controller is robust enough to make the system tracks the desired speed, even without considering road grade and change of overall vehicle mass. Furthermore, an automatic air conditioner was running during the entire driving test, which often causes an abrupt change of engine accessory torque. For instance, spikes in the speed error plot in Fig. 8.8 around 240 and 360 seconds may result from several disturbance and discrete change of vehicle variables, e.g., abrupt change of accessory torque and/or road grade. However, once the torque converter is locked, the tracking error becomes less than 0.5 m/s, which is on the order of the wheel speed sensor resolution.

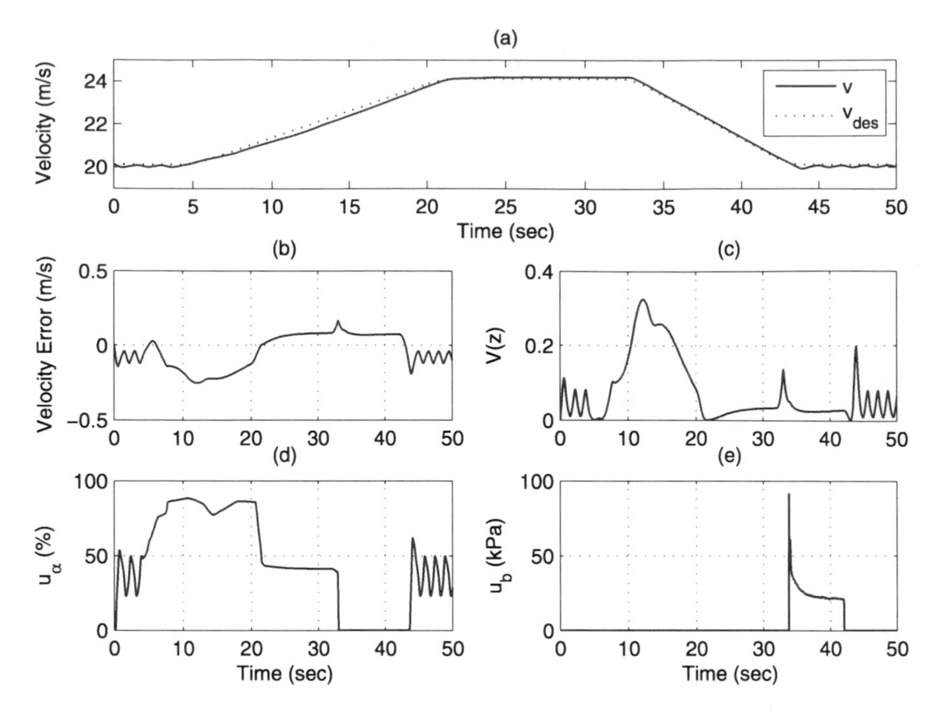

Fig. 8.7 Velocity tracking performance (**a**) and (**b**), quadratic function level (**c**), and control inputs (**d**) and (**e**)

8.3 Active Fault Tolerant Control

Before considering possible faults, it is assumed that the following measurement information and control input are reliable.

- Since hardware redundancy already exists in terms of engine speed and four wheel speed measurements, it can be assumed that the engine speed, wheel speed, and velocity measurements are available without fault as long as there is no slip and the torque converter is locked as assumed in Sect. 8.2.1.
- Generally speaking, the air brake system is a dual brake system and has emergency/parking brake mechanism. So the failure of brake pressure generation happens very rarely [57]. Therefore, it can be assumed that there is no actuator failure in (8.8).
- The road grade is known a prior via a digital map.

There are two types of faults considered here: sensor and parametric fault. Mathematically, the measurement output y with possible sensor faults can be described as

$$y = \left[I - F_s(\nu)\right][v \; v - v_{des} \; \omega_e \; T_e \; P_b \; \theta]^T \in \Re^6 \tag{8.20}$$

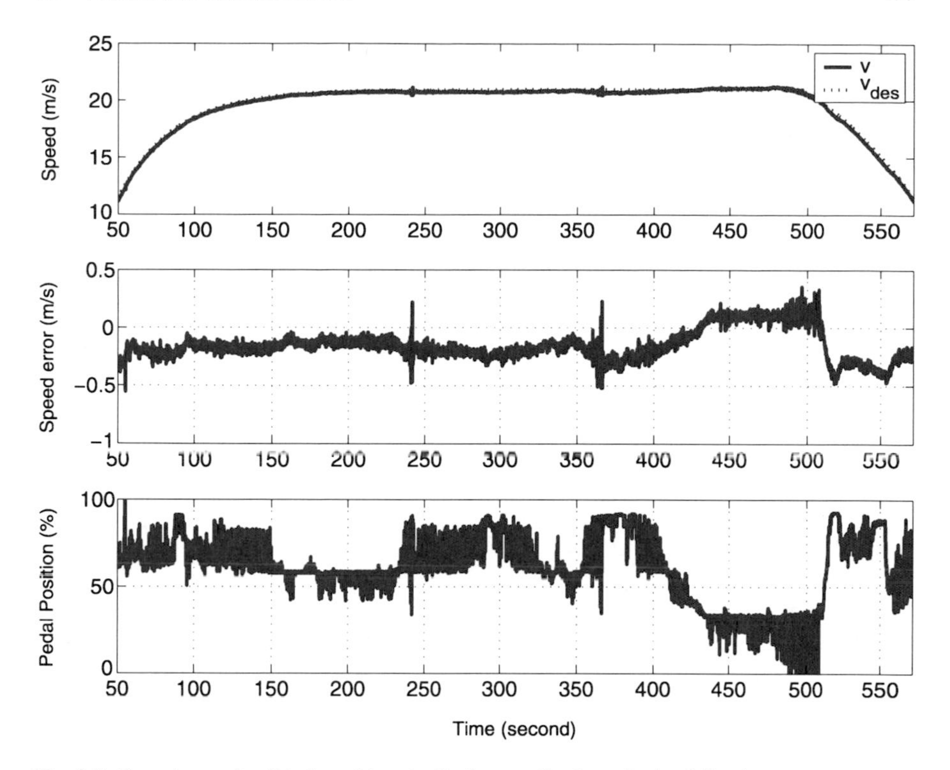

Fig. 8.8 Experimental validation of longitudinal controller for velocity following

where $F_s(\nu) = \mathrm{diag}(\nu_1, \nu_2, \nu_3, \nu_4, \nu_5, \nu_6) \in \Re^{6\times 6}$ and $\nu_i \in [0, 1]$. If the parametric fault related with actuators is only considered, a set of parameters can be written as

$$F_p(\mu)\Theta = \left[\frac{\tau_e}{1-\mu_1} \; (1-\mu_2)K_b \; \frac{\tau_b}{1-\mu_3}\right]^T \in \Re^3 \tag{8.21}$$

where $F_p(\mu) = \mathrm{diag}[1/(1-\mu_1), (1-\mu_2), 1/(1-\mu_3)] \in \Re^{3\times 3}$ and $\mu_i \in [0, 1]$. It is noted that the parametric faults in the engine or brake actuator result in an increase of the time constant and thus cause a slow time response. This is the reason that $1/(1-\mu_1)$ and $1/(1-\mu_3)$ instead of $(1-\mu_i)$ are multiplied by τ_e and τ_b, respectively.

8.3.1 Error Dynamics for Faulty System

Using the above assumptions, consider only ν_1, ν_3, ν_6 in F_s of (8.20) for more clarity of derivation. However, any combination of possible faults can be considered in the derivation of the error dynamics of the faulty system. For the given assumptions, the equations of motion for engine control in (8.6) and (8.7) are

$$\begin{cases} \dot{v} = \frac{\bar{T}_e + (T_e - T_{edes}) + (T_{edes} - \bar{T}_e)}{J_{eq}} - f_1(v, \omega_e, \theta) + \Delta f_{1m}, \\ \dot{T}_e = \frac{1-\mu_1}{\tau_e}(-T_e + u_e) + \Delta f_{2m}. \end{cases} \tag{8.22}$$

Using the definitions of S_{2e}, ξ_{2e}, the low-pass filter error of $\xi_{2e} := T_{edes} - \bar{T}_e$ and the synthetic engine torque $\bar{T}_e$ in (8.10), the faulty system in (8.22) can be written as

$$\begin{cases} \dot{v} = \dot{v}_{des} - K_{1e} y_2 + (S_{2e} + \xi_{2e})/J_{eq} + \Delta f_{1m}, \\ \dot{T}_e = (1 - \mu_1)(\dot{T}_{edes} - K_{2e} S_{2e}) + \Delta f_{2m} + \Delta f_{2s} \end{cases} \tag{8.23}$$

where

$$\Delta f_{2s} = (1 - \mu_1)\nu_4(K_{2e} - 1/\tau_e)T_e.$$

If the DSC law for engine control in (8.11) is applied and the filter error dynamics is added, the augmented error dynamics for the faulty system is as follows:

$$\begin{cases} \dot{S}_{1e} = -K_{1e}(1 - \nu_2)S_{1e} + (S_{2e} + \xi_{2e})/J_{eq} + \Delta f_{1m}, \\ \dot{S}_{2e} = -K_{2e}(1 - \mu_1)S_{2e} + \mu_1 \xi_{2e}/\tau_{2e} + \Delta f_{2m} + \Delta f_{2s}, \\ \dot{\xi}_{2e} = -\xi_{2e}/\tau_{2e} + J_{eq}[K_{1e}(1 - \nu_2)\dot{S}_{1e} - \ddot{v}_{des} - \dot{f}_1 - K_{1e}\dot{\nu}_2 S_{1e}]. \end{cases} \tag{8.24}$$

It is noted that more possible faults may be considered for the derivation of the error dynamics while only three possible faults such as ν_2, ν_4, and μ_1 are considered above.

Similarly, when the brake pressure is greater than the push-out pressure (P_o), the longitudinal equation for brake control is presented as follows:

$$\begin{cases} \dot{v} = \frac{-(1-\mu_2) R_g K_b [\bar{P}_b - P_o + (P_b - P_{bdes}) + (P_{bdes} - \bar{P}_b)]}{J_{eq}} - f_1 + \Delta f_{1m}, \\ \dot{P}_b = \frac{1-\mu_3}{\tau_b}[-P_b + u_b] + \Delta f_{3m}. \end{cases}$$

The error dynamics for braking control is given by

$$\begin{cases} \dot{S}_{1b} = -K_{1b}(1 - \mu_2)(1 - \nu_2)S_{1b} - (1 - \mu_2)\frac{R_g K_b}{J_{eq}}(S_{2b} + \xi_{2b}) \\ \qquad + \Delta f_{1m} + \Delta f_{1p}, \\ \dot{S}_{2b} = -K_{2b}(1 - \mu_3)S_{2b} + \mu_3 \xi_{2b}/\tau_{2b} + \Delta f_{3m} + \Delta f_{3s}, \\ \dot{\xi}_{2b} = -\frac{\xi_{2b}}{\tau_{2b}} - \frac{J_{eq}}{R_g K_b}[K_{1b}(1 - \nu_2)\dot{S}_{1b} - \ddot{v}_{des} - \dot{f}_1 - K_{1b}\dot{\nu}_2 S_{1b})] \end{cases} \tag{8.25}$$

where $\xi_{2b} := P_{bdes} - \bar{P}_b$,

$$\Delta f_{1p} = -\mu_2(f_1 + \dot{v}_{des}), \quad \text{and} \quad \Delta f_{3s} = (1 - \mu_3)\nu_5(K_{2b} - 1/\tau_b)P_b.$$

Finally, combining (8.24) with (8.25), we obtain the hybrid error dynamics as follows: for $i = e, b$,

$$
\begin{aligned}
T_i \dot{z}_i &= \bar{A}_i^f z_i + [e_1\ e_2\ \bar{B}_{di}] \begin{bmatrix} \Delta f_{1i} \\ \Delta f_{2i} \\ \dot{f}_{1i} \end{bmatrix} + \bar{B}_{di} \ddot{v}_{des} \\
&= \bar{A}_i^f z_i + \bar{B}_{wi} w_i + \bar{B}_{di} \ddot{v}_{des}
\end{aligned}
\tag{8.26}
$$

where the vectors are defined as

$$
\begin{aligned}
z_i &= [S_{1i}\ S_{2i}\ \xi_{2i}]^T \in \Re^3, \qquad e_1 = [1\ 0\ 0]^T \in \Re^3, \qquad e_2 = [0\ 1\ 0]^T \in \Re^3, \\
\bar{B}_{di} &= [0\ 0\ -a_i]^T \in \Re^3, \quad a_i = \begin{cases} J_{eq}, & \text{if } i = e, \\ -J_{eq}/(R_g K_b), & \text{if } i = b, \end{cases} \\
\Delta f_{1i} &:= \begin{cases} \Delta f_{1m}, & \text{for } i = e, \\ \Delta f_{1m} + \Delta f_{1p}, & \text{for } i = b, \end{cases} \qquad \Delta f_{2i} := \begin{cases} \Delta f_{2m} + \Delta f_{2s}, & \text{for } i = e, \\ \Delta f_{3m} + \Delta f_{3s}, & \text{for } i = b, \end{cases} \\
\dot{f}_{1i} &:= \dot{f}_1 + K_{1i} \dot{v}_2 S_{1i},
\end{aligned}
$$

and the matrices are

$$
\begin{aligned}
T_i &= \begin{bmatrix} 1 & 0 & 0 \\ 0 & 1 & 0 \\ -a_i(1-\nu_2)K_{1i} & 0 & 1 \end{bmatrix}, \\
\bar{A}_e^f &= \begin{bmatrix} -K_{1i}(1-\nu_2) & 1/a_i & 1/a_i \\ 0 & -K_{2i}(1-\mu_1) & \mu_1/\tau_{2e} \\ 0 & 0 & -1/\tau_{2i} \end{bmatrix}, \\
\bar{A}_b^f &= \begin{bmatrix} -K_{1i}(1-\mu_2)(1-\nu_2) & (1-\mu_2)/a_i & (1-\mu_2)/a_i \\ 0 & -K_{2i}(1-\mu_3) & \mu_3/\tau_{2b} \\ 0 & 0 & -1/\tau_{2i} \end{bmatrix}.
\end{aligned}
$$

After multiplying T_i^{-1} to both sides in (8.26), the augmented closed-loop error dynamics for the faulty system is rewritten as

$$
\dot{z}_i = A_i^f z_i + B_{wi}^f w_i + B_{di} \ddot{v}_{des} \tag{8.27}
$$

where the system matrices are

$$
\begin{aligned}
A_i^f &= T_i^{-1} \bar{A}_i^f, \\
B_{wi}^f &= T_i^{-1} \bar{B}_{wi} = T_i^{-1} [e_1\ e_2\ \bar{B}_{di}] = \begin{bmatrix} 1 & 0 & 0 \\ 0 & 1 & 0 \\ a_i(1-\nu_2)K_{1i} & 0 & -a_i \end{bmatrix} = [B_{ui}\ e_2\ B_{di}].
\end{aligned}
$$

8.3.2 Sensor Fault Handling

As discussed in Sect. 8.1, a set of longitudinal controllers are proposed to compensate for various types and degrees of faults in the transit bus. For the sensor fault, it is assumed that it is detected and isolated in the fault detection and diagnostic system in Fig. 8.1. As suggested in Sect. 7.2.3, the fault will be determined whether it is tolerable or not via a fault classification algorithm. If it is classified as an intolerable fault, a nonlinear observer is used to handle the sensor fault.

Three possible sensor faults (ν_2, ν_4, and ν_5 in $F_s(\nu)$) are considered and the corresponding error dynamics is derived in the previous section. However, while the measurements y_4 and y_5 are the state of the vehicle model in the longitudinal direction, $y_2 = v - v_{des}$ represents a relation between preceding and following vehicle if there is a preceding vehicle with $v_{des} = v_{prec}(t)$. Therefore, if a model-based observer technique is used, the estimation of $v - v_{des}$ is in general based on a kinematic model between two vehicles [40, 108]. Furthermore, the sensor fault of range rate has a high impact on the closed-loop system in terms of stability and performance. More specifically, as seen in (8.17), the corresponding sensor fault ν_2 is involved directly in the system matrices A_i^f and B_{wi}. Due to the importance of this measurement, multiple sensors such as radar, lidar, and/or vision sensor are used to handle the sensor fault of y_2 [40, 108]. Therefore, as long as the estimation of $v - v_{des}$ converges to a true value whatever the sensor fusion algorithm used, the error dynamics in (8.27) becomes

$$\dot{z}_i = A_i z_i + B_{wi} w_i + B_{di} \ddot{v}_{des}$$

where A_i is defined in (8.17) which is derived for a non-faulty system and

$$B_{wi} = \begin{bmatrix} 1 & 0 & 0 \\ 0 & 1 & 0 \\ a_i K_{1i} & 0 & -a_i \end{bmatrix}.$$

If there is no parametric fault, the corresponding error dynamics is equivalent to (8.17), thus stability and performance in the term of quadratic boundedness can be analyzed via convex optimization. The consideration of a parametric fault will be discussed in next section.

Next, consider the sensor fault in y_4 or y_5 under the assumption of no fault in y_2. As discusses in Sect. 8.1, estimated values instead of measurement will be used in the control law based on a nonlinear observer. Before designing the observer, the equations of motion in (8.6), (8.10), and (8.8) can be written in state space form

$$\begin{cases} \dot{x}_{1i} = x_{2i}/a_i + f_{1i} + \Delta f_{1m}, \\ \dot{x}_{2i} = -x_{2i}/\tau_i + u_i/\tau_i + \Delta f_{im} \end{cases} \tag{8.28}$$

where

$$x_{1i} = v, \qquad x_{2i} = \begin{cases} T_e, & \text{for } i = e, \\ P_b, & \text{for } i = b, \end{cases}$$

$$f_{1i} = \begin{cases} -f_1, & \text{for } i = e, \\ \frac{R_g K_b P_o}{J_{eq}} - f_1, & \text{for } i = b, \end{cases} \qquad \Delta f_{im} = \begin{cases} \Delta f_{2m}, & \text{for } i = e, \\ \Delta f_{3m}, & \text{for } i = b. \end{cases}$$

Then, (8.28) can be written in matrix form of (8.1) as follows: for $i = e, b$,

$$\begin{aligned} \dot{x}_i &= A_{si} x_i + B_{ui} u_i + B_f f_{1i} + \Delta f_i, \\ y &= C x_i \end{aligned} \tag{8.29}$$

where

$$A_{si} = \begin{bmatrix} 0 & 1/a_i \\ 0 & -1/\tau_i \end{bmatrix}, \qquad B_{ui} = \begin{bmatrix} 0 \\ 1/\tau_i \end{bmatrix}, \qquad B_f = \begin{bmatrix} 1 \\ 0 \end{bmatrix},$$

$$\Delta f_i = \begin{bmatrix} \Delta f_{1m} \\ \Delta f_{im} \end{bmatrix}, \qquad C = [1\ 0]$$

and the pair (C, A) is observable. The nonlinear observer proposed in (8.2) is

$$\dot{\hat{x}}_i = A_{si}\hat{x}_i + B_{ui}u_i + B_f f_{1i}(\hat{x}_i) + L(y - C\hat{x}_i). \tag{8.30}$$

After defining the estimation error $e_i = x_i - \hat{x}_i$, the error dynamics of the observer is obtained by subtracting (8.30) from (8.29) as follows:

$$\begin{aligned} \dot{e}_i &= (A_{si} - LC)e_i + B_f\big[f_{1i}(x_i) - f_{1i}(\hat{x}_i)\big] + \Delta f_i \\ &= A_{obi}e_i + B_f\phi_i + \Delta f_i \end{aligned} \tag{8.31}$$

where

$$\phi_i = f_{1i}(x_i) - f_{1i}(\hat{x}_i) = -\frac{R_g h C_a}{J_{eq}}\big(v^2 - \hat{v}^2\big) = -\frac{R_g h C_a}{J_{eq}}(v + \hat{v})e_{1i}$$

and e_{1i} is the first element of the estimation error vector e_i.

If the measurements y_4 and y_5 are replaced by the estimation of the nonlinear observer in (8.30), the commanded engine torque and brake pressure in (8.11) and (8.13) are rewritten, respectively, as

$$\begin{aligned} u_e &= \hat{T}_e + \tau_e\big(\dot{T}_{edes} - K_{2e}\hat{S}_{2e}\big) = \hat{x}_{2e} - \tau_e\big(\xi_{2e}/\tau_{2e} + K_{2e}\hat{S}_{2e}\big), \\ u_b &= \hat{P}_b + \tau_b\big(\dot{P}_{bdes} - K_{2b}\hat{S}_{2b}\big) = \hat{x}_{2b} - \tau_b\big(\xi_{2b}/\tau_{2b} + K_{2b}\hat{S}_{2b}\big) \end{aligned}$$

where $\hat{S}_{2e} = \hat{T}_e - T_{edes}$ and $\hat{S}_{2b} = \hat{P}_b - P_{bdes}$. Then, the equation of motion for engine control in (8.22) is

$$\begin{cases} \dot{v} = \frac{\bar{T}_e + (T_e - \hat{T}_e) + (\hat{T}_e - T_{edes}) + (T_{edes} - \bar{T}_e)}{J_{eq}} - f_1 + \Delta f_{1m}, \\ \dot{T}_e = \frac{d}{dt}(e_{2e} + \hat{T}_e) = \frac{1}{(1-\mu_1)\tau_e}(-T_e + u_e) + \Delta f_{2m}. \end{cases}$$

The augmented error dynamics including filter error dynamics and observer error dynamics in (8.31) is summarized as

$$\begin{cases} \dot{S}_{1e} = -K_{1e}S_{1e} + (e_{2e} + S_{2e} + \xi_{2e})/J_{eq} + \Delta f_{1m}, \\ \dot{\hat{S}}_{2e} + \dot{e}_{2e} = -K_{2e}\hat{S}_{2e} + \Delta f_{2m}, \\ \dot{e}_e = A_{obi}e_e + B_f\phi_e + \Delta f_e, \\ \dot{\xi}_{2e} = -\xi_{2e}/\tau_{2e} + J_{eq}[K_{1e}\dot{S}_{1e} - \ddot{v}_{des} - \dot{f}_1] \end{cases} \tag{8.32}$$

and can be written in matrix form as follows:

$$T_e\dot{z}_e^s = \bar{A}_e^s z_e^s + \bar{B}_{we}g_e + B_{de}\ddot{v}_{des} \tag{8.33}$$

where $z_e^s = [S_{1e}\ \hat{S}_{2e}\ e_{1e}\ e_{2e}\ \xi_{2e}]^T \in \Re^5$, $g_e = [\Delta f_{1m}\ \Delta f_{2m}\ \phi_e\ \dot{f}_1]^T \in \Re^4$,

$$T_e = \begin{bmatrix} 1 & 0 & 0 & 0 & 0 \\ 0 & 1 & 0 & 1 & 0 \\ 0 & 0 & 1 & 0 & 0 \\ 0 & 0 & 0 & 1 & 0 \\ -a_e K_{1e} & 0 & 0 & 0 & 1 \end{bmatrix},$$

$$\bar{A}_e = \begin{bmatrix} -K_{1e} & 1/a_e & 0 & 1/a_e & 1/a_e \\ 0 & -K_{2e} & 0 & 0 & 0 \\ 0 & 0 & 0 & 1/a_e & 0 \\ 0 & 0 & 0 & -1/\tau_e & 0 \\ 0 & 0 & 0 & 0 & -1/\tau_e \end{bmatrix},$$

$$\bar{B}_{we} = \begin{bmatrix} 1 & 0 & 0 & 0 \\ 0 & 1 & 0 & 0 \\ 1 & 0 & 1 & 0 \\ 0 & 1 & 0 & 0 \\ 0 & 0 & 0 & -a_e \end{bmatrix}, \quad \text{and} \quad B_{de} = \begin{bmatrix} 0 \\ 0 \\ 0 \\ 0 \\ -a_e \end{bmatrix}.$$

Similarly, the error dynamics for brake control are derived and the overall error dynamics can be summarized as

$$T_i \dot{z}_i^s = \bar{A}_i^s z_i^s + \bar{B}_{wi} g_i + B_{di} \ddot{v}_{des} \quad \text{for } i = e, b. \tag{8.34}$$

After multiplying both sides of (8.34) by T_i^{-1} such that

$$T_i^{-1} = \begin{bmatrix} 1 & 0 & 0 & 0 & 0 \\ 0 & 1 & 0 & -1 & 0 \\ 0 & 0 & 1 & 0 & 0 \\ 0 & 0 & 0 & 1 & 0 \\ a_i K_{1i} & 0 & 0 & 0 & 1 \end{bmatrix},$$

the augmented error dynamics for both engine and brake control is

$$\dot{z}_i^s = A_i^s z_i^s + B_{wi} g_i + B_{di} \ddot{v}_{des} \tag{8.35}$$

where $A_i^s = T_i^{-1} \bar{A}_i^s$ and $B_{wi} = T_i^{-1} \bar{B}_{wi}$. Furthermore, if the nonlinear function g_i is decomposed into the vanishing and nonvanishing perturbation term such that

$$g_i = p_i + q_i$$

where p_i goes to zero and q_i does not as z_i^s goes to zero, the above error dynamics is rewritten as

$$\dot{z}_i^s = A_i^s z_i^s + B_{wi} p_i + [B_{wi}\ B_{di}] \begin{bmatrix} q_i \\ \ddot{v}_{des} \end{bmatrix}.$$

If the vanishing perturbation term p_i is norm-bounded or diagonal norm-bounded, quadratic boundedness can be analyzed using Theorem 4.6. Furthermore, the eigenvalues of A_i^s for DSC and the nonlinear observer can be assigned independently using the separation principle stated in Chap. 4.

Remark 8.2 If v_{des} in $S_{1i} = v - v_{des}$ is known, e.g., there is no preceding vehicle and the control objective is to make $v \to v_{des}$, $\hat{S}_{1i} = \hat{v} - v_{des}$ may be used as suggested in Chap. 4 and Sect. 8.1. Then, the first line of (8.32) can be written as

$$\dot{\hat{S}}_{1e} + \dot{\hat{e}}_{1e} = -K_{1e}S_{1e} + (e_{2e} + S_{2e} + \xi_{2e})/J_{eq} + \Delta f_{1m}.$$

Then, T_i in (8.34) is as derived in (4.42)

$$T_i = \begin{bmatrix} \mathbf{I}_2 & \mathbf{I}_2 & 0_2 \\ \mathbf{0}_2 & \mathbf{I}_2 & 0_2 \\ -K & 0_2^T & 1 \end{bmatrix} \in \Re^{5\times 5}$$

where $K = [a_i K_{1i}\ 0]$. Thus, the augmented error dynamics, which is slightly different from (8.35), can be derived.

Example 8.2 (Sensor fault handling for longitudinal control of a transit bus) Consider the longitudinal control problem in Example 8.1. For the clarity of analysis, the model uncertainty in Example 8.1 is not considered for the design of the nonlinear observer and controller, but included for simulation to give more realistic results.

Observer Design Without consideration of model uncertainty, the system equation in (8.29) is written in NLDI form on a domain $\mathscr{D} = \{x_i \in \Re^2 | v \le v_{\max}\}$ for $i = e, b$ as follows:

$$\begin{aligned} \dot{x}_i &= A_{si}x_i + B_{ui}u_i + B_f f_{1i}, \\ y &= Cx_i \end{aligned}$$

where f_{1i} is locally Lipschitz on $\mathscr{D}$ such that

$$\left| f_{1i}(x_i) - f_{1i}(\hat{x}_i) \right| = \left| -\frac{R_g h C_a}{J_{eq}}(v + \hat{v})(v - \hat{v}) \right| \le \gamma |v - \hat{v}|$$

and the corresponding Lipschitz constant is $\gamma := |2R_g hC_a v_{\max}/J_{eq}| - 0.0104$ for the system parameters given in Table 8.2 and $v_{\max} = 25$.

As proposed in Algorithm 4.2 in Chap. 4, we solve the following COP for the switched nonlinear system to check whether the calculated stability margin is greater than the Lipschitz constant. For a fixed $\varepsilon \in [10^{-1},\ 10^2]$ and $i = e, b$,

$$\begin{aligned} &\text{maximize} && \beta \\ &\text{subject to} && P > 0, \quad \beta > 0, \\ &&& \begin{bmatrix} P^T A_i + A_i^T P + \beta I - \frac{1}{\varepsilon} C^T C & P \\ P & -I \end{bmatrix} < 0. \end{aligned} \tag{8.36}$$

Figure 8.9 shows that the stability margin, $\gamma_{\max}$, is greater than the given Lipschitz constant for all $\varepsilon_k \in [10^{-1},\ 10^2]$. Therefore, we can go to Step 3 in Algorithm 4.2 without a coordinate transformation. If COP (4.27) for the switched nonlinear system is solved for the given γ, the corresponding observer gain matrix is $L = [9.7315\ \ -0.004]^T \in \Re^2$. Then, the eigenvalues of A_{obi} in (8.31) are assigned as

$$\lambda(A_{obi}) = \begin{bmatrix} \{-9.7315,\ -33.333\} & \text{for } i = e, \\ \{-9.7315,\ -7.6923\} & \text{for } i = b. \end{bmatrix}$$

For the observer gain calculated above, the estimation values are compared with true ones in Fig. 8.10. While the estimation performance of v and P_b is very good, there is a larger error for estimation of T_e. This may result from not considering a black-box model of the engine (see Fig. 8.5). It is also noted that the estimation of velocity is switched between two observers based on the switching command of engine and brake control. In addition, the estimation of engine torque is zero during brake control and vice versa.

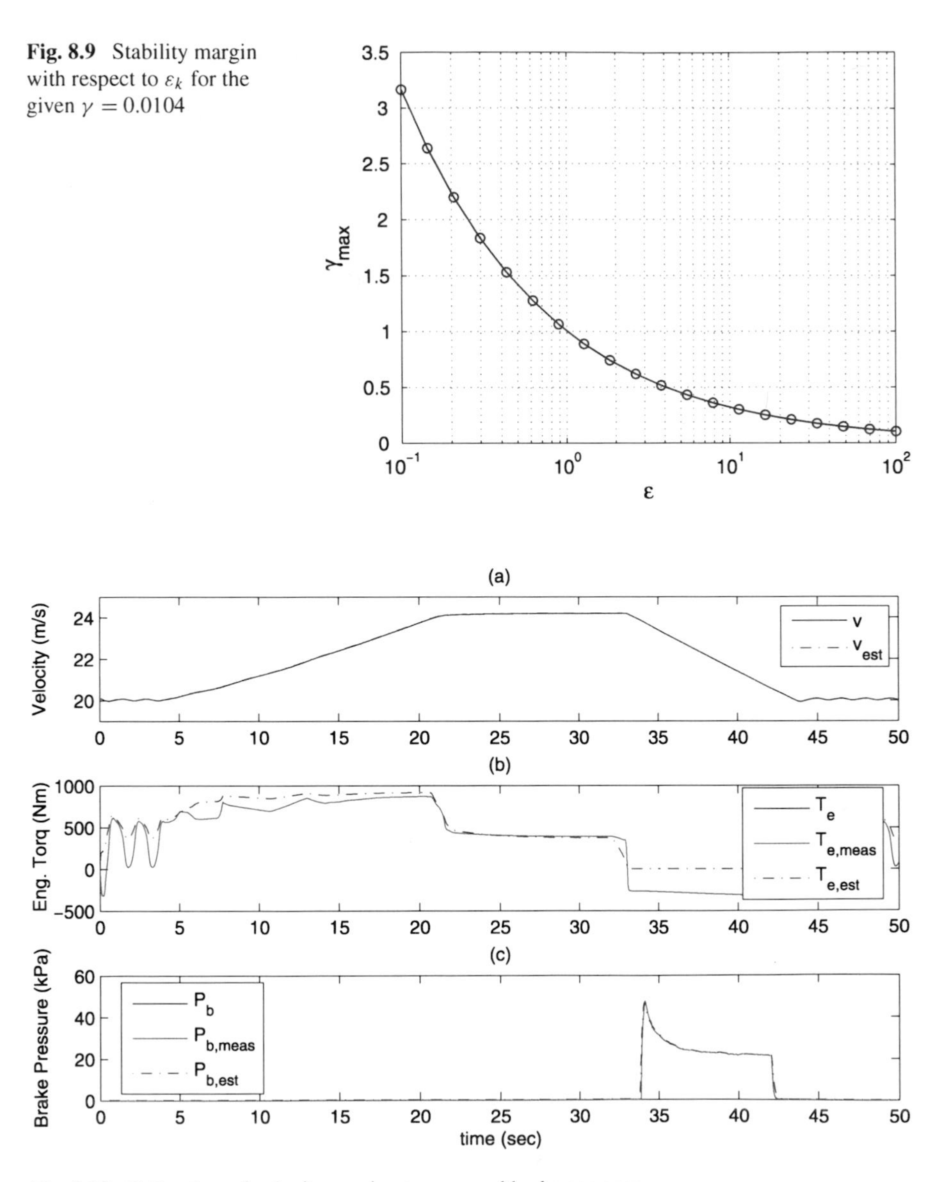

Fig. 8.9 Stability margin with respect to ε_k for the given $\gamma = 0.0104$

Fig. 8.10 Estimation of velocity, engine torque, and brake pressure

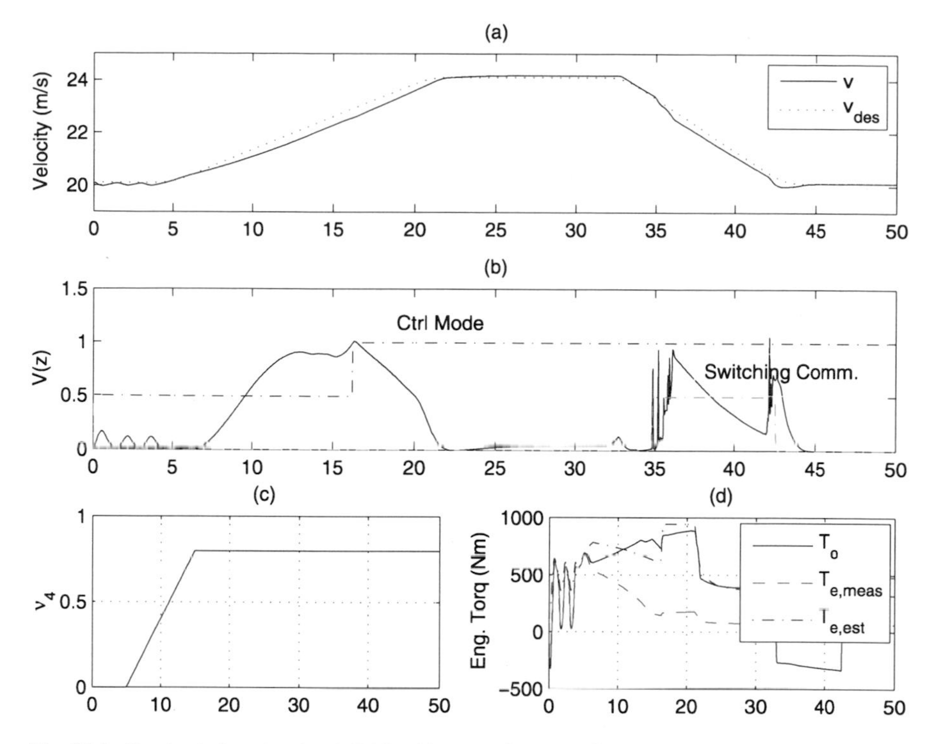

Fig. 8.11 Quadratic function level $V(z)$ with controller reconfiguration: sensor fault case

Sensor Fault Handling A set of controllers which includes the default DSC controller designed in Example 8.1 and the observer-based DSC is used for sensor fault handling. If the sensor fault is classified an intolerable fault via fault classification and isolated via FDD, the default controller is switched to the observer-based controller as a controller reconfiguration scheme.

Suppose there is a sensor fault $\nu_4(t)$ in the measurement of T_e. When ν_4 occurs at 5 s and increases up to 0.8 at 15 s as shown in Fig. 8.11(c), the quadratic Lyapunov-like function, $V(z)$, calculated in Example 8.1 is shown in Fig. 8.11(b). When the function level reaches 1 at about 16 s, the controller is switched to the observer-based controller (see controller mode line in Fig. 8.11(b)) and it is shown that the quadratic function level does not increase and stays below 1. Without switching to the observer-based controller, a larger velocity tracking error is expected and the system may become unstable.

8.3.3 *Trajectory Reconfiguration for Longitudinal Control*

Since there are too many possible parametric faults to be considered, the selection of the parametric fault may depend on the impact on system stability and the severity of performance degradation. As discussed in Sect. 7.2.3, if a parametric fault is classified as tolerable, i.e., a Lyapunov-like function level is below 1, it implies that the current nonlinear controller can compensate for the specific parametric fault without switching to other controllers. In other words, it is said that the corresponding parametric fault has a minor impact on stability and performance degradation.

In general, a parametric fault related with actuators has a direct impact on actuator performance so its consideration is quite important. The parametric fault can be also called an actuator fault since the actuator fault can be modeled mathematically as the parametric fault. For instance, if parametric fault in (8.21) is considered, μ_1 represents a multiplicative fault for τ_e and results in performance degradation of engine torque generation. Similarly, μ_2 generates performance degradation of brake torque generation even though there is no fault in a pneumatic brake actuator to produce brake pressure, i.e., $\mu_3 = 0$.

Under the assumption of no sensor fault, the error dynamics in (8.26) is simplified as follows: for $i = e$,

$$\begin{aligned} T_i \dot{z}_i &= \bar{A}_i^f z_i + [e_1 \; e_2 \; B_{di}] \begin{bmatrix} \Delta f_{1m} \\ \Delta f_{2m} \\ \dot{f}_1 \end{bmatrix} + B_{di} \ddot{v}_{des} \\ &= \bar{A}_i^f z_i + \bar{B}_{\tilde{w}e} \tilde{w}_e + B_{di} \ddot{v}_{des} \end{aligned} \tag{8.37a}$$

and for $i = b$,

$$\begin{aligned} T_i \dot{z}_i &= \bar{A}_i^f z_i + [e_1 \; e_1 \; e_2 \; B_{di}] \begin{bmatrix} \Delta f_{1m} \\ \Delta f_{1p}(\mu_2) \\ \Delta f_{3m} \\ \dot{f}_1 \end{bmatrix} + B_{di} \ddot{v}_{des} \\ &= \bar{A}_i^f z_i + \bar{B}_{\tilde{w}b} \tilde{w}_b + B_{di} \ddot{v}_{des} \end{aligned} \tag{8.37b}$$

where all definitions of vectors and the system matrices T_i, $\bar{A}_i^f$, and B_{di} are given in (8.26). It is interesting to remark that the multiplicative fault, μ_2 in (8.1), results in expansion of $\tilde{w}_b$ and the corresponding $\bar{B}_{\tilde{w}b}$ due to the addition of Δf_{1p}. After multiplying T_i^{-1} on both sides of (8.37a), (8.37b), the switched error dynamics is written as

$$\dot{z}_i = A_i^f z_i + \tilde{B}_{wi} \tilde{w}_i + B_{di} \ddot{v}_{des} \quad \text{for } i = e, b. \tag{8.38}$$

Once the error dynamics is derived, the next step is to decompose the perturbation term $\tilde{w}_i$ into vanishing and nonvanishing terms as done through all previous chapters. As shown in Examples 8.1 and 8.2, the model uncertainty Δf_m can be decomposed into the vanishing and nonvanishing perturbation terms on a domain $\mathscr{D}$, e.g.,

$$\mathscr{D} = \left\{ x_i \in \Re^2 \,||v| \le v_{\max}, |\dot{v}| \le a_{\max}, |\ddot{v}| \le b_{\max} \right\} \tag{8.39}$$

where $x_e = [v \; T_e]^T$, $x_b = [v \; P_b]^T$, and $v_{\max}, a_{\max}, b_{\max} > 0$. Therefore, let

$$\Delta f_{jm} := p_i + q_i \quad \text{for } j = 1, 2, 3$$

where p_i represents the vanishing perturbation term and q_i stands for the nonvanishing term. While the upper bounds of the vanishing and nonvanishing terms depend on what model uncertainty is used, $\dot{f}_1$ is obtained in a generic way as follows:

$$\dot{f}_1 = J_{11}\dot{v} = J_{11}(\dot{S}_{1i} + \dot{v}_{des}) = J_{11}(c_{zi}z_i + \dot{v}_{des} + \Delta f_{1i}),$$
$$|\dot{f}_1| = |J_{11}||c_{zi}z_i + \dot{v}_{des} + \Delta f_{1i}| \le \gamma|c_{zi}z_i + \dot{v}_{des} + \Delta f_{1i}|$$

where J_{11} is derived in Example 8.1, γ is a Lipschitz constant for f_1 satisfying $|J_{11}| = |\partial f_1/\partial v| \le \gamma$, $c_{zi} = [-K_{1i}\ 1/a_i\ 1/a_i]$, and

$$\Delta f_{1i} = \begin{cases} \Delta f_{1m}, & \text{for } i = e, \\ \Delta f_{1m} + \Delta f_{1p}, & \text{for } i = b. \end{cases}$$

Similarly, using derivation of Δf_{1p} in (8.25), its upper bound can be written as

$$\left|\Delta f_{1p}(\mu_2)\right| := |t_1 + t_2| = \left|-\mu_2(f_1 + \dot{v}_{des})\right| \le \mu_2|f_1| + \mu_2|\dot{v}_{des}|$$

where $\mu_2 \in [0, 1]$,

$$|t_1| \le \mu_2|f_1|, \quad \text{and} \quad |t_2| \le \mu_2|\dot{v}_{des}|.$$

Since $|f_1|$ and $|\dot{v}_{des}|$ are bounded on the domain $\mathscr{D}$, Δf_{1p} can be considered as a nonvanishing perturbation.

After w_i in (8.38) is decomposed into the vanishing and nonvanishing terms, (8.38) can be rewritten as for $i = e$

$$\dot{z}_i = A_i^f z_i + B_{wi}\begin{bmatrix} p_1 \\ p_2 \\ J_{11}p_1 \\ J_{11}c_{zi}z_i \end{bmatrix} + [B_{ui}\ e_2\ B_{di}\ B_{di}\ B_{di}]\begin{bmatrix} q_1 \\ q_2 \\ J_{11}q_1 \\ J_{11}\dot{v}_{des} \\ \ddot{v}_{des} \end{bmatrix} \tag{8.40a}$$

and for $i = b$,

$$\dot{z}_i = A_i^f z_i + B_{wi}\begin{bmatrix} p_1 \\ p_3 \\ J_{11}p_1 \\ J_{11}c_{zi}z_i \end{bmatrix}$$
$$+ [B_{ui}\ B_{ui}\ B_{ui}\ e_2\ B_{di}\ B_{di}\ B_{di}\ B_{di}\ B_{di}]\begin{bmatrix} q_1 \\ t_1 \\ t_2 \\ q_3 \\ J_{11}q_1 \\ J_{11}t_1 \\ J_{11}t_2 \\ J_{11}\dot{v}_{des} \\ \ddot{v}_{des} \end{bmatrix} \tag{8.40b}$$

where $B_{wi} = [B_{ui}\ e_2\ B_{di}\ B_{di}] \in \Re^{3\times 4}$ and $B_{ui} = T_i^{-1}e_1 \in \Re^3$. Combining (8.40a) with (8.40b), the switched error dynamics is summarized as

$$\dot{z}_i = A_i^f z_i + B_{wi}w_i + B_{ri}r_i \quad \text{for } i = e, b. \tag{8.41}$$

It is noted that the dimension of w_i and r_i relies on what type of fault or uncertainty is considered. For instance, if the model uncertainty Δf_{2m} considered in Example 8.1 is the vanishing perturbation such as $\Delta f_{2m} = p_2$, the corresponding q_2 can be omitted in r_e.

After the decomposition of w_i as done in (8.41), the last step to derive the error dynamics for analysis and design is to obtain the upper bound of w_i. Note that it is not necessary to derive an explicit form of w_i, but it is required to estimate the linear upper bound of w_i. If w_i is norm-bounded componentwise (see also in Example 8.3 below), the error dynamics in (8.41) is classified as DNLDI. Then, the quadratic boundedness or input-output stability in the term of $\mathscr{L}_2$ gain can be discussed (refer to Chaps. 2 and 3).

As proposed in Sect. 8.1.2, if the parametric fault is classified to be intolerable and isolatable in the actuator or related system, the two-step trajectory reconfiguration procedure is applied: First, the stabilization step to check the closed-loop system to be stabilized with the assumption of no nonvanishing perturbation, i.e., $r_i = 0$ is performed. Then, if w_i is diagonally norm-bounded, the error dynamics in (8.27) is written as

$$\dot{z}_i = A_i^f z_i + B_{wi} w_i, \quad |w_i| \le |C_{zi} z_i|.$$

Using the result of Theorem 2.3, the existence of a quadratic Lyapunov function for the error dynamics above can be tested for the given μ_i via convex optimization.

Next, the reconfiguration step is to determine the magnitude of r_i to guarantee quadratic boundedness satisfying the performance specifications using the remaining capability of the actuators. Suppose the desired trajectory $r_{des} = [v_{des}\ \dot{v}_{des}\ \ddot{v}_{des}]^T \in \Re^3$ is feasible on the domain $\mathscr{D}$ in (8.39). Since f_1 is locally Lipschitz on $\mathscr{D}$, f_1 and $J_{11} = \partial f_1/\partial v$ are bounded such that $|f_1| \le \delta$ and $|J_{11}| \le \gamma$ for the positive constants δ and γ. Therefore, the nonvanishing perturbation, r_i in (8.41), is norm-bounded on $\mathscr{D}$ as

$$\begin{aligned} r_e^T r_e &= (1 + J_{11}^2) q_1^T q_1 + J_{11}^2 \dot{v}_{des}^2 + \ddot{v}_{des}^2 \le (1+\gamma^2)\gamma_0 v_{\max}^2 + \gamma^2 a_{\max}^2 + b_{\max}^2 \\ r_b^T r_b &= r_e^T r_e + (1 + J_{11}^2)(t_1^2 + t_2^2) \\ &\le (1+\gamma^2)\gamma_0 v_{\max}^2 + \gamma^2 a_{\max}^2 + b_{\max}^2 + \mu_2^2 (1+\gamma^2)(\delta^2 + a_{\max}^2) := r_{\max}^2 \\ \Longrightarrow &\quad \|r_i\| \le r_{\max}(\mu_2, v_{\max}, a_{\max}, b_{\max}) \end{aligned} \tag{8.42}$$

where the magnitude of $r_{\max}$ can be modified by redefining the domain $\mathscr{D}$ with a change of $v_{\max}$, $a_{\max}$, and $b_{\max}$. As discussed in Sect. 8.1.2, if a line search of $r_{\max}$ is not available online, then $r_{\max}$ can be determined in advance for specific magnitudes of actuator faults. Then, if the size of the occurred fault is less than one of the predetermined fault, the pre-calculated $r_{\max}$ can be used.

Example 8.3 (Trajectory reconfiguration for longitudinal control of a transit bus) Consider Example 8.1 again with parametric faults, μ_i in (8.21). Suppose all μ_i are detected and isolated via fault detection and diagnosis. As done in Example 8.1, the model uncertainty Δf_{jm} in w_i in (8.41) is decomposed into vanishing and nonvanishing terms such that

$$\begin{aligned}
\Delta f_{1m} &= p_1 + q_1, \quad |p_1| \le \gamma_0 \big|[1\ 0\ 0] z_i\big|, |q_1| \le \gamma_0 |v_{des}|, \\
\Delta f_{2m} &= p_2, \quad |p_2| \le \gamma_1 \big|[0\ K_{2e}\ 1/\tau_{2e}] z_e\big|, \\
\Delta f_{3m} &= p_3, \quad |p_3| \le \gamma_1 \big|[0\ K_{2b}\ 1/\tau_{2b}] z_b\big|
\end{aligned}$$

where γ_0 and γ_1 are defined in Example 8.1. Since Δf_{2m} and Δf_{3m} are bounded by only vanishing linear terms, the error dynamics in (8.41) can be simplified with $q_2 = q_3 = 0$ to

$$\dot{z}_i = A_i^f z_i + B_{wi} w_i + B_{ri} r_i \quad \text{for } i = e, b. \tag{8.43}$$

where

$$\begin{aligned}
B_{re} &= [B_{ui}\ B_{di}\ B_{di}\ B_{di}] \in \Re^{3\times 4}, \\
r_e &= [q_1\ J_{11} q_1\ J_{11} \dot{v}_{des}\ \ddot{v}_{des}]^T \in \Re^4, \\
B_{rb} &= [B_{ui}\ B_{ui}\ B_{ui}\ B_{di}\ B_{di}\ B_{di}\ B_{di}\ B_{di}] \in \Re^{3\times 8}, \quad \text{and} \\
r_b &= [q_1\ t_1\ t_2\ J_{11} q_1\ J_{11} t_1\ J_{11} t_2\ J_{11} \dot{v}_{des}\ \ddot{v}_{des}]^T \in \Re^8.
\end{aligned}$$

Furthermore, the vanishing perturbation w_i is diagonally norm-bounded, i.e., there exist C_{zi} for $i = e, b$ such that

$$|w_i| = \left|\begin{bmatrix} p_1 \\ p_i \\ J_{11} p_1 \\ J_{11} c_{zi} z_i \end{bmatrix}\right| \le \left|\begin{bmatrix} \gamma_0 [1\ 0\ 0] z_i \\ \gamma_1 [0\ K_{2i}\ 1/\tau_{2e}] z_i \\ \gamma\gamma_0 [1\ 0\ 0] z_i \\ \gamma [-K_{1i}\ 1/a_i\ 1/a_i] z_i \end{bmatrix} z_i\right| := |C_{zi} z_i|$$

where $p_e = p_2$, $p_b = p_3$, and γ is defined in Example 8.1. Therefore, the switched error dynamics is summarized as

$$\begin{aligned}
\dot{z}_i &= A_i^f z_i + B_{wi} w_i + B_{ri} r_i \quad \text{for } i = e, b, \\
|w_i| &\le |C_{zi} z_i|
\end{aligned}$$

which are in switched DNLDI form (refer to Chap. 2). Finally, if the coordinate transformation is used such that $\bar{z}_i = D z_i$ as done in Example 8.1, the error dynamics is

$$\begin{aligned}
\dot{\bar{z}}_i &= A_i^f z_i + D B_{wi} w_i + D B_{ri} r_i \quad \text{for } i = e, b, \\
|w_i| &\le |C_{zi} D^{-1} \bar{z}_i|.
\end{aligned} \tag{8.44}$$

Case I: Tolerable Parametric Fault Consider μ_1 and μ_3 with the assumption of $\mu_2 = 0$. The stabilization step is to check whether the error dynamics in (8.44) with $r_i = 0$ is quadratically stable with respect to μ_1 and μ_3. When LMI (2.49) in Theorem 2.3 is solved for the given controller gain in Example 8.1, the solution of LMI (2.49) is feasible for any $\mu_1, \mu_3 \in [0, 1)$. Thus, the error dynamics is quadratically stable for any μ_1, μ_3. Next, for the given μ_1 and μ_3, quadratic boundedness is

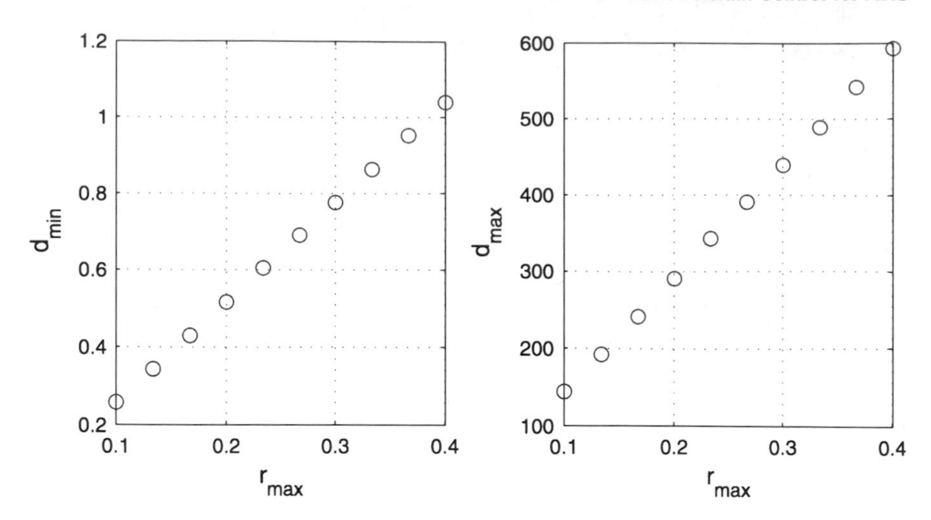

Fig. 8.12 Minimum and maximum diameter of ellipsoid with respect to $r_{\max}$ with $\mu_1 = \mu_3 = 0.8$

investigated with respect to the magnitude of r_i satisfying $\|r_i\| \leq r_0$ where r_0 is chosen for the design of a longitudinal controller under normal driving condition (refer to Example 8.1). That is, we need to determine $r_{\max}$ satisfying $\|r_i\| \leq r_{\max} \leq r_0$ to guarantee quadratic boundedness. It is also interesting to remark that $d_{\min}$ represents the upper bound of S_{1i} in z_i of (8.44). Therefore, it is expected that the maximum deviation of the tracking error is about 0.5 m/s on $\mathscr{D}$ in the presence of the parametric fault $\mu_1 = \mu_2 = 0.8$.

Let $v_{\max} = 25$ m/s, $a_{\max} = 0.8$ m/s^2, and $b_{\max} = 0.4$ m/s^3 given for the domain $\mathscr{D}$ in (8.39) and $\mu_1 = \mu_3 = 0.8$ which represents that the time constants of the engine and brake actuators become five time larger. For the given μ_1 and μ_3, LMI (2.61) of Algorithm 2.2 in Chap. 2 is solved for 10 equally spaced points $r_{\max}[i] \in [0.1,\ r_0]$ where $r_0 = 0.4003$ and $i = 1, \ldots, 10$. More specifically, the quadratic bound of (8.44) is calculated iteratively with respect to α for a fixed $r_{\max}[i]$ as done in Example 8.1. Figure 8.12 shows the minimum and maximum diameter of the ellipsoidal error bound calculated for each $r_{\max}[i]$. It is shown that the size of the error bound is almost proportional to the magnitude of $r_{\max}$. Roughly speaking, we can say that the smaller the exogenous input is given, the smaller the error bound is expected.

Suppose the desired velocity is given in Fig. 8.13(a). Then, r_0 is about 0.2 with $|\dot{v}_{des}| \leq 0.4$ and $|\ddot{v}_{des}| \leq 0.2$. When $\mu_1 = 0.8$ and $\mu_2 = 0.8$ occur, respectively, at 5 and 15 (s), time responses of a quadratic function level and velocity tracking error are shown in Figs. 8.13(b) and (d). Using Fig. 8.12, the minimum diameter of the error bound for $r_{\max} = 0.2$ is about 0.5 and thus the maximum tracking error is expected to be about 0.25 m/s. Figure 8.13(d) shows that the tracking error is bounded by about 0.25 m/s.

As shown in Fig. 8.13(b), the quadratic function level does not exceed 1 even in the presence of parametric fault. Thus, the occurred fault is declared as the tolerable fault via fault classification in Sect. 7.2.3. It implies that the default longitudinal controller designed in Example 8.1 is inherently robust enough to handle this type and

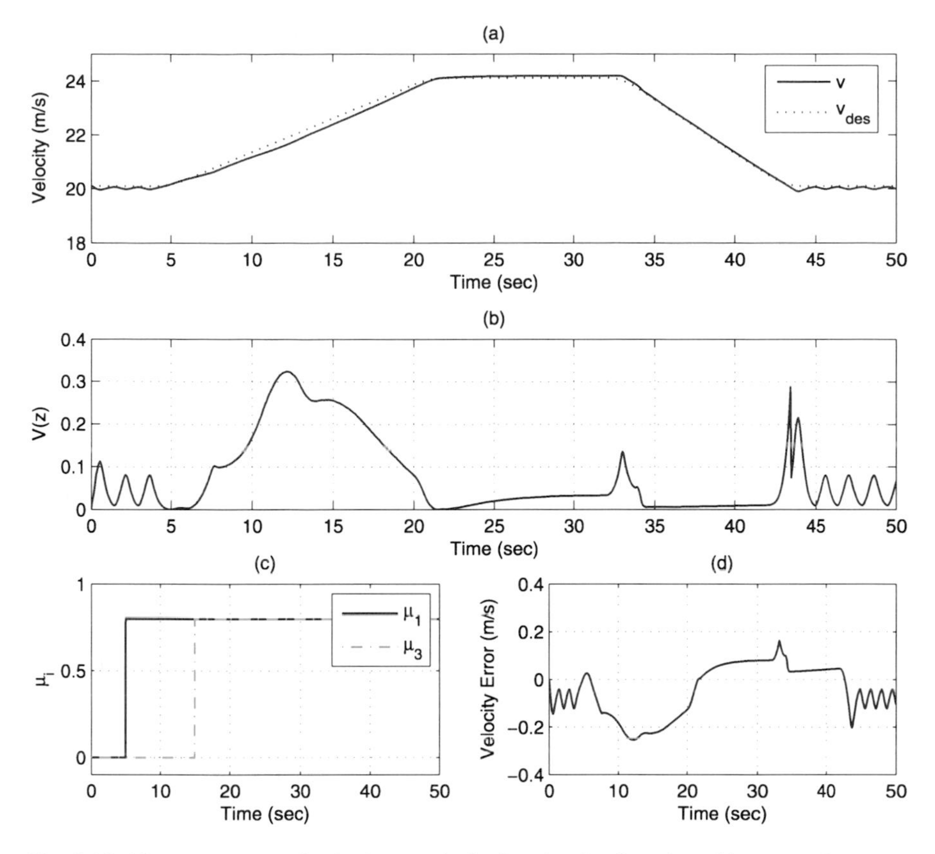

Fig. 8.13 Time responses of velocity, quadratic function level, and tracking error for $\mu_1 = \mu_3 = 0.8$

degree of fault. Hence, the fault tolerant control system does not need any controller reconfiguration and just provides a warning signal to the driver.

Case II: Trajectory Reconfiguration Consider μ_2 with the assumption of $\mu_1 = \mu_3 = 0$. The solution of LMI (2.49) for the error dynamics in (8.44) is feasible for any $\mu_2 \in [0, 1)$ and thus they are quadratically stable for any μ_2. For the given domain $\mathscr{D}$ above and $\mu_2 = 0.7$, the minimum and maximum diameter of the ellipsoidal error bound for 10 equally spaced points $r_{\max}[i] \in [0.1, r_0]$ where $r_0 = 0.6944$ and $i = 1, \dots, 10$ is shown in Fig. 8.14. It is noted that the value of r_0 increases due to inclusion of μ_2 (see also in (8.42)). As shown in the earlier case, similarly it is shown that the size of the error bound becomes smaller as the magnitude of the exogenous input r_i is smaller. However, if Fig. 8.14 is compared with Fig. 8.12, it is recognized that the change of $d_{\min}$ with respect to $r_{\max}$ is steeper when μ_2 is considered. It implies that μ_2 has a more serious impact on performance degradation than μ_1 and μ_3 for the given default controller.

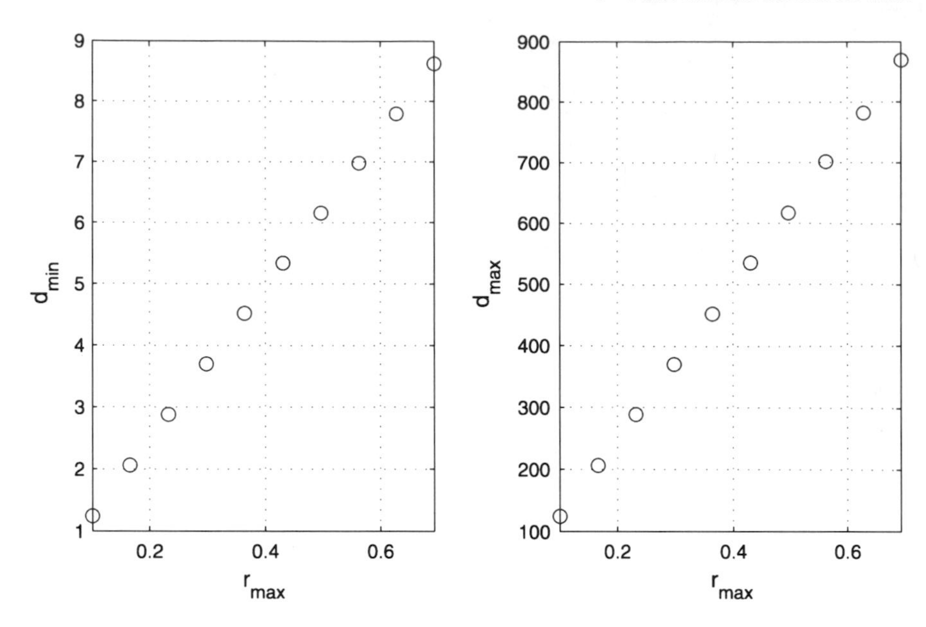

Fig. 8.14 Minimum and maximum diameter of ellipsoid with respect to $r_{\max}$ with $\mu_2 = 0.7$

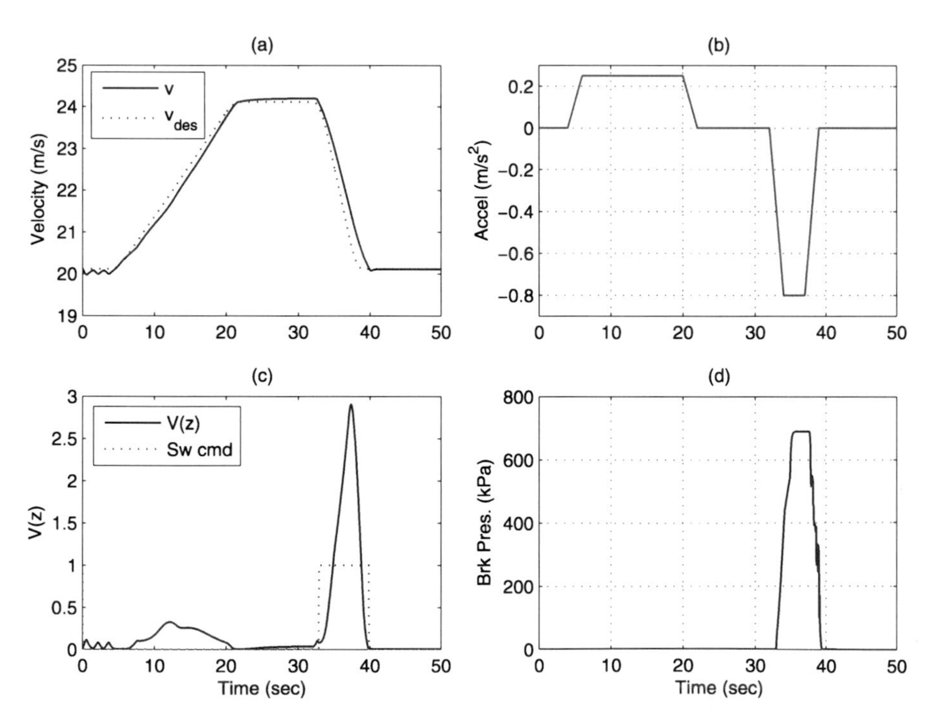

Fig. 8.15 Velocity tracking (**a**), desired acceleration (**b**), quadratic function level (**c**), and brake pressure (**d**)

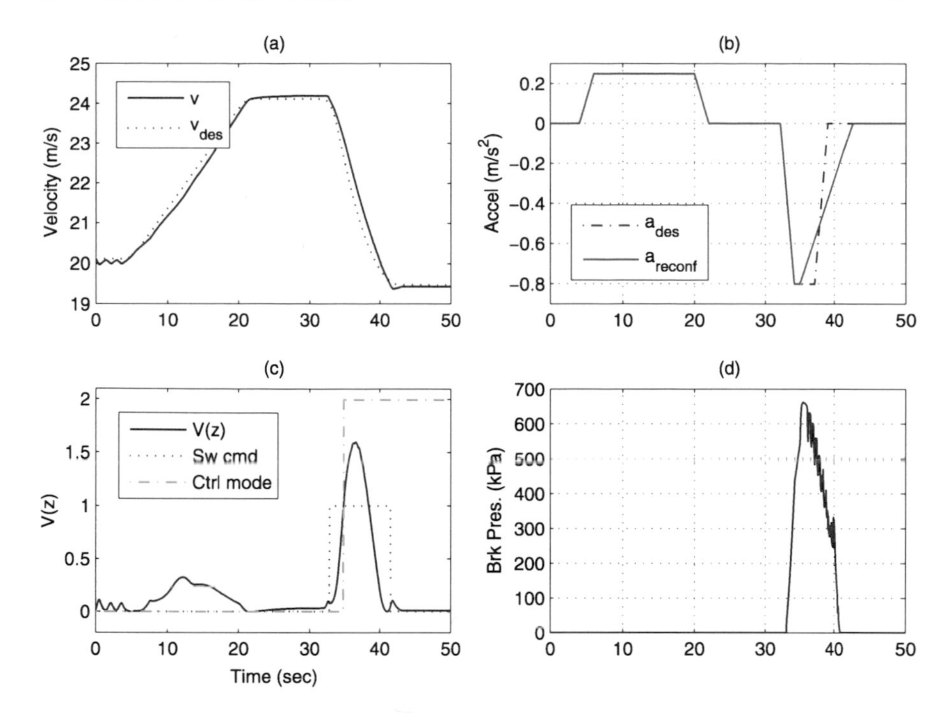

Fig. 8.16 Velocity tracking (**a**), desired acceleration (**b**), quadratic function level (**c**), and brake pressure (**d**) with trajectory reconfiguration

Suppose the desired velocity is modified with a larger deceleration as in Figs. 8.15(a) and (b), i.e., $|\dot{v}_{des}| \leq 0.8$ and $|\ddot{a}_{des}| \leq 0.4$. The time responses of a quadratic function level and brake pressure in the presence of 70% multiplicative parametric fault in K_b ($\mu_2 = 0.7$) are shown in Figs. 8.15(c) and (d). As expected through the performance degradation analysis above, a larger tracking error is generated and the quadratic function level exceeds 1. Thus, the occurred fault can be classified an intolerable one and we need the trajectory reconfiguration proposed in Sect. 8.1.2.

Although there may be many possible methods to reconfigure v_{des}, $\dot{v}_{des}$, and $\ddot{v}_{des}$ to reduce the magnitude of r_i in (8.44), only $\ddot{v}_{des}$ will be modified because γ and γ_0 are small and thus the magnitude of $\ddot{v}_{des}$ becomes dominant in r_i. That is, if the occurred parametric fault is classified an intolerable one, the inequality constraint of $\ddot{v}_{des}$, $|\ddot{v}_{des}| \leq 0.4$, is changed to $|\ddot{v}_{des}| \leq 0.1$. When the same parametric fault is classified the intolerable one at about 34 seconds, the default controller is switched to the DSC with trajectory reconfiguration—see 'ctrl mode' in Fig. 8.16(c). Then, after the new inequality constraint of $\ddot{v}_{des}$ is applied, the desired acceleration profile is modified as shown in Fig. 8.16(b). Sequentially, the maximum of the quadratic function level is reduced, thus the tracking performance is improved as shown in Fig. 8.16(c). In addition, the saturation of brake pressure is prevented as seen in Fig. 8.16(d). It is noted that the maximum of the quadratic function level can be reduced more if the inequality constraint of $|\dot{v_{des}}| \leq a_{\max}$ is also modified. For instance, if $a_{\max} = 0.8$ is changed to 0.4 in Fig. 8.16(b), this trajectory reconfiguration may result in a smaller tracking error but require more time to reduce the speed.

It is interesting to remark that the combined effect of multiple parametric faults may be investigated in the same framework of analysis. For instance, if both Case I and Case II happen at the same time, the stabilization and trajectory reconfiguration step can be conducted. As mentioned earlier, μ_2 has a more serious impact on the performance degradation. Therefore, if the same trajectory reconfiguration in Case II is applied, the similar time response of the tracking error is shown even though the coupled effect on the quadratic function level is more pronounced.

Chapter 9
Biped Robot Control

Increased attention has recently been paid to biped locomotion for a wide range of topics from model formulation, trajectory generation, environment/object recognition, controls, and the realization of biped robots [18, 104, 112]. From a control design perspective, the challenges in biped robots arise from the complexity in mechanisms, the variety of operating conditions/constraints (e.g., multiple contact points with the environment, various walking speeds and step size, underactuation, etc.), as well as uncertainties like disturbances and noise. While many control approaches for biped locomotion can be found in the literature, most of them can be roughly classified into PID and nonlinear control in the continuous time domain [35]. However, the PID controller has been designed either heuristically or by trial and error due to the complexity and nonlinearity of the associated dynamic model.

Among nonlinear control techniques applied for biped walking, most of them can be classified into feedback linearization [35, 72], computed torque control [69], and sliding mode control [63, 101]. To prove stable walking of the nonlinear controller based on biped robot models, the Poincaré method was introduced to analyze the closed-loop system of the biped model formulated as a nonlinear system with impulse effects [35]. Another approach was to use a Lyapunov function candidate and to show the existence of the controller gains to guarantee the stability under appropriate assumptions [17]. However, due to the mathematical complexity of the dynamic model, none of these control approaches has produced a closed-loop system with provable stability properties. The complexity of nonlinear control for biped walking challenges us to come up with systematic design procedures to meet both stability and performance. Furthermore, few studies of the controller design for biped walking with a variable step size (or variable walking speed) have been found in the literature.

In this chapter, the results of Chaps. 2 and 6 are applied to biped robot control for stable biped walking and it is shown that the tracking error is bounded in the sense of quadratic boundedness. More precisely speaking, the design methodology of multi-input multi-output (MIMO) nonlinear systems is used for a biped robot model. Then, the closed-loop system with provable stability is derived and is reformulated in the form of convex optimization problems to estimate an ellipsoidal upper bound of tracking error during biped walking with a variable step size.

B. Song, J.K. Hedrick, *Dynamic Surface Control of Uncertain Nonlinear Systems*, Communications and Control Engineering,
DOI 10.1007/978-0-85729-632-0_9,

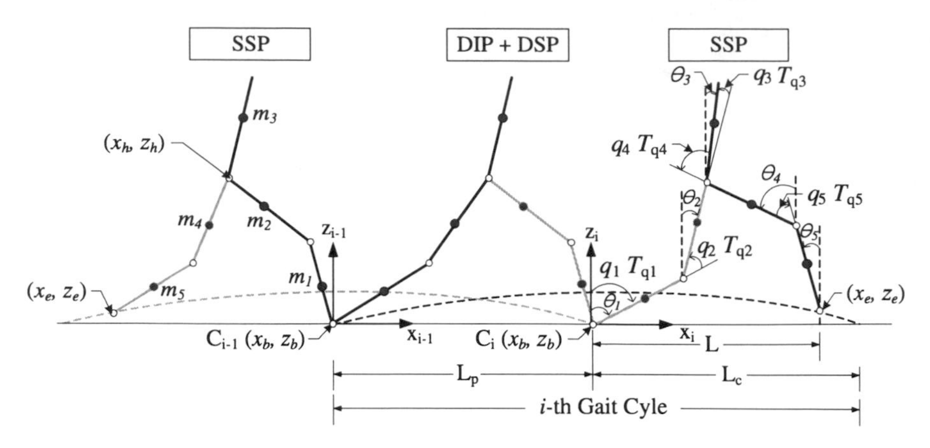

Fig. 9.1 Biped walking of a planar five-link model

The remainder of the chapter is organized into four sections. While a complete model of a planner 5-link biped is described in Sect. 9.1, a two-phase trajectory generation approach for biped walking with variable steps is proposed in Sect. 9.2. In Sect. 9.3, DSC is designed for MIMO nonlinear systems and its stability and performance will be estimated using the piecewise augmented error dynamics in the framework of convex optimization. Finally, the ellipsoidal error bound will be estimated numerically and the zero moment point (ZMP) is calculated for showing stable walking via simulations in Sect. 9.4.

9.1 Hybrid Biped Model

Among many dynamic models to describe planar biped locomotion found in the literature, most of them include both a single-leg support phase (SSP) and an impact (or impulse) phase model which is an instantaneous period switched between a series of SSP (see Fig. 9.1) [35, 101]. A double-leg support phase (DSP) model has been usually neglected due to its complexity (see [101] and references therein). In [17, 63], a three-phase model including SSP, DSP, and double impact phase (DIP) models was proposed and it was claimed that DSP is necessary to realize the stable motion over a wide range of walking speeds (see Fig. 9.1). The DIP model is usually dependent on assumed contact conditions between leg ends and ground. For instance, if the contact is assumed as compliant contact, i.e., the normal forces of two legs on the ground are modeled via a nonlinear spring-damper, and the tangential forces as dynamic friction, the dynamics of DSP can be derived [16, 60]. However, in reality a walking surface is quite stiff and it may not be possible to choose reasonable parameters for a compliant contact model. An alternative model is to assume DIP as a rigid contact, thus resulting in an algebraic relation for computing the ground reaction forces [18].

For the stable biped walking with a variable size step, the three-phase dynamic model is used. The DSP will be considered for both trajectory generation discussed in Sect. 9.2 and controller design in Sect. 9.3, and the DIP is assumed as rigid contact. Furthermore, it is noted that the degree of freedom is 5 for SSP and 3 for DSP, respectively, due to kinematic constraints when a planar five-link biped is considered [63]. Consequently the biped model can be described in the coordinates of either θ or q as shown in Fig. 9.1. First, using Lagrange's equation [91, 101], an equation of motion for SSP is derived in a standard second order system as follows:

$$D(\theta)\ddot{\theta} + H(\theta, \dot{\theta})\dot{\theta} + G(\theta) = T_\theta \tag{9.1}$$

where $D(\theta) \in \Re^{5\times5}$ is the positive definite and symmetric inertia matrix, $H(\theta, \dot{\theta}) \in \Re^{5\times5}$ is related to centrifugal and Coriolis terms, $G(\theta) \in \Re^5$ represents gravity terms, $\theta \in \Re^5$ is the generalized coordinate, and $T_\theta \in \Re^5$ is the torque. Furthermore, the matrices D, H, and G are given in [63] as

$$\begin{aligned} D_{ij} &= f_{ij}\cos(\theta_i - \theta_j), \\ H_{ij} &= f_{ij}\sin(\theta_i - \theta_j)\dot{\theta}_j, \\ G_i &= g_i \sin(\theta_i) \end{aligned}$$

where $i, j = 1, \ldots, 5$, f_{ij} and g_i are

$$f_{ij} = \begin{cases} I_i + m_i d_i^2 + a_i(\sum_{j=i+1}^5 m_j) l_i^2, & j = i, \\ a_i m_j d_j l_i + a_i a_j (\sum_{k=j+1}^5 m_k) l_i l_j, & j > i, \\ f_{ji}, & j < i, \end{cases}$$

$$g_i = m_i d_i g + a_i \left(\sum_{j=i+1}^{5} m_j \right) l_i g.$$

All parameters and variables are described in Table 9.1.

The equation of motion for DSP with two holonomic constraints is written as follows [17, 63]:

$$D(\theta)\ddot{\theta} + H(\theta, \dot{\theta})\dot{\theta} + G(\theta) = J^T(\theta)\lambda + T_\theta, \tag{9.2a}$$

$$\Phi(\theta) = \begin{bmatrix} x_e - x_b - L_c \\ z_e - z_b \end{bmatrix} = 0 \tag{9.2b}$$

where λ is a 2×1 vector of Lagrange multipliers and J is the 2×5 Jacobian matrix $J = \partial\Phi/\partial\theta \in \Re^{2\times5}$. Furthermore, a dynamic system with holonomic constraints can be formulated without the terms of constraint forces via a set of independent generalized coordinates [34]. Therefore, if the transformation is defined from θ to the independent generalized coordinates $p = [x_h \; z_h \; \theta_3]^T \in \Re^3$, i.e.,

$$p(\theta) = \begin{bmatrix} l_1 \sin\theta_1 + l_2 \sin\theta_2 \\ l_1 \cos\theta_1 + l_2 \cos\theta_2 \\ \theta_3 \end{bmatrix},$$

the equation of motion for DSP is derived as [63]

$$\ddot{p} = B\dot{p} + C(T_\theta - H\dot{\theta} - G) \tag{9.3}$$

Table 9.1 Variables and parameters for a biped robot

Variable	Description (unit)
m_i	mass of the ith link (kg)
l_i	length of the ith link (m)
d_i	distance between the mass center of the ith link (m)
I_i	moment of inertia of the ith link (kg m^2)
C_i	coordinate of the supporting limb tip in the ith gait cycle
θ_i	angle of line i with respect to the vertical axis through joint i (rad)
q_i	relative angle between jointed links (rad)
a_i	a number defined by $a_i = 0$ if $i = 3$ and $a_i = 1$ otherwise
τ_s	swing period, i.e. Single-leg Support Phase (SSP) period (second)
τ_t	gait period, i.e. SSP + Double-leg Support Phase (DSP) period (second)
L	the distance between the tips of two lower limbs (m)
L_p, L_c	previous and current desired step size, respectively (m)
(x_b, z_b)	position of the supporting limb tip with respect to C_i
(x_e, z_e)	position of the swing limb tip with respect to C_i
(x_h, z_h)	position of the hip with respect to C_i

where $B = L_{11} - N_1 N_2^{-1} L_{21}$, $C = M_1 - N_1 N_2^{-1} M_2$, and

$$L = \begin{bmatrix} L_{11} & L_{12} \\ L_{21} & L_{22} \end{bmatrix} = \dot{S} S^{-1}, \quad S = \begin{bmatrix} R \\ J \end{bmatrix} \in \Re^{5\times 5}, \quad R = \frac{\partial p}{\partial \theta} \in \Re^{3\times 5},$$

$$M = \begin{bmatrix} M_1 \\ M_2 \end{bmatrix} = S D^{-1}, \qquad N = \begin{bmatrix} N_1 \\ N_2 \end{bmatrix} = M J^T.$$

When a swing limb tip touches the ground with a collision, the joint angular velocity will be changed with a sudden jump at the beginning of DSP under the assumption of rigid contact. The DIP model describes this impact process switching from a SSP to a DSP model. By use of principles of conservation of impulse and momentum, the DIP model is described as [63]

$$\begin{bmatrix} W(\theta^-) & J^T(\theta^-) \\ J(\theta^-) & 0 \end{bmatrix} \begin{bmatrix} \dot{\theta}^+ \\ P_l \end{bmatrix} = \begin{bmatrix} W(\theta^-)\dot{\theta}^- \\ 0 \end{bmatrix} \tag{9.4}$$

where $P_l \in \Re^2$ is the impulse forces at the contact point of the rear limb and $W(\theta^-) \in \Re^{5\times 5}$ is

$$W_{ij} = \begin{cases} I_i - m_i d_i (a_i l_i - d_i), & j = i, \\ [a_i m_j d_j l_i + a_j m_i (a_i l_i - d_i) l_j + a_i a_j (\sum_{k=j+1}^{i-1} m_k) l_i l_j] \cos(\theta_i - \theta_j), & j < i, \\ 0, & j > 1. \end{cases}$$

From an implementation point of view, the angles at joints may be measured in terms of the relative joint angles, and the actuator joint torques rather than the segmental torques are applied. Therefore, using the linear transformation function $T_{q\theta} : \theta \to q$ and defining p_q as

$$T_{q\theta} = \begin{bmatrix} 1 & 0 & 0 & 0 & 0 \\ 1 & -1 & 0 & 0 & 0 \\ 0 & 1 & -1 & 0 & 0 \\ 0 & 0 & 1 & 1 & 0 \\ 0 & 0 & 0 & 1 & -1 \end{bmatrix} \quad \text{and} \quad p_q(q) = \begin{bmatrix} l_1 \sin(q_1) + l_2 \sin(q_1 - q_2) \\ l_1 \cos(q_1) + l_2 \cos(q_1 - q_2) \\ q_1 - q_2 - q_3 \end{bmatrix},$$

both (9.1) and (9.3) can be rewritten with respect to relative joint angles (q_i) as follows:

$$\text{For SSP,} \quad \begin{cases} \ddot{q} = D_q^{-1}(q)\{T_q - H_q(q, \dot{q}) - G_q(q)\}, \\ S_{s\to d} = \{q \mid z_e(q) = 0\}, \\ \mathscr{T}_{s\to d} : \begin{bmatrix} \dot{q}^+ \\ P_l \end{bmatrix} = \begin{bmatrix} W_q(q^-) & J_q^T(q^-) \\ J_q(q^-) & 0 \end{bmatrix}^{-1} \begin{bmatrix} W_q(q^-)\dot{q}^- \\ 0 \end{bmatrix}, \end{cases} \tag{9.5a}$$

$$\text{For DSP,} \quad \begin{cases} \ddot{p_q} = B_q \dot{p}_q + C_q\{T_q - H_q(q, \dot{q}) - G_q(q)\}, \\ S_{d\to s} = \{q \mid z_b(q) > 0\}, \\ \mathscr{T}_{d\to s} : \begin{bmatrix} (x_e^+, z_e^+) \\ (x_b^+, z_b^+) \end{bmatrix}_{C_i^+} = \begin{bmatrix} (x_b^-, z_b^-) \\ (x_e^-, z_e^-) \end{bmatrix}_{C_{i-1}^-} \end{cases} \tag{9.5b}$$

where $S_{s\to d}$ and $S_{d\to s}$ are the switching conditions for transition from SSP to DSP and vice versa. $\mathscr{T}_{s\to d}$ is the transition function from SSP to DSP and the impulse model described in (9.4) is applied to change the angular velocity when $q \in S_{s\to d}$. The transition function from DSP to SSP, $\mathscr{T}_{d\to s}$, is applied when one full gait cycle is completed. Since the leg is switched after the full gait cycle (see switching from $i-1$ to ith gait cycle in Fig. 9.1), the positions of the leg tips corresponding (x_e, z_e) and (x_b, z_b) should also be switched with respect to the coordinates C_i^+ during the ith gait cycle.

9.2 Trajectory Generation

Once desired trajectories of (x_e, z_e) and (x_h, z_h) are assigned, the corresponding relative angular positions can be generated by using both kinematic constraints and inverse kinematics [21]. Suppose (x_e, z_e) and (x_h, z_h) are defined for SSP and DSP as follows:

$$x_h = -\frac{L_c + L_p}{4}\cos\left(\frac{\pi t}{\tau_t}\right) + \frac{L_c - L_p}{4}, \qquad z_h = 0.5, \tag{9.6a}$$

$$x_e = \begin{cases} -\frac{L_p + L_c}{2}\cos\left(\frac{\pi t}{\tau_s}\right) + \frac{L_c - L_p}{2}, & \text{for SSP,} \\ L_c, & \text{for DSP,} \end{cases} \tag{9.6b}$$

$$z_e = \begin{cases} \frac{L_h}{2}\{1 - \cos\left(\frac{2\pi}{\tau_s} t\right)\}, & \text{for SSP,} \\ 0, & \text{for DSP.} \end{cases} \tag{9.6c}$$

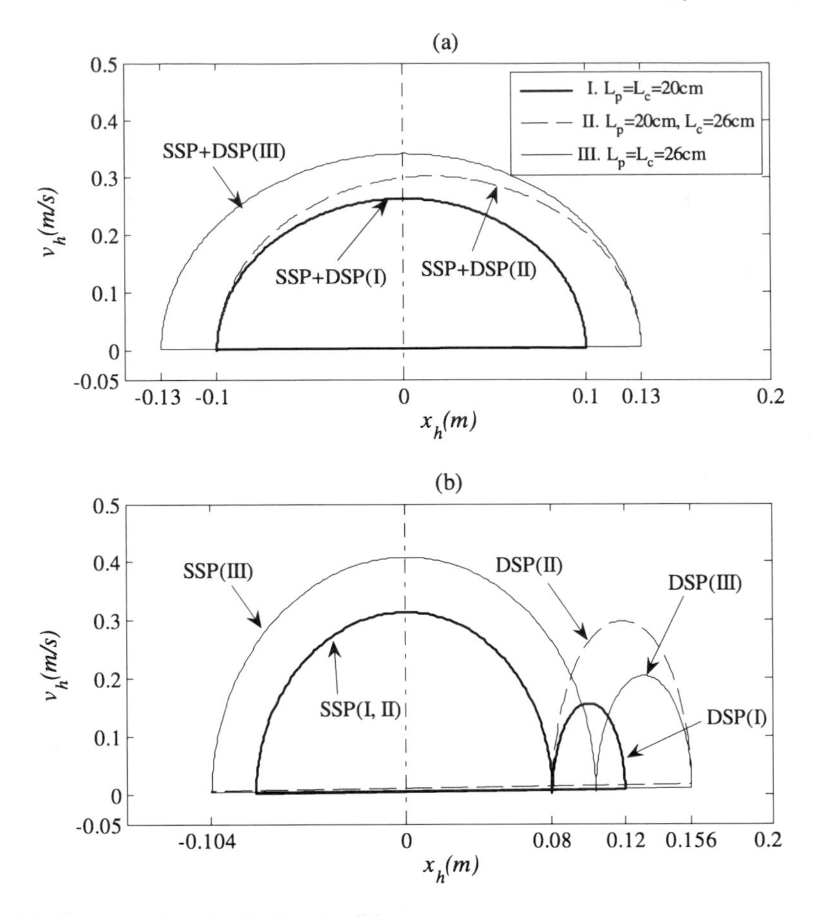

Fig. 9.2 Desired trajectories for biped walking

When the design parameters are given as $L_c = 0.26$, $L_p = 0.2$, $L_h = 0.03$, $\tau_t = 1.2$, and $\tau_s = 0.8$, the phase portrait with respect to x_h and v_h is shown in Fig. 9.2(a). That is, a step size is given as 0.2 m for the first gait cycle (see line I of Fig. 9.2(a)). Then, the step size is changed to 0.26 m and the corresponding trajectory for the second gait cycle is shown as line II in the figure and represents a transient walking period. Finally the trajectory for the third gait cycle with an increased step size is shown as line III in Fig. 9.2(a). Since (9.6a), (9.6b), (9.6c) generates the trajectory regardless of a type of the dynamic model, it is named as a one-phase trajectory generation method. When the one-phase method is used, the zero moment point (ZMP) varies from -0.03 to 0.13 and the magnitude of ZMP variation increases when the step size is changed (see Fig. 9.3, dotted line (a)). It is noted that ZMP is set to zero when the ZMP is placed between two limb tips during DSP.

To minimize the magnitude of ZMP variation and avoid a minus value of ZMP of the one-phase trajectory generation method, two trajectory generation profiles are considered for SSP and DSP, respectively, while the vertical position of the hip is

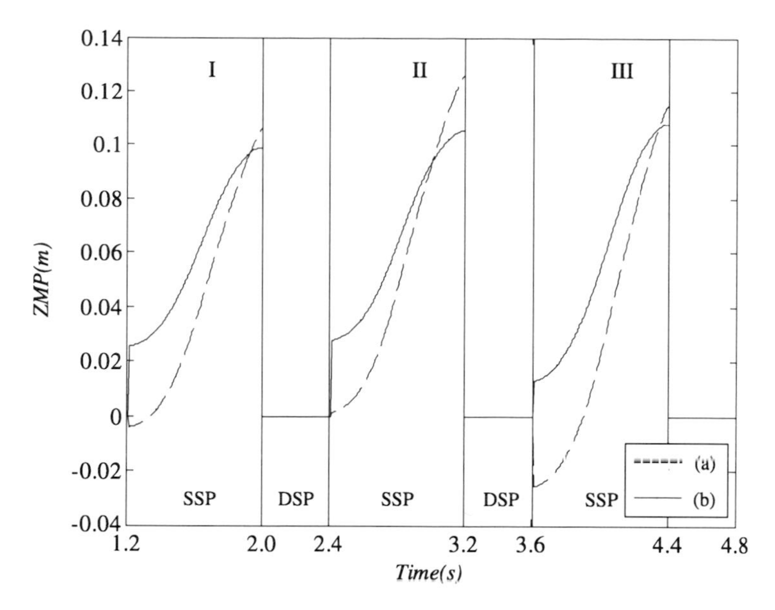

Fig. 9.3 ZMP comparison of two trajectory generation methods: (**a**) one-phase, (**b**) two-phase method

kept constant, e.g., $z_h = 0.5$. The corresponding x_h based on cosine functions are written as follows:

$$x_h = \begin{cases} -L_p R_s \cos(\frac{\pi t}{\tau_s}), & \text{for SSP,} \\ -\frac{L_c - R_s(L_c + L_p)}{2} \cos\frac{\pi(t-\tau_s)}{\tau_t - \tau_s} + \frac{L_c - R_s(L_c - L_p)}{2}, & \text{for DSP} \end{cases} \tag{9.7}$$

where R_s is a proportional ratio of displacement of x_h to a current step size in SSP, e.g., if $R_s = 0.5$, it means that x_h positions in a half of a current step size at the end of SSP. If R_s is assigned as 0.5, the generated trajectory is quite similar with that of the one-phase method in (9.6a) and there is usually a large variation of ZMP during SSP as shown in Fig. 9.2. Therefore, in the two-phase trajectory generation approach, we let x_h move less than a half of a current step size during SSP, thus enabling us to minimize the variation of ZMP. On the other hand, x_h is moving further for better stable walking during DSP where the corresponding ZMP is placed between two supporting limb tips (see Fig. 9.2). When $R_s = 0.4$, the phase portraits of x_h and v_h are shown in Fig. 9.2(b), and it is composed of two half circles: one is for SSP and the other is for DSP. This is the reason the proposed method is named as a two-phase trajectory generation method. Figure 9.3 shows that the magnitude of ZMP variation is minimized and a negative value of ZMP is avoided via the two-phase method.

Using both kinematical constraints and inverse kinematics, we can derive the desired relative angular positions based on (9.6c) and (9.7) as follows:

$$q_1 = \cos^{-1}\left(\frac{x_h}{k_h}\right) + \cos^{-1}\left(\frac{l_1^2 + k_h^2 - l_2^2}{2l_1 k_h}\right),$$

$$
\begin{aligned}
q_2 &= \pi - \cos^{-1}\left(\frac{l_1^2 + l_2^2 - k_h^2}{2l_1 l_2}\right), \\
q_3 &= \cos^{-1}\left(\frac{x_h}{k_h}\right) - \cos^{-1}\left(\frac{l_2^2 + k_h^2 - l_1^2}{2l_2 k_h}\right), \\
q_4 &= \cos^{-1}\left(\frac{x_e - x_h}{k_e}\right) + \cos^{-1}\left(\frac{l_4^2 + k_e^2 - l_5^2}{2l_4 k_e}\right), \\
q_5 &= \pi - \cos^{-1}\left(\frac{l_4^2 + l_5^2 - k_e^2}{2l_4 l_5}\right)
\end{aligned}
\tag{9.8}
$$

where $k_h = \sqrt{x_h^2 + z_h^2}$ and $k_e = \sqrt{(x_e - x_h)^2 + (z_e - z_h)^2}$. It is noted that x_h, k_h, and k_e are dependent on the status of the biped walking phase and thus two different desired trajectories are generated for the design of a motion control in the next section.

9.3 Motion Control for SSP and DSP

9.3.1 Application of Dynamic Surface Control

The design methodology of DSC developed in Chap. 6 will be applied to the biped robot model described in (9.5a), (9.5b). First, the equation of motion for SSP in (9.5a) can be written in state-space form as follows:

$$
\begin{aligned}
\dot{p}_{1S} &= p_{2S}, \\
\dot{p}_{2S} &= D_q^{-1}(T_q - N_q)
\end{aligned}
\tag{9.9}
$$

where $p_{1S} := q \in \Re^5$, $p_{2S} := \dot{q} \in \Re^5$, and $N_q := H_q(q, \dot{q}) + G_q(q)$. Define the first error vector as $S_{1S} := p_{1S} - r_S \in \Re^5$ where $r_S \in \Re^5$ is the desired trajectory for SSP in (9.8). After differentiating the error vector S_{1S}, the synthetic input, $\bar{p}_{2s}$, is defined as

$$
\begin{aligned}
\dot{S}_{1S} &= \dot{p}_{1S} - \dot{r}_S = p_{2S} - \dot{r}_S, \\
\bar{p}_{2S} &= \dot{r}_S - \Lambda_{1S} S_{1S}
\end{aligned}
\tag{9.10a}
$$

where Λ_{1S} is a 5×5 diagonal gain matrix, $\Lambda_{1S} = \mathrm{diag}(K_{11}, \ldots, K_{51})$. The $\bar{p}_{2S}$ forces the error vector to stay in an arbitrary error bound for an appropriate Λ_{1S}.

Next, a set of first-order low-pass filters is introduced to calculate the desired value which will be used to define the second error vector (refer to Chap. 6).

$$
T_S \dot{p}_{fS} + p_{fS} = \bar{p}_{2S}, \quad p_{fS}(0) := \bar{p}_{2S}(0)
\tag{9.10b}
$$

where T_S is a 5×5 diagonal matrix containing time constants as

$$
T_S := \mathrm{diag}(\tau_{1S}, \ldots, \tau_{5S}).
$$

Next, the second error surface vector can be defined as $S_{2S} := p_{2S} - p_{fS}$. After differentiating the surface vector, the control torque, which enables the error surface vector stays in an arbitrary bound, is designed as follows:

$$\begin{aligned} \dot{S}_{2S} &= \dot{p}_{2S} - \dot{p}_{fS} = D_q^{-1}(T_q - N_q) - \dot{p}_{fS}, \\ T_q &= N_q + D_q(\dot{p}_{fS} - \Lambda_{2S} S_{2S}). \end{aligned} \tag{9.10c}$$

Finally, using (9.10b), (9.10c) can be rewritten as

$$T_q = N_q + D_q\{T_{2S}^{-1}(\bar{p}_{2S} - p_{fS}) - \Lambda_{2S} S_{2S}\} \tag{9.10d}$$

where $\Lambda_{2S} := \mathrm{diag}(\lambda_{21}^S, \ldots, \lambda_{25}^S)$ is a 5×5 diagonal gain matrix.

In order to design a motion controller for DSP, similarly (9.5b) can be written in form of a state-space representation as follows:

$$\begin{aligned} \dot{p}_{1D} &= p_{2D}, \\ \dot{p}_{2D} &= B_q p_{2D} + C_q(T_q - N_q) \end{aligned} \tag{9.11}$$

where $p_{1D} := p_q \in \Re^3$ and $p_{2D} := \dot{p}_q \in \Re^3$. First, define the first error vector, $S_{1D} := p_{1D} - q_D \in \Re^3$ where q_D is the desired trajectory for DSP. Then, after differentiating it, similarly we can derive the synthetic input for the equations for DSP as follows:

$$\begin{aligned} \dot{S}_{1D} &= \dot{p}_{1D} - \dot{r}_D = p_{2D} - \dot{r}_D, \\ \bar{p}_{2D} &= \dot{r}_D - \Lambda_{1D} S_{1D}, \end{aligned} \tag{9.12a}$$

$$T_{2D}\dot{p}_{fD} + p_{fD} = \bar{p}_{2D}, \quad p_{2d}(0) := \bar{p}_{2D}(0) \tag{9.12b}$$

where $\Lambda_{1D} := \mathrm{diag}(\lambda_{11}^D, \lambda_{12}^D, \lambda_{13}^D)$ is the 3×3 diagonal controller gain matrix and T_{2D} is a 3×3 diagonal matrix containing time constants as $T_{2D} := \mathrm{diag}(\tau_{21}^D, \tau_{22}^D, \tau_{23}^D)$. After defining the second error surface vector $S_{2D} := p_{2D} - p_{2d}$ and differentiating the vector, the control torque can be obtained as follows:

$$\dot{S}_{2D} = \dot{p}_{2D} - \dot{p}_{fD} = B_q p_{2D} + C_q(T_q - N_q) - \dot{p}_{fD}, \tag{9.12c}$$

$$\begin{aligned} T_q &= N_q + C_q^-(-B_q p_{2D} + \dot{p}_{fD} - \Lambda_{2D} S_{2D}) \\ &= N_q + C_q^-\{-B_q p_{2D} + T_{2D}^{-1}(\bar{p}_{2D} - p_{fD}) - \Lambda_{2D} S_{2D}\} \end{aligned} \tag{9.12d}$$

where C_q^- is a 5×3 pseudo-inverse matrix as $C_q^- = C_q^T(C_q C_q^T)^{-1}$ and $\Lambda_{2D} := \mathrm{diag}(\lambda_{21}^D, \lambda_{22}^D, \lambda_{23}^D)$ is a 3×3 gain matrix.

9.3.2 Augmented Error Dynamics

Before discussing both stability and performance of the proposed control, we will derive the error dynamics of the closed-loop system including low-pass filter dynamics. If p_{fi} and $\bar{p}_{2i}$ are subtracted and added in the first row in (9.9) and (9.11), and the control torque calculated in (9.10d) and (9.12d) is applied to the second row in (9.9) and (9.11), respectively, the overall closed-loop dynamics including the low-pass filter dynamics can be written without loss of generality as follows:

$$\begin{cases} \dot{p}_{1i} = (p_{2i} - p_{fi}) + (p_{fi} - \bar{p}_{2i}) + \bar{p}_{2i}, \\ \dot{p}_{2i} = \dot{p}_{fi} - \Lambda_{2i} S_{2i}, \\ T_{2i}\dot{p}_{fi} + p_{fi} = \bar{p}_{2i} \end{cases} \tag{9.13}$$

where the subscript i stands for the phase of the model, i.e., $i = S$ and $i = D$ represent SSP and DSP, respectively. After defining the filter error $\xi_{2i} := p_{fi} - \bar{p}_{2i}$ and using (9.10a) and (9.12a), the augmented error dynamics can be written as

$$\begin{cases} \dot{p}_{1i} = S_{2i} + \xi_{2i} + \dot{r}_i - \Lambda_{1i} S_{1i}, \\ \dot{p}_{2i} = \dot{p}_{fi} - \Lambda_{2i} S_{2i}, \\ \dot{\xi}_{2i} = \dot{p}_{fi} - \dot{\bar{p}}_{2i} = -T_{2i}^{-1}\xi_{2i} - \ddot{r}_i + \Lambda_{1i}\dot{S}_{1i}. \end{cases} \tag{9.14}$$

Equation (9.14) can be rewritten in terms of error vectors as follows:

$$\begin{cases} \dot{S}_{1i} = S_{2i} + \xi_{2i} - \Lambda_{1i} S_{1i}, \\ \dot{S}_{2i} = -\Lambda_{2i} S_{2i}, \\ \dot{\xi}_{2i} - \Lambda_{1i}\dot{S}_{1i} = -T_{2i}^{-1}\xi_{2i} - \ddot{r}_i. \end{cases} \tag{9.15}$$

Then, (9.15) can be written in matrix form as

$$\begin{bmatrix} I & 0 & 0 \\ 0 & I & 0 \\ -\Lambda_{1i} & 0 & I \end{bmatrix} \begin{bmatrix} \dot{S}_{1i} \\ \dot{S}_{2i} \\ \dot{\xi}_{2i} \end{bmatrix} = \begin{bmatrix} -\Lambda_{1i} & I & I \\ 0 & -\Lambda_{2i} & 0 \\ 0 & 0 & -T_{2i}^{-1} \end{bmatrix} \begin{bmatrix} S_{1i} \\ S_{2i} \\ \xi_{2i} \end{bmatrix} + \begin{bmatrix} 0 \\ 0 \\ -I \end{bmatrix} \ddot{r}_i.$$

Finally, the augmented error dynamics can be written as follows:

$$\begin{bmatrix} \dot{S}_{1i} \\ \dot{S}_{2i} \\ \dot{\xi}_{2i} \end{bmatrix} = \begin{bmatrix} -\Lambda_{1i} & I & I \\ 0 & -\Lambda_{2i} & 0 \\ -\Lambda_{1i}^2 & \Lambda_{1i} & \Lambda_{1i} - T_{2i}^{-1} \end{bmatrix} \begin{bmatrix} S_{1i} \\ S_{2i} \\ \xi_{2i} \end{bmatrix} + \begin{bmatrix} 0 \\ 0 \\ -I \end{bmatrix} \ddot{r}_i.$$

The error dynamics can be represented as

$$\dot{z}_i = A_i z_i + B_i d_i \quad \text{for } i = S, D \tag{9.16}$$

where $z_i := [S_{1i}^T \; S_{2i}^T \; \xi_{2i}^T]^T$ and $d_i := \ddot{r}_i$. It is noted that the dimensions of z_i and d_i are dependent on the index i, i.e., $z_i \in \Re^{15}$ and $d_i \in \Re^5$ for $i = S$ and $z_i \in \Re^9$ and $d_i \in \Re^3$ for $i = D$.

Remark 9.1 While the augmented error dynamics in (9.16) is the same as (6.54), they are a piecewise linear system subject to an exogenous input signal (d_i). That is, two linear systems have different orders (one is a 15th order and the other is a 9th order system). When the DIP model is considered, the augmented error dynamics can be written as

$$\dot{z}_i = A_i z_i + B_i d_i,$$
$$\mathcal{T}_i^j : \begin{cases} S_{2D}^+ = p_{2D}^+ - p_{fD} = \dot{p}_q(q, \dot{q}^+) - p_{fD}, & \text{if } i = S,\ j = D, \\ S_{2S}^+ = p_{2S} - p_{fS} = \dot{q} - p_{fS}, & \text{if } i = D,\ j = S \end{cases} \tag{9.17}$$

where $\mathcal{T}_i^j$ is the transition function from i to j. As described in (9.5a), (9.5b) the angular velocities are changed instantaneously to $\dot{q}^+$ when SSP is switched into DSP. Thus S_{2D}^+ is changed instantaneously.

9.3.3 Piecewise Quadratic Boundedness

Suppose the exogenous input signal (d_i) is bounded as $d_i^T d_i \leq \gamma_i$ and the upper bound γ_i is known in the trajectory generation procedure. Then, (9.16) can be rewritten as

$$\dot{z}_i = A_i z_i + \bar{B}_i u_i \quad \text{for } i = S, D \tag{9.18}$$

where $u_i \in \Re^5$ is the unit-peak input such as $u_i^T u_i \leq 1$, i.e., $u_i := d_i / \sqrt{\gamma_i}$, and $\bar{B}_i := \sqrt{\gamma_i} B_i$. Then, (9.18) can be classified a polytopic linear differential inclusion (PLDI). As discussed in Chap. 2, quadratic boundedness can be estimated via convex optimization once the error dynamics are given in one of the PLDI, NLDI, and DNLDI forms. Furthermore, as discussed in Chap. 7, if the switched error dynamics with same system order for a switched nonlinear system are obtained, simultaneous quadratic stability or boundedness can be applied for analysis (refer to vehicle control example in Chap. 7). However, the error dynamics given in (9.18) has two linear error dynamics with different orders. Therefore, the definitions of reachable set and quadratic boundedness are modified as follows:

Definition 9.1 Suppose the diagonal gain matrices Λ_{1i}, Λ_{2i} and T_{2i} for DSC are given. A set, R_{up} described in (9.19) below is called *piecewise reachable sets* of the augmented error dynamics given by (9.18)

$$R_{up}^i = \{z_i(T) | z_i, u_i, \text{ satisfy (9.18)}, z_i(0) = 0, u_i^T u_i \leq 1, T \geq 0\}. \tag{9.19}$$

Definition 9.2 If there exists an ellipsoid, $\varepsilon_i = \{z_i | z_i^T P_i z_i \leq 1, P_i \succ 0\}$, contains the reachable sets of the augmented error dynamics described in (9.19), the error dynamics is said to be *quadratically piecewise bounded with Lyapunov matrix* P_i, i.e., if there exists $P_i \succ 0$ such that

$$\frac{dV(z_i)}{dt} = \frac{d}{dt}(z_i^T P_i z_i) = (A_i z_i + \bar{B}_i u_i)^T P_i z_i + z_i^T P_i (A_i z_i + \bar{B}_i u_i) < 0 \tag{9.20}$$

for all z_i and u_i satisfying $\{z_i | z_i^T P_i z_i > 1\}$ and $u_i^T u_i < 1$, respectively. Then the system in (9.5a), (9.5b) is called *quadratically piecewise trackable* via DSC for the given gain matrices Λ_{1i}, Λ_{2i} and T_{2i}.

Using the inequality condition (9.20), the following theorem can be stated for the piecewise quadratically boundedness of the augmented error dynamics in (9.18).

Theorem 9.1 *Suppose the matrices A_i and $\bar{B}_i$ are given. The system given in* (9.5a), (9.5b) *is piecewise quadratically tractable via DSC for the given set of diagonal* *gain matrices* Λ_{1i}, Λ_{2i}, *and a filter time constant matrix* T_{2i} *if there exist* $P_i \succ 0$ *and* $\alpha \geq 0$ *such that*

$$\begin{bmatrix} A_i^T P_i + P_i A_i + \alpha P_i & P_i \bar{B}_i \\ \bar{B}_i^T P_i & -\alpha I \end{bmatrix} \leq 0. \tag{9.21}$$

Then, the ellipsoid, $\varepsilon_i = \{z_i | z_i^T P_i z_i \leq 1\}$, *contains the piecewise reachable sets of the augmented error dynamics in* (9.18).

The proof is omitted due to the similarity of the results in Chap. 2. If all orders of the piecewise augmented error dynamics are the same, the sufficient condition for simultaneous quadratic tracking can be obtained: if there exist $P \succ 0$ and $\alpha \geq 0$ satisfying LMI (9.21). For the given error dynamics in (9.18), the order of the error vectors for SSP and DSP is 15 and 9, respectively, i.e., $z_S \in \Re^{15}$, $z_D \in \Re^9$. Therefore, only piecewise quadratic tracking can be verified by solving LMI (9.21).

Remark 9.2 When the DIP model is considered, there is an instantaneous change of S_{2i} right after SSP is switched to DSP as described in (9.17). If the initial value of z_i is placed outside the ellipsoidal error bound, i.e., $(z_i^+)^T P_D z_i^+$, z_i goes into the ellipsoidal error bound as long as Theorem 9.1 is satisfied. If z_i stays in the ellipsoidal error bound, z_i remains in the bound as long as the error dynamics is quadratically bounded.

Remark 9.3 LMI (9.21) can be solved in the framework of convex optimization. The smallest ellipsoidal error bound whose size may be defined by either a volume or a maximum diameter of the ellipsoid can be calculated numerically [12]. If the maximum diameter of the ellipsoid is minimized, the algorithm to solve LMI (9.21) is the following:

$$\begin{aligned} \text{For a fixed } \alpha > 0, \quad & \text{maximize} \quad \lambda_{\min}(P_i) \\ & \text{subject to} \quad P_i > 0, \quad \text{LMI (9.21)}. \end{aligned} \tag{9.22}$$

Through the line search of α, the ellipsoidal error bound can be estimated numerically.

9.4 Simulation Results

In order to validate the proposed controller and its performance estimation based on the ellipsoidal error bound, a three-phase biped model described in Sect. 9.1 is used for simulation and the corresponding system parameters are given as shown in

Table 9.2 System parameters for simulation

Link no.	m_i	I_i	l_i	d_i
1	6.3	0.0024	0.30	0.150
2	8.6	0.0031	0.29	0.145
3	9.3	0.0607	0.35	0.175
4	8.6	0.0031	0.29	0.145
5	6.3	0.0024	0.30	0.150

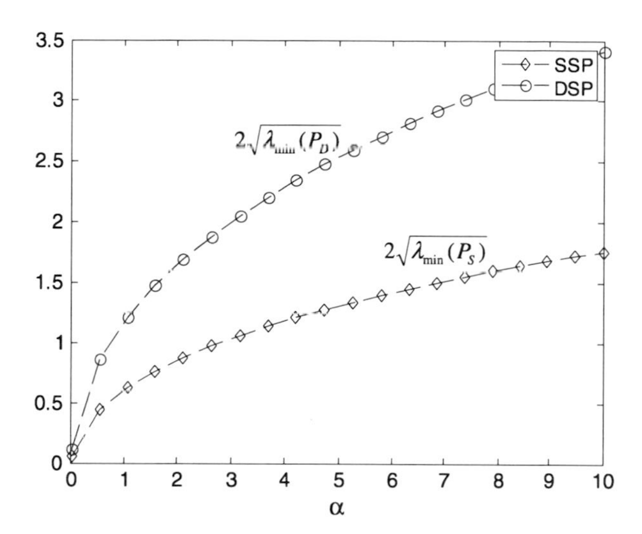

Fig. 9.4 Diameter of the ellipsoidal error bound vs. α

Table 9.2. When the controller gains, Λ_{1i}, Λ_{2i}, and τ_{2i}, are assigned as follows:

$$\text{for } i = S, \quad \begin{cases} \Lambda_{1S} = \text{diag}(60, 60, 80, 160, 160), \\ \Lambda_{2S} = \text{diag}(80, 80, 150, 200, 200), \\ T_{2S} = 0.02 \times I, \end{cases}$$

$$\text{for } i = D, \quad \begin{cases} \Lambda_{1D} = \text{diag}(60, 60, 100), \\ \Lambda_{2D} = \text{diag}(80, 80, 200), \\ T_{2D} = 0.02 \times I, \end{cases}$$

the matrices A_i and B_i in (9.18) are obtained. Furthermore, it can be checked whether A_i is Hurwitz, which is a necessary condition for the existence of a solution in Theorem 9.1.

Next, when $\gamma_S = 253.4$ and $\gamma_D = 94.85$ coming from the desired trajectory in (9.8), an ellipsoidal error bound is estimated by solving the LMI problem (9.22) where the smallest ellipsoid is defined by minimizing the largest diameter. When α is given from 0.01 to 10 for line searching, the maximum diameter of the ellipsoid, $2\sqrt{\lambda_{\min}(P_i)}$, is calculated as shown in Fig. 9.4. This calculation is performed using both CVX [32] and SeDuMi for semidefinite programming [93]. Since the largest diameter increases as α becomes larger, the smallest ellipsoidal error bound is chosen

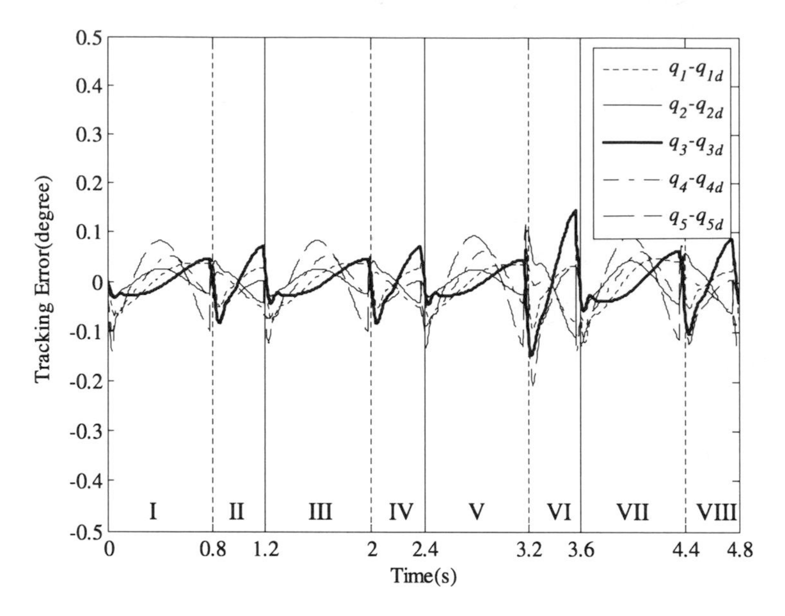

Fig. 9.5 Angular position errors at joints, $q_i - q_{id}$

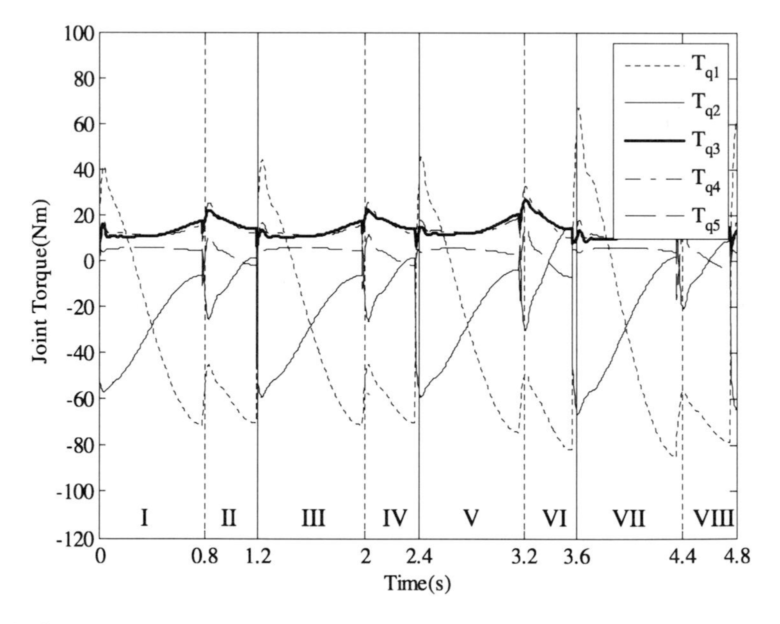

Fig. 9.6 Control torque at joints, T_{qi}

for $\alpha = 0.01$. Thus, once $P_S \in \Re^{15\times 15}$ and $P_D \in \Re^{9\times 9}$ are calculated, the piecewise quadratic function level, $z_i^T P_i z_i$, will be calculated later.

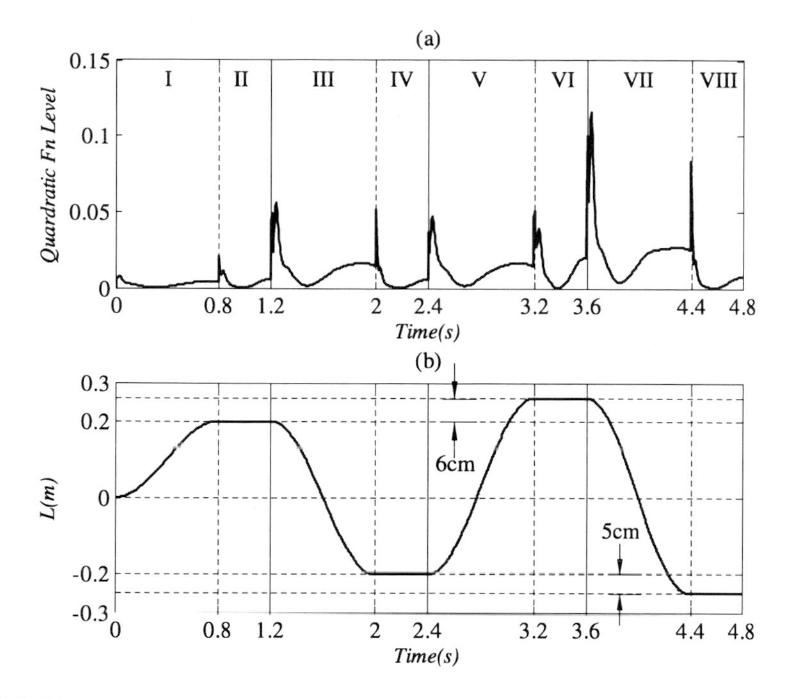

Fig. 9.7 Piecewise quadratic function level (a) and displacement between the tips of two lower limbs (b)

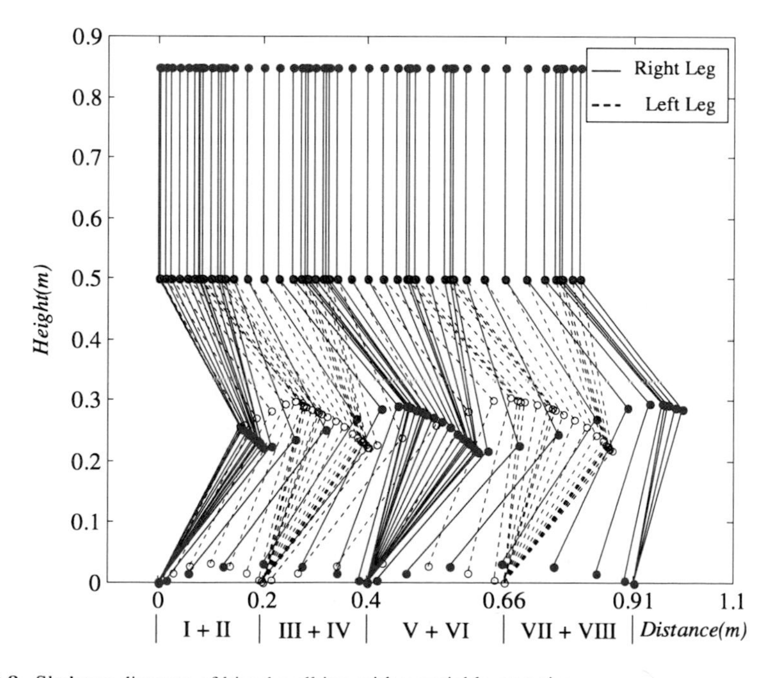

Fig. 9.8 Skeleton diagram of biped walking with a variable step size

Figure 9.5 shows time responses of the tracking angular position error at each joint, and the maximum error is less than 0.17 degree. The corresponding control torque is shown in Fig. 9.6, and the maximum torque is about 80 Nm, which is realizable in real applications. In the time period I of the figure, the step size is given to be 20 cm (see Fig. 9.7(b)) and a left leg is moving forward from standing (see Fig. 9.8). DSP is maintained for 0.4 second in the time period II, and a full swing of a right leg with a size of 40 cm is done in the time period III. After DSP period (IV), the step size of the left leg is increased to 26 cm and the total swing length of the left leg is 46 cm at the end of V. Finally, the total swing length of the right leg becomes 51 cm at the end of VII (see Figs. 9.7 and 9.8).

Finally, using the piecewise quadratic function, $z_i^T P_i z_i$, calculated above, the time response of the quadratic function is shown in Fig. 9.7(a). As shown in the figure, the simulation result that the function level is less than one validates the piecewise quadratic tracking that the precalculated piecewise ellipsoidal error bound contains the simulated errors.

Appendix
Proofs

A.1 Proof of Lemma 2.1

Lemma 2.1 *For the given class of nonlinear system* (2.3), *the augmented closed-loop error dynamics with DSC is*

$$\begin{aligned}\dot{z} &= A_{cl}z + B_w w + B_r r,\\ \|w\| &\le \gamma \|C_z z\|\end{aligned} \tag{A.1}$$

where

$$\begin{aligned}
\dot{f} &= [\dot{f}_1 \ \cdots \ \dot{f}_{n-1}]^T = \frac{\partial f}{\partial x}\dot{x} = JC_z z + J_1 \dot{x}_{1d} = w + J_1 \dot{x}_{1d} \in \Re^{n_w},\\
w &= J(x)C_z z, \quad J = \frac{\partial f}{\partial x}, \ J_1 \textit{ is the first column of } J,\\
r &= \begin{bmatrix} J_1 \dot{x}_{1d} \\ \ddot{x}_{1d} \end{bmatrix} \in \Re^{n_w+1} := \Re^{n_r},\\
B_r &= \begin{bmatrix} \mathbf{0}_{n\times n_w} & 0_n \\ T_\xi^{-1} & -T_\xi^{-1} b_r \end{bmatrix} = [B_w \ B_e] \in \Re^{n_z \times n_r}, \quad b_r = [1 \ 0 \ \cdots \ 0] \in \Re^{n_w},
\end{aligned}$$

and $C_z = [\underline{A}_{11} \ T_\xi]$ *where the notation* $\underline{A}_{11}$ *is the matrix where the last row of the matrix* A_{11} *is eliminated.*

Proof As derived in (2.18) of Sect. 2.4, the closed-loop error dynamics is

$$\dot{z} = A_{cl}z + B_w \dot{f} + B_d \ddot{x}_{1d}. \tag{A.2}$$

Furthermore, $\dot{f}$ in the above equation can be expressed as a function of z and $\dot{x}_{1d}$ as follows:

$$\begin{aligned}
\dot{f}_1 &= \frac{\partial f_1}{\partial x_1}\dot{x}_1 = \frac{\partial f_1}{\partial x_1}(x_2 + f_1) = \frac{\partial f_1}{\partial x_1}\big[(x_2 - x_{2d}) + (x_{2d} - \bar{x}_2) + \bar{x}_2 + f_1\big]\\
&= J_{11}\{S_2 + \xi_2 + \dot{x}_{1d} - f_1 - K_1 S_1 + f_1\}\\
&= J_{11}\{S_2 + \xi_2 - K_1 S_1\} + J_{11}\dot{x}_{1d} = J_{11}c_{z1}z + J_{11}\dot{x}_{1d},
\end{aligned} \tag{A.3}$$

B. Song, J.K. Hedrick, *Dynamic Surface Control of Uncertain Nonlinear Systems*, Communications and Control Engineering,
DOI 10.1007/978-0-85729-632-0,

where $c_{z1} = [-K_1\ 1\ 0\ \cdots\ 0\ 1\ 0\ \cdots\ 0] \in \Re^{n_z}$. For $2 \le i \le n-1$,

$$\begin{aligned}\dot{f}_i &= \sum_{j=1}^{i} \frac{\partial f_i}{\partial x_j}\dot{x}_j = \sum_{j=1}^{i} \frac{\partial f_i}{\partial x_j}(x_{j+1} + f_j) = \sum_{j=1}^{i} J_{ij}(S_{i+1} + \xi_{i+1} + \dot{x}_{id} - K_i S_i) \\ &= \sum_{j=1}^{i} J_{ij}(x) c_{zj} z + J_{i1}\dot{x}_{1d} = J_{i1}\dot{x}_{1d} + \bar{J}_i(x) C_z z \end{aligned} \tag{A.4}$$

where

$$c_{zj} = \left[\cdots -K_j\ 1\ 0\ \cdots\ 0 - \frac{1}{\tau_j}\ 1\ 0\ \cdots\ 0 \right] \in \Re^{1\times n_z},$$

$$\begin{aligned} C_z &= \begin{bmatrix} c_{z1} \\ c_{z2} \\ \vdots \\ c_{z(n-1)} \end{bmatrix} = \left[\begin{array}{ccccc|cccccc} -K_1 & 1 & 0 & \cdots & 0 & 1 & 0 & \cdots & 0 & 0 \\ 0 & -K_2 & 1 & \cdots & 0 & -1/\tau_2 & 1 & \cdots & 0 & 0 \\ \vdots & \vdots & \vdots & \ddots & \vdots & \vdots & \vdots & \ddots & \vdots & \vdots \\ 0 & 0 & 0 & \cdots & 1 & 0 & 0 & \cdots & -1/\tau_{n-1} & 1 \end{array}\right] \\ &= [\underline{A}_{11}\ T_\xi] \in \Re^{(n-1)\times n_z}, \end{aligned} \tag{A.5}$$

$$J = \begin{bmatrix} \frac{\partial f_1}{\partial x} \\ \frac{\partial f_2}{\partial x} \\ \vdots \\ \frac{\partial f_{n-1}}{\partial x} \end{bmatrix} = \begin{bmatrix} \bar{J}_1 \\ \bar{J}_2 \\ \vdots \\ \bar{J}_{n-1} \end{bmatrix} = \begin{bmatrix} \frac{\partial f_1}{\partial x_1} & 0 & 0 & \cdots & 0 \\ \frac{\partial f_2}{\partial x_1} & \frac{\partial f_2}{\partial x_2} & 0 & \cdots & 0 \\ \vdots & \vdots & \vdots & \ddots & \vdots \\ \frac{\partial f_{n-1}}{\partial x_1} & \frac{\partial f_{n-1}}{\partial x_2} & \frac{\partial f_{n-1}}{\partial x_3} & \cdots & \frac{\partial f_{n-1}}{\partial x_{n-1}} \end{bmatrix}$$
$$\in \Re^{(n-1)\times(n-1)},$$

where $\bar{A}_{11}$ is the matrix where the last row of the matrix A_{11} is eliminated, A_{11} and T_ξ are defined in (2.17), and $\bar{J}_i$ is the ith row of J.

Combining (A.3) and (A.4), we can write

$$\dot{f} = J\dot{x} = J(C_z z + e_1\dot{x}_{1d}) = JC_z z + J_1\dot{x}_{1d}, \tag{A.6}$$

where $e_1 = [1\ 0\ \cdots\ 0] \in \Re^{n-1}$ and $J_1 = Je_1 \in \Re^{n-1}$ is the first column of J. After combining (A.2) and (A.6), we can write the augmented error dynamics as

$$\begin{aligned}\dot{z} &= A_z z + B_w(JC_z z + J_1\dot{x}_{1d}) + B_d\ddot{x}_{1d} = A_z z + B_w JC_z z + [B_w\ B_d]\begin{bmatrix} J_1\dot{x}_{1d} \\ \ddot{x}_{1d} \end{bmatrix} \\ &= A_z z + B_w w + B_r r \end{aligned}$$

where

$$w = JC_z z \in \Re^{n_w}, \qquad r = \begin{bmatrix} J_1\dot{x}_{1d} \\ \ddot{x}_{1d} \end{bmatrix} \in \Re^{n_w+1} := \Re^{n_r}.$$

Since w in the error dynamics is norm-bounded such that

$$\|w\| = \|JC_z z\| \le \|J\|\|C_z z\| \le \gamma\|C_z z\|$$

where the last inequality comes from the assumption of f in (2.4), the augmented error dynamics in (A.1) is derived. □

A.2 Proof of Lemma 3.1

Lemma 3.1 *For the given uncertain nonlinear system* (3.1), *the augmented closed-loop error dynamics with DSC is*

$$\dot{z} = A_{cl}z + B_{wu}w_u + B_{ru}r_u \tag{A.7}$$

where

$$w_u = \begin{bmatrix} p \\ Jp \\ w \end{bmatrix} \in \Re^{n_u+n_w+n_w} := \Re^{n_{wu}}, \qquad r_u = \begin{bmatrix} q \\ Jq \\ r \end{bmatrix} \in \Re^{n_u+n_w+n_r} := \Re^{n_{ru}},$$

$$\Delta f = p + q \in \Re^{n_u},$$

$$B_{wu} = [B_u \ B_w \ B_w] \in \Re^{n_z \times n_{wu}}, \qquad B_{ru} = [B_u \ B_w \ B_r] \in \Re^{n_z \times n_{ru}},$$

B_u *is defined in* (3.3), *and all other vectors and matrices are defined in Lemma* 2.1. *Furthermore, there are* C_{zi} *such that*

$$|w_{ui}| \le \|C_{zi}z\|.$$

Proof Due to similarity, the derivation of the following error dynamics in (3.3) is omitted, but the readers may refer to Sect. 2.4 for the detailed derivation:

$$\dot{z} = A_{cl}z + B_w\dot{f} + B_{ue}r_e \tag{A.8}$$

where $z = [S\ \xi]^T \in \Re^{n_z}$, $r_e = [\Delta f\ \ddot{x}_{1d}]^T \in \Re^{n_u+1}$, and

$$B_{ue} = \begin{bmatrix} \mathbf{I}_{n(n-1)} & 0_n \\ T_\xi^{-1}K & -T_\xi^{-1}b_r \end{bmatrix} = [B_u \ B_e], \quad B_u = \begin{bmatrix} \mathbf{I}_{n(n-1)} \\ T_\xi^{-1}K \end{bmatrix} \in \Re^{n_z \times n_u}.$$

Then, we need to decompose $\dot{f}$ and Δf into vanishing and nonvanishing perturbation terms.

First, $\dot{f}$ in (A.8) can be regarded as a function of z and $\dot{x}_{1d}$

$$\begin{aligned} \dot{f}_1 &= \frac{\partial f_1}{\partial x_1}\dot{x}_1 = \frac{\partial f_1}{\partial x_1}(x_2 + f_1 + \Delta f_1) \\ &= \frac{\partial f_1}{\partial x_1}\big[(x_2 - x_{2d}) + (x_{2d} - \bar{x}_2) + \bar{x}_2 + f_1 + \Delta f_1\big] \\ &= J_{11}\{S_2 + \xi_2 + \dot{x}_{1d} - f_1 - K_1S_1 + f_1 + \Delta f_1\} \\ &= J_{11}\{S_2 + \xi_2 - K_1S_1\} + J_{11}\dot{x}_{1d} + J_{11}\Delta f_1 \\ &= J_{11}c_{z1}z + J_{11}\dot{x}_{1d} + J_{11}\Delta f_1. \end{aligned} \tag{A.9}$$

For $2 \le i \le n-1$,

$$\begin{aligned} \dot{f}_i &= \sum_{j=1}^{i} \frac{\partial f_i}{\partial x_j}\dot{x}_j = \sum_{j=1}^{i} \frac{\partial f_i}{\partial x_j}(x_{j+1} + f_j + \Delta f_j) \\ &= \sum_{j=1}^{i} J_{ij}(S_{j+1} + \xi_{j+1} + \dot{x}_{jd} - K_jS_j + \Delta f_j) \end{aligned}$$

$$= \sum_{j=1}^{i} J_{ij}(c_{zj}z + \Delta f_j) + J_{i1}\dot{x}_{1d} = J_{i1}\dot{x}_{1d} + \bar{J}_i(C_z z + \Delta f) \tag{A.10}$$

where C_z is defined in (A.5). Therefore, as derived in (A.6), similarly

$$\dot{f} = J\dot{x} = J(C_z z + e_1\dot{x}_{1d} + \Delta f) = JC_z z + J\Delta f + J_1\dot{x}_{1d} \tag{A.11}$$

where J and J_1 are defined in Lemma 2.1.

Second, the existence of upper bound of Δf is stated as follows:

Claim There exist C_{ui} and D_{ui} such that

$$|\Delta f_i| \le \|C_{ui}z\| + \|D_{ui}r_d\|$$

where $r_d = [x_{1d}\ \dot{x}_{1d}]^T \in \Re^2$.

Proof

$$|x_1| = |S_1 + x_{1d}| \le |S_1| + |x_{1d}| := |c_{x1}z| + |d_{x1}r_d|,$$

$$\begin{aligned}
|x_2| &= |S_2 + \xi_2 + \bar{x}_2| \le |S_2 + \xi_2 - K_1S_1| + |\dot{x}_{1d}| + \big|f_1(x_1)\big| \\
&\le |c_{z1}z| + |\dot{x}_{1d}| + m_{11}|x_1| \\
&\le |c_{z1}z| + |\dot{x}_{1d}| + m_{11}\big(|c_{x1}z| + |x_{1d}|\big) \\
&= \left\| \begin{bmatrix} c_{z1} \\ m_1 c_{x1} \end{bmatrix} z \right\|_1 + \left\| \begin{bmatrix} m_{11} & 0 \\ 0 & 1 \end{bmatrix} \begin{bmatrix} x_{1d} \\ \dot{x}_{1d} \end{bmatrix} \right\|_1 := \|C_{x2}z\|_1 + \|D_{x2}r_d\|_1,
\end{aligned}$$

$$\begin{aligned}
|x_3| &= |S_3 + \xi_3 + \bar{x}_3| \le \left| S_3 + \xi_3 - \frac{\xi_2}{\tau_2} - K_2S_2 \right| + \big|f_2(x_1, x_2)\big| \\
&\le |c_{z2}z| + m_{21}|x_1| + m_{22}|x_2| \\
&\le |c_{z2}z| + m_{21}\big(|c_{x1}z| + |d_{x1}r_d|\big) + m_{22}\big(\|C_{x2}z\|_1 + \|D_{x2}r_d\|_1\big) \\
&= \left\| \begin{bmatrix} c_{z2} \\ m_{22}C_{x2} \\ m_{21}c_{x1} \end{bmatrix} z \right\|_1 + \left\| \begin{bmatrix} m_{22}D_{x2} \\ m_{21}d_{x1} \end{bmatrix} r_d \right\|_1 := \|C_{x3}z\|_1 + \|D_{x3}r_d\|_1
\end{aligned}$$

where $r_d = [x_{1d}\ \dot{x}_{1d}]^T \in \Re^2$, $c_{x1} = [1\ 0\ \cdots\ 0] \in \Re^{1\times n_z}$, $d_{x1} = [1\ 0]$, and c_{zi} is the i-th row of C_z in (2.19). By induction, for $3 \le i \le n-1$,

$$|x_{i+1}| = |S_{i+1} + \xi_{i+1} + \bar{x}_{i+1}| = \left| S_{i+1} + \xi_{i+1} - \frac{\xi_i}{\tau_i} - K_iS_i - f_i(x_1, \ldots, x_i) \right|$$

$$\begin{aligned}
&\le |c_{zi}z| + \sum_{j=1}^{i} m_{ij}|x_j| \\
&= \left\| \begin{bmatrix} c_{zi} \\ m_{ii}C_{xi} \\ \vdots \\ m_{i2}C_{x2} \\ m_{i1}c_{x1} \end{bmatrix} z \right\|_1 + \left\| \begin{bmatrix} m_{ii}D_{x(i-1)} \\ \vdots \\ m_{i2}D_{x2} \\ m_{i1}d_{x1} \end{bmatrix} r_d \right\|_1 := \|C_{x(i+1)}z\|_1 + \|D_{x(i+1)}r_d\|_1
\end{aligned}$$

where $C_{xi} \in \Re^{n_C \times n_z}$ and $D_{xi} \in \Re^{n_D \times n_{r_d}}$. Therefore,

$$
\begin{aligned}
|\Delta f_i| &\le \sum_{j=1}^{i} n_{ij}|x_j| = n_{i1}|x_1| + \cdots + n_{ij}|x_j| \\
&\le \left\| \begin{bmatrix} n_{i1}c_{x1} \\ n_{i2}C_{x2} \\ \vdots \\ n_{ii}C_{xi} \end{bmatrix} z \right\|_1 + \left\| \begin{bmatrix} n_{i1}d_{x1} \\ n_{i2}D_{x2} \\ \vdots \\ n_{ii}D_{xi} \end{bmatrix} r_d \right\|_1 := \|\bar{C}_{ui}z\|_1 + \|\bar{D}_{ui}r\|_1 \\
&\le \sqrt{n_C}\|\bar{C}_{ui}z\|_2 + \sqrt{n_D}\|\bar{D}_{ui}r_d\|_2 := \|C_{ui}z\|_2 + \|D_{ui}r\|_2
\end{aligned}
$$

where n_C and n_D are the number of row of C_{ui} and D_{ui}, respectively. The last inequality comes from equivalence of matrix norm. □

Using the result of the claim, Δf_i can be written as a function p_i and q_i such that

$$
\Delta f_i = p_i + q_i, \quad |p_i| \le \|C_{ui}z\|, \ |q_i| \le \|D_{ui}r_d\|. \tag{A.12}
$$

Then, using (A.11) and (A.12), the augmented error dynamics (A.8) is written as

$$
\begin{aligned}
\dot{z} &= A_{cl}z + B_w(JC_z z + J\Delta f + J_1\dot{x}_{1d}) + B_u\Delta f + B_e\ddot{x}_{1d}, \\
\dot{z} &= A_{cl}z + (B_wJC_z z + B_wJp + B_u p) + (B_wJq + B_wJ_1\dot{x}_{1d} + B_e\ddot{x}_{1d}) \\
&= A_{cl}z + [B_u\ B_w\ B_w]\begin{bmatrix} p \\ Jp \\ JC_z z \end{bmatrix} + [B_u\ B_w\ B_w\ B_e]\begin{bmatrix} q \\ Jq \\ J_1\dot{x}_{id} \\ \ddot{x}_{1d} \end{bmatrix}.
\end{aligned}
$$

Furthermore, using the definition of w, r, and B_r in Lemma 2.1, the above equation can be rewritten as

$$
\begin{aligned}
\dot{z} &= A_{cl}z + [B_u\ B_w\ B_w]\begin{bmatrix} p \\ Jp \\ w \end{bmatrix} + [B_u\ B_w\ B_r]\begin{bmatrix} q \\ Jq \\ r \end{bmatrix} \\
&:= A_{cl}z + B_{uw}w_u + B_{ur}r_u.
\end{aligned}
$$

Moreover, the vanishing perturbation term $w_u \in \Re^{n_{wu}}$ is diagonally norm-bounded as follows:

$$
\begin{cases}
|p_i| \le \|C_{ui}z\|, \\
|\bar{J}_i p| \le \|\bar{J}_i\|\|p\| \le \gamma_i\|C_{ui}z\| = \|\tilde{C}_{ui}z\|, \\
|w_i| \le \gamma_i\|c_{zi}z\| = \|\tilde{c}_{zi}z\|
\end{cases}
$$

where $\tilde{c}_{zi} = \gamma_i c_{zi}$, $\gamma_i = \max(\gamma_{i1}, \ldots, \gamma_{i(n-1)})$ and c_{zi} is defined in (A.5). The first inequality comes from the result of the above claim and the last inequality comes from the existence of γ_{ij} such that $\|J_{ij}\| := \|\partial f_i/\partial x_j\| \le \gamma_{ij}$ in the assumption of (3.1). Therefore, there exist C_{zi} such that

$$
|w_{ui}| \le \|C_{zi}z\| \tag{A.13}
$$

for $i = 1, \ldots, n_{wu}$. □

A.3 Proof of Lemma 3.2

Lemma 3.2 *For the given autonomous nonlinear system* (3.16) *with mismatched uncertainties, the augmented error dynamics with DSC is written as*

$$\dot{z} = A_{cl}z + B_w w + B_{rd}r_d + B_{nd}n_d \tag{A.14}$$

where z, A_{cl}, and B_w are defined in (2.18),

$$r_d = \begin{bmatrix} r \\ D_2\Phi d \\ Jd \end{bmatrix} \in \Re^{n_{rd}}, \qquad n_d = \begin{bmatrix} d \\ D_1 d \end{bmatrix} \in \Re^{n_{nd}},$$

$$B_{rd} = [B_r \ B_w \ B_w] \in \Re^{n_z \times n_{rd}}, \quad \text{and} \quad B_{nd} = [B_u \ B_w] \in \Re^{n_z \times n_{nd}}$$

where r, B_r, and J are defined in Lemma 2.1, *B_u is defined in* (3.3),

$$d = [\Delta f_1 - h_1 \ \cdots \ \Delta f_{n-1} - h_{n-1}]^T \in \Re^{n-1} := \Re^{n_u},$$

$$h = [h_1 \ \cdots \ h_{n-1}]^T \in \Re^{n_u}, \quad h_i = \frac{S_i \rho_i^2}{2\varepsilon},$$

$$D_1 = \operatorname{diag}\left(\begin{bmatrix} \frac{\rho_1^2}{2\varepsilon} & \cdots & \frac{\rho_{n-1}^2}{2\varepsilon} \end{bmatrix}\right) \in \Re^{n_u \times n_u}, \quad \text{and}$$

$$D_2 = \operatorname{diag}\left(\begin{bmatrix} \frac{S_1 \rho_1}{\varepsilon} & \cdots & \frac{S_{n-1}\rho_{n-1}}{\varepsilon} \end{bmatrix}\right) \in \Re^{n_u \times n_u}.$$

Furthermore, if all assumptions in (3.16) *are satisfied for all $x \in \mathscr{D}$, x_{1d} is the feasible output trajectory in $\mathscr{D}$, and z is bounded on the compact and convex set Ω_c, there exist $\tilde{C}_{wi}$ and $n_0 > 0$ for a given ε such that*

$$|w_i| \le \left\|\tilde{C}_{wi}(\varepsilon) z\right\| \quad \text{and} \quad z^T E n_d \le n_0(\varepsilon)$$

where $E = [\mathbf{I}_{n_z \times n_u} \ \mathbf{I}_{n_z \times n_u}] \in \Re^{n_z \times n_{nd}}$.

Proof As derived in (3.3) of Sect. 3.1, similarly the closed error dynamics is

$$\begin{aligned} \dot{z} &= A_{cl}z + B_w g + [B_u \ B_e]\begin{bmatrix} d \\ \ddot{x}_{1d} \end{bmatrix}, \\ \dot{z} &= A_{cl}z + B_w g + B_e \ddot{x}_{1d} + B_u d \end{aligned} \tag{A.15}$$

where

$$g = [\dot{f}_1 + \dot{h}_1 \ \cdots \ \dot{f}_{n-1} + \dot{h}_{n-1}]^T \in \Re^{n-1} := \Re^{n_w}.$$

Since $\dot{x}_i$ is written in a function of z, $\dot{x}_{1d}$, and d as

$$\begin{cases} \dot{x}_1 = S_2 + \xi_2 - K_1 S_1 + \dot{x}_{1d} + d_1, \\ \dot{x}_i = S_{i+1} + \xi_{i+1} - K_i S_i + \dot{x}_{id} + d_i \quad \text{for } i = 2, \ldots, n-1, \\ \dot{x}_n = -K_n S_n + \dot{x}_{nd}, \end{cases}$$

the nonlinear system with DSC is written as

$$\dot{x} = \tilde{C}_z z + e_1 \dot{x}_{1d} + \tilde{d} \tag{A.16}$$

where $\tilde{C}_z = [A_{11}\ \underline{T}_\xi] \in \Re^{n\times n_z}$, $\tilde{d} = [d_1\ \cdots\ d_{n-1}\ 0]^T = \left[\begin{smallmatrix} d \\ 0 \end{smallmatrix}\right] \in \Re^n$, and

$$\underline{T}_\xi = \begin{bmatrix} 1 & 0 & \cdots & 0 & 0 \\ -\frac{1}{\tau_2} & 1 & \cdots & 0 & 0 \\ 0 & -\frac{1}{\tau_3} & \cdots & 0 & 0 \\ \vdots & \vdots & \ddots & \vdots & \vdots \\ 0 & 0 & \cdots & -\frac{1}{\tau_{n-1}} & 1 \\ 0 & 0 & \cdots & 0 & -\frac{1}{\tau_n} \end{bmatrix} = \begin{bmatrix} T_\xi \\ -\frac{1}{\tau_n} e_{n_\xi}^T \end{bmatrix} \in \Re^{n\times n_\xi}$$

where e_i is the unit vector with one in the ith element. Similarly $\dot{S}_i$ is written as a function of z and d as

$$\begin{aligned} &\dot{S}_i = S_{i+1} + \xi_{i+1} - K_i S_i + d_i \quad \text{for } i = 1, \ldots, n-1 \\ &\implies \quad \dot{S}_{1:n-1} = A_s z + d \end{aligned} \tag{A.17}$$

where $S_{1:n-1} = [S_1\ \cdots\ S_{n-1}]^T \in Re^{n-1}$, $A_s = [\underline{A}_{11}\ \mathbf{I}_{n-1}] \in \Re^{(n-1)\times n_z}$ and $\underline{A}_{11}$ is defined in Lemma 2.1.

Since the ith component of $\dot{h}$ is written as

$$\dot{h}_i = \frac{\dot{S}_i \rho_i^2}{2\varepsilon} + \frac{S_i \rho_i}{\varepsilon} \sum_{j=1}^{n} \frac{\partial \rho_i}{\partial x_j} \dot{x}_j = \frac{\rho_i^2}{2\varepsilon} \dot{S}_i + \frac{S_i \rho_i}{\varepsilon} \bar{\Phi}_i \dot{x}$$

where $\bar{\Phi}_i$ is the ith row of the matrix Φ, the element of g in (A.15) can be decomposed into the vanishing and nonvanishing terms as follows:

$$\begin{aligned} g_1 = \dot{f}_1 + \dot{h}_1 &= \sum_{j=1}^{n} \frac{\partial f_1}{\partial x_j} \dot{x}_j + \frac{\rho_1^2}{2\varepsilon} \dot{S}_1 + \frac{S_1 \rho_1}{\varepsilon} \bar{\Phi}_1 \dot{x} \\ &= \left(\bar{J}_1 + \frac{S_1 \rho_1}{\varepsilon} \bar{\Phi}_1 \right) \dot{x} + \frac{\rho_1^2}{2\varepsilon} \dot{S}_1 \\ &= \frac{\rho_1^2}{2\varepsilon} (\bar{A}_{s1} z + d_1) + \left(\bar{J}_1 + \frac{S_1 \rho_1}{\varepsilon} \bar{\Phi}_1 \right) (C_z z + e_1 \dot{x}_{1d} + d) \\ &= \left[\frac{\rho_1^2}{2\varepsilon} \bar{A}_{s1} + \left(\bar{J}_1 + \frac{S_1 \rho_1}{\varepsilon} \bar{\Phi}_1 \right) C_z \right] z + \frac{\rho_1 \Phi_{11} \dot{x}_{1d}}{\varepsilon} S_1 \\ &\quad + \frac{\rho_1^2}{2\varepsilon} d_1 + J_{11} \dot{x}_{1d} + \left(\bar{J}_1 + \frac{S_1 \rho_1}{\varepsilon} \bar{\Phi}_1 \right) d \\ &= w_1 + \frac{\rho_1^2}{2\varepsilon} d_1 + J_{11} \dot{x}_{1d} + \left(\bar{J}_1 + \frac{S_1 \rho_1}{\varepsilon} \bar{\Phi}_1 \right) d, \end{aligned}$$

where $\bar{A}_{s1}$ is the first row of A_s and the vanishing perturbation term p_1 is

$$w_1 = \left[\frac{\rho_1^2}{2\varepsilon} \bar{A}_{s1} + \left(\bar{J}_1 + \frac{S_1 \rho_1}{\varepsilon} \bar{\Phi}_1 \right) C_z \right] z + \frac{\rho_1 \Phi_{11} \dot{x}_{1d}}{\varepsilon} e_1^T z.$$

For $2 \le i \le n-1$,

$$
\begin{aligned}
g_i = \dot{f}_i + \dot{h}_i &= \sum_{j=1}^{n} \frac{\partial f_i}{\partial x_j}\dot{x}_j + \frac{\dot{S}_i \rho_i^2}{2\varepsilon} + \frac{S_i \rho_i}{\varepsilon}\sum_{j=1}^{n}\frac{\partial \rho_i}{\partial x_j}\dot{x}_j \\
&= \left[\frac{\rho_i^2}{2\varepsilon}\bar{A}_{si} + \left(\bar{J}_i + \frac{S_i\rho_i}{\varepsilon}\bar{\Phi}_i\right)C_z\right]z + \frac{\rho_i \Phi_{i1}\dot{x}_{1d}}{\varepsilon}S_i \\
&\quad + \frac{\rho_i^2}{2\varepsilon}d_i + J_{i1}\dot{x}_{1d} + \left(\bar{J}_i + \frac{S_i\rho_i}{\varepsilon}\bar{\Phi}_i\right)d \\
&= w_i + \frac{\rho_i^2}{2\varepsilon}d_i + J_{i1}\dot{x}_{1d} + \left(\bar{J}_i + \frac{S_i\rho_i}{\varepsilon}\bar{\Phi}_i\right)d,
\end{aligned}
$$

where $J_{ij} = \partial f_i/\partial x_j$, $\Phi_{ij} = \partial \rho_i/\partial x_j$, and

$$
w_i = \left[\frac{\rho_i^2}{2\varepsilon}\bar{A}_{si} + \left(\bar{J}_i + \frac{S_i\rho_i}{\varepsilon}\bar{\Phi}_i\right)C_z\right]z + \frac{\rho_i \Phi_{i1}\dot{x}_{1d}}{\varepsilon}e_i^T z. \tag{A.18}
$$

Then, g can be written in vector form as

$$
g = w + J_1\dot{x}_{1d} + D_1 d + (J + D_2\Phi)\tilde{d} \tag{A.19}
$$

where $w = [w_1 \ \cdots \ w_{n-1}]^T \in \Re^{n_u}$.

Therefore, (A.15) can be rewritten as

$$
\begin{aligned}
\dot{z} &= A_{cl}z + B_w w + [B_w \ B_e \ B_w \ B_w]\begin{bmatrix} J_1\dot{x}_{1d} \\ \ddot{x}_{1d} \\ D_2\Phi\tilde{d} \\ J\tilde{d}\end{bmatrix} + [B_d \ B_w]\begin{bmatrix} d \\ D_1 d\end{bmatrix} \\
\Longrightarrow \quad \dot{z} &= A_{cl}z + B_w w + B_{rd}r_d + B_{nd}n_d
\end{aligned}
$$

which is written in (A.14).

Next, consider the boundedness of w and n. Using (A.18), the diagonal norm-bound of w is written as follows:

$$
\begin{aligned}
|w_i| &= \left|\left[\frac{\rho_i^2}{2\varepsilon} \ \ \bar{J}_i \ \ \frac{S_i\rho_i}{\varepsilon}\bar{\Phi}_i \ \ \frac{\rho_i}{\varepsilon}\Phi_{i1}\dot{x}_{1d}\right]\begin{bmatrix}\bar{A}_{si}z \\ C_z z \\ C_z z \\ e_i^T z\end{bmatrix}\right| := \left|a_i^T C_{wi} z\right| & \text{(A.20)} \\
&\le \|a_i\|\|C_{wi}z\| & \text{(A.21)}
\end{aligned}
$$

where the inequality comes from Cauchy–Schwartz inequality and

$$
a_i = \begin{bmatrix} \frac{\rho_i^2}{2\varepsilon} \\ \bar{J}_i^T \\ \frac{S_i\rho_i}{\varepsilon}\bar{\Phi}_i^T \\ \frac{\rho_i}{\varepsilon}\Phi_{i1}\dot{x}_{1d}\end{bmatrix} \in \Re^{2n_u+2}, \qquad C_{wi} = \begin{bmatrix}\bar{A}_{si} \\ C_z \\ C_z \\ e_i^T\end{bmatrix} \in \Re^{(2n_u+2)\times n_z}.
$$

It is noted that a_i includes all variables and C_{wi} is given for a fixed set of controller gains. Based on the assumptions in (3.16), $\dot{x}_{1d}$ is a feasible trajectory on the convex set $\mathscr{D}_i$, J_{ij}, Φ_{ij}, and ρ_i are bounded on $\mathscr{D}$, and z is bounded on the convex set $\Omega_c = \{z \in \Re^{n_z} | V(z) \leq c\}$ where $V(z)$ is a continuously differentiable, positive definite function. Therefore, since all elements of a_i are bounded on $\mathscr{D}_i \times \Omega_c$, there exists a positive constant μ_i such that $\|a_i\|_2 \leq \mu_i$ and thus there exists $\tilde{C}_{wi}$ such that

$$|w_i| \leq \mu_i \|C_{wi} z\| := \|\tilde{C}_{wi} z\|$$

where $\tilde{C}_{wi} = \mu_i C_{wi}$.

Finally, using Young's inequality, the inequality condition of d can be derived as follows:

$$\begin{aligned} &\frac{S_i^2 \rho_i^2}{2\varepsilon} + \frac{\varepsilon}{2} \geq |S_i| \rho_i \geq |S_i||\Delta f_i| \geq S_i \Delta f_i \\ &\Longrightarrow \quad S_i \Delta f_i - \frac{S_i^2 \rho_i^2}{2\varepsilon} = S_i d_i \leq \frac{\varepsilon}{2} \\ &\Longrightarrow \quad \sum_{i=1}^{n_u} S_i d_i = z^T \mathbf{I}_{n_z \times n_u} d \leq \frac{n_u \varepsilon}{2}. \end{aligned}$$

Then,

$$z^T En = z^T \begin{bmatrix} \mathbf{I}_{n_u} & \mathbf{I}_{n_u} \\ \mathbf{0} & \mathbf{0} \end{bmatrix} \begin{bmatrix} \mathbf{I}_{n_u} \\ D_1 \end{bmatrix} d = \sum_{i=1}^{n_u} \left(1 + \frac{\rho_i^2}{2\varepsilon}\right) S_i d_i \leq \frac{n_u \varepsilon}{2} + \sum_{i=1}^{n_u} \frac{\rho_i^2}{4}.$$

Since ρ_i is continuous and Φ_{ij} is bounded on $\mathscr{D}$, there exists ρ_{max} such that

$$\max_i \left|\rho_i(x)\right| \leq \rho_{\max}$$

for all i and $x \in \mathscr{D}$. Therefore, there exists $n_0 > 0$ for a given $\varepsilon > 0$ such that

$$z^T En \leq \frac{n_u \varepsilon}{2} + \frac{n_u \rho_{\max}^2}{4} := n_0(\varepsilon). \tag{A.22}$$

□

A.4 Proof of Theorem 5.3

Theorem 5.3 *Suppose there exist* $\{P_1, \sigma_1, \lambda_1\}$ and $\{P_2, \sigma_2, \lambda_2\}$ *satisfying all LMIs in Theorem* 5.1 *where* $\lambda_1 > \lambda_2$ *(see also in Fig.* 5.1*). Furthermore, if* P_1, σ_1, λ_1, and λ_2 *satisfy*

$$\begin{bmatrix} A_0^T P + P A_0 + (\lambda_1 - \lambda_2) c c^T + \sigma \gamma^2 C_z^T C_z & P B_u & P B_w \\ B_u^T P & -\frac{1}{\lambda_1 - \lambda_2} & 0 \\ B_w^T P & 0 & -\sigma I \end{bmatrix} < 0 \tag{A.23}$$

where $A_0 := A_{op} + \lambda_1 B_u c^T$. *Then, there exist* $\tilde{P}(\lambda)$ and $\tilde{\sigma}(\lambda)$ *satisfying Theorem* 5.1 *for any* $A_0(\lambda) = A_{op} + \lambda B_u c^T$, $\lambda \in [\lambda_2, \lambda_1]$.

Proof Suppose $\tilde{P}_t = (1-t)P_1 + tP_2 > 0$ for $t \in [0, 1]$, which is a convex combination of P_1 and P_2. Similarly, let $\tilde{\sigma}_t = (1-t)\sigma_1 + t\sigma_2 \geq 0$. Moreover, denote

$\varepsilon_\lambda \triangleq \mathscr{E}[\tilde{P}_t] = \{z \in \Re^{n_z} | z^T \tilde{P}_t z \le 1, \forall t \in [0,1]\}$. First, we will show the existence of $\tilde{P}_t$ for the inequality (5.15) for an arbitrary $\lambda^* \in [\lambda_2, \lambda_1]$. Using the Schur complement of (5.15), we have

$$\begin{aligned}\mathscr{F}[\tilde{P}_t, \lambda^*] &\triangleq z^T \tilde{P}_t z - \frac{\lambda^{*2}}{u_0^2}|c^T z|^2 = (1-t)z^T P_1 z + t z^T P_2 z - \frac{\lambda^{*2}}{u_0^2}|c^T z|^2 \\ &\ge \frac{|c^T z|^2}{u_0^2}\{(1-t)\lambda_1^2 + t\lambda_2^2 - \lambda^{*2}\}\end{aligned}$$

for $\forall z \in \varepsilon_{\lambda^*} \subset \mathscr{E}(P)$ and the last inequality comes from the inequality condition (5.15) such that $z^T P_1 z \ge \frac{\lambda_1^2 |c^T z|^2}{u_0^2}$ holds for $\forall z \in \mathscr{E}(P_1)$ and $z^T P_2 z \ge \frac{\lambda_2^2 |c^T z|^2}{u_0^2}$ for $\forall z \in \mathscr{E}(P_2)$. Therefore, there exists a $\tilde{P}$ for any $t \in (0, t_1]$ such that

$$\mathscr{F}\big[\tilde{P}(t), \lambda^*\big] \ge 0, \quad \forall\, z \in \mathscr{E}(\tilde{P})$$

where $t_1 = \frac{\lambda_1^2 - \lambda^{*2}}{\lambda_1^2 - \lambda_2^2}$. The rest of proof is to show that the existence of $\tilde{P}_t$ satisfies the inequality (5.14) for λ^*. Using the Schur complement, denote

$$\begin{aligned}\mathscr{L}[\tilde{P}, \tilde{\sigma}, \lambda^*] \triangleq{}& z^T\big(A_{cl}^T \tilde{P} + \tilde{P} A_{cl}\big)z + 2(\lambda^* - 1)z^T \tilde{P} B_u K z \\ &+ 2z^T \tilde{P} B_w w - \tilde{\sigma}\big\{\|w\|^2 - \gamma^2 \|C_z z\|^2\big\}\end{aligned}$$

for $\forall z \in \varepsilon_{\lambda^*}$. Then

$$\mathscr{L}[\tilde{P}, \tilde{\sigma}, \lambda^*] = (1-t)\mathscr{L}[P_1, \sigma_1, \lambda^*] + t\mathscr{L}[P_2, \sigma_2, \lambda^*].$$

Claim (see Claim 1 in [71]) Given any $\delta(\lambda) \in \Re$ such that $|\delta|^2 \le (\lambda_1 - \lambda_2)^2$,

$$2\delta z^T P B_u c^T z \le (\lambda_1 - \lambda_2) z^T P B_u B_u^T P z + (\lambda_1 - \lambda_2) z^T c c^T z$$

for all $z \in \Re^{n_z}$.

Using the claim, we have

$$\begin{aligned}\mathscr{L}[P_1, \sigma_1, \lambda^*] &= \mathscr{L}[P_1, \sigma_1, \lambda_1] + 2(\lambda^* - \lambda_1)z^T P_1 B_u c^T z \\ &\le \mathscr{L}[P_1, \sigma_1, \lambda_1] + (\lambda_1 - \lambda_2)z^T P_1 B_u B_u^T P_1 z + (\lambda_1 - \lambda_2)z^T c c^T z.\end{aligned}$$

By the inequality condition (5.18),

$$\mathscr{L}[P_1, \sigma_1, \lambda_1] + (\lambda_1 - \lambda_2)z^T P_1 B_u B_u^T P_1 z + (\lambda_1 - \lambda_2)z^T c c^T z < 0.$$

Therefore,

$$\mathscr{L}[P_1, \sigma_1, \lambda^*] < 0$$

for all $z \in \mathscr{E}(P_1)$. Furthermore, since $\mathscr{E}[P_1]$ and $\mathscr{E}[P_2]$ are compact, there exist positive scalars $\bar{r}$ and $\bar{p}$ such that

$$\mathscr{L}[P_2, \sigma_2, \lambda^*] \le \bar{r} \quad \text{and} \quad \mathscr{L}[P_1, \sigma_1, \lambda^*] \le -\bar{p}.$$

Therefore, there exist $\tilde{P}_t$ for any $t \in (0, t_2]$ such that

$$\mathscr{L}[\tilde{P}_t, \tilde{\sigma}, \lambda^*] < 0, \quad \forall z \in \varepsilon_\lambda \subset \mathscr{E}(P_1)$$

where $t_2 = \frac{\bar{p}}{\bar{r}+\bar{p}}$. The above inequality is equivalent to the matrix inequality (5.14) in Theorem 5.1 for all $\lambda \in [\lambda_1, \lambda_2]$.

Finally, let $t^* = \min(t_1, t_2)$. Then, for any $t \in (0, t^*]$, there exist $\tilde{P}_t(\lambda)$ and $\tilde{\sigma}_t(\lambda)$ satisfying LMI (5.14) and (5.15) for any $\lambda \in [\lambda_1, \lambda_2]$. □

References

1. Aboky, C., Sallet, G., Vivalda, J.C.: Observers for Lipschitz nonlinear systems. Int. J. Control **75**(3), 204–212 (2002)
2. Ahmed-Ali, T., Lamnabhi-Lagarrigue, F.: Sliding observer-controller design for uncertain triangular nonlinear systems. IEEE Trans. Autom. Control **44**(6), 1244–1249 (1999)
3. Arcak, M., Kokotović, P.: Nonlinear observers: a circle criterion design and robustness analysis. Automatica **37**(12), 1923–1930 (2001)
4. Aubin, J.P., Cellina, A.: Differential Inclusions. A Series of Comprehensive Studies in Mathematics, vol. 264. Springer, Berlin (1984)
5. Bae, H., Ryu, J., Gerdes, C.: Road grade and vehicle parameter estimation for longitudinal control using gps. In: Proceedings of IEEE Intelligent Transportation Systems Conference, pp. 166–171 (2001)
6. Barmish, B.R.: Necessary and sufficient conditions for quadratic stability of an uncertain system. J. Optim. Theory Appl. **46**(4), 399–408 (1985)
7. Basseville, M., Nikiforov, I.V.: Detection of Abrupt Changes. Prentice Hall, New York (1993)
8. Blanchini, F.: Set invariance in control. Automatica **35**, 1747–1767 (1999)
9. Blanchini, F., Miani, S.: Constrained stabilization via smooth Lyapunov functions. Syst. Control Lett. **35**, 155–163 (1998)
10. Bow, S.-T.: Pattern Recognition and Image Preprocessing. Dekker, New York (1992)
11. Boyd, S., Balaktrishnan, V., Kabamba, P.: A bisection method for computing the h_∞ norm of a transfer matrix and relative problems. Math. Control Signals Syst. **2**, 207–219 (1989)
12. Boyd, S., El Ghaoui, L., Feron, E., Balakrishnan, V.: Linear Matrix Inequalities in System and Control Theory. SIAM, Philadelphia (1994)
13. Boyd, S., Vandenberghe, L.: Convex Optimization. Cambridge University Press, Cambridge (2004)
14. Brockman, M., Corless, M.: Quadratic boundedness of nonlinear dynamical systems. In: Proceedings of 34th IEEE Conference on Decision and Control, New Orleans, LA, pp. 504–509 (1995)
15. Brockman, M., Corless, M.: Quadratic boundedness of nominally linear systems. Int. J. Control **71**(6), 1105–1117 (1998)
16. Canudas-De-Wit, C., Olsson, H., Astrom, K.J., Lischinsky, P.: A new model for control of system with friction. IEEE Trans. Autom. Control **40**(3), 419–425 (1995)
17. Chemori, A., Loria, A.: Control of a planar underactuated biped on a complete walking cycle. IEEE Trans. Autom. Control **49**(5), 838–843 (2004)
18. Chevallereau, C., Abba, G., Aoustin, Y., Plestan, F., Westervelt, E.R., Canudas-De-Wit, C., Grizzle, J.W.: Rabbit: a testbed for advanced control theory. IEEE Control Syst. Mag. **23**, 57–79 (2003)

B. Song, J.K. Hedrick, *Dynamic Surface Control of Uncertain Nonlinear Systems*,
Communications and Control Engineering,
DOI 10.1007/978-0-85729-632-0,

19. Cho, D., Hedrick, J.K.: Automotive power modeling for control. J. Dyn. Syst. Meas. Control **111**, 568–576 (1989)
20. Choi, S.: Design of a robust controller for automotive engines: theory and experiment. Ph.D. thesis, University of California at Berkeley (1993)
21. Choi, J.-W., Song, B., Park, T.W.: A study on modeling of gait cycle and its zmp variation for biped. In: Proceedings of KSMTE Autumn Conference, Suwon, Korea, pp. 57–62 (2005)
22. Corless, M., Leitmann, G.: Continuous state feedback guaranteeing uniform ultimate boundedness for uncertain dynamical systems. IEEE Trans. Autom. Control **26**(5), 1139–1144 (1981)
23. Corless, M., Leitmann, G.: Bounded controllers for robust exponential convergence. J. Optim. Theory Appl. **76**(1), 1–12 (1993)
24. Emelyanov, S.V., Korovin, S.K., Nersisyan, A.L., Nisenzov, Y.: Output feedback stabilization of uncertain plants. Int. J. Control **55**(1), 61–81 (1992)
25. Frank, P.M.: Fault diagnosis in dynamic systems using analytical and knowledge-based redundancy—a survey and some new results. Automatica **26**(3), 459–474 (1990)
26. Frank, P.M., Ding, X.: Survey of robust residual generation and evaluation methods in observer-based fault detection systems. J. Process Control **7**(6), 403–424 (1997)
27. Gavel, D.T., Siljak, D.D.: Decentralized adaptive control: structural conditions for stability. IEEE Trans. Autom. Control **34**(4), 413–426 (1989)
28. Gerdes, J.C.: Decoupled design of robust controllers for nonlinear systems: As motivated by and applied to coordinated throttle and brake control for automated highways. Ph.D. thesis, U.C. Berkeley, March 1996
29. Gerdes, J.C., Hedrick, J.K.: "loop-at-a-time" design of dynamic surface controllers for nonlinear systems. J. Dyn. Syst. Meas. Control **124**, 104–110 (2002)
30. Gertler, J.J.: Survey of model-based failure detection and isolation in complex plants. IEEE Control Syst. Mag. **8**(6), 3–11 (1988)
31. Gong, Z., Wen, C., Mital, D.P.: Decentralized robust controller design for a class of interconnected uncertain systems: with unknown bound of uncertainty. IEEE Trans. Autom. Control **41**(6), 850–854 (1996)
32. Grant, M., Boyd, S.: CVX: Matlab software for disciplined convex programming, version 1.21. http://cvxr.com/cvx/ (2011)
33. Green, J.H., Hedrick, J.K.: Nonlinear speed control of automotive engines. In: Proceedings of the American Control Conference, San Diego, CA, pp. 2891–2897 (1990)
34. Greenwood, D.T.: Principles of Dynamics, 2nd edn. Prentice Hall, New York (1988)
35. Grizzle, J.W., Abba, G., Plestan, F.: Asymptotically stable walking for biped robots: analysis via systems with impulse effects. IEEE Trans. Autom. Control **46**(1), 51–64 (2001)
36. Hedrick, J.K., Won, M.C., Choi, S.-B.: Fuel injection control of automotive engines. In: Proceedings of IEEE CDC, Buena Vista, FL (1994)
37. Hedrick, J.K., Yip, P.P.: Multiple sliding surface control: theory and application. J. Dyn. Syst. Meas. Control **122**, 586–593 (2000)
38. Henrion, D., Tarbouriech, S.: Lmi relaxations for robust stability of linear systems with saturating controls. Automatica **35**, 1599–1604 (1999)
39. Howell, A., Hedrick, J.K.: Nonlinear observer design via convex optimization. In: American Control Conference, pp. 2088–2093 (2002)
40. Howell, A., Song, B., Hedrick, J.K.: Cooperative range estimation and sensor diagnostics for vehicle control. In: Proceedings of ASME IMECE, p. DSC TOC (2003)
41. Hu, T., Lin, Z.: On enlarging the basin of attraction for linear systems under saturated linear feedback. Syst. Control Lett. **40**, 59–69 (2000)
42. Ioannou, P.A.: Decentralized adaptive control of interconnected systems. IEEE Trans. Autom. Control **31**(4), 291–298 (1986)
43. Isermann, R.: Process fault detection based on modeling and estimation methods—a survey. Automatica **20**(4), 387–404 (1984)
44. Isidori, A.: Nonlinear Control Systems, 3rd edn. Springer, Berlin (1995)
45. Jain, S., Khorrami, F.: Decentralized adaptive control of a class of large-scale interconnected nonlinear systems. IEEE Trans. Autom. Control **42**(2), 136–154 (1997)

46. Jain, S., Khorrami, F.: Decentralized adaptive output feedback design for large-scale nonlinear systems. IEEE Trans. Autom. Control **42**(5), 729–735 (1997)
47. Jankovic, M.: Adaptive nonlinear output feedback tracking with a partial high-gain observer and backstepping. IEEE Trans. Autom. Control **42**(1), 106–113 (1997)
48. Jiang, Z.-P.: Decentralized and adaptive nonlinear tracking of large-scale systems via output feedback. IEEE Trans. Autom. Control **45**(11), 2122–2128 (2000)
49. Johansson, M., Rantzer, A.: Computation of piecewise quadratic Lyapunov functions for hybrid systems. IEEE Trans. Autom. Control **43**(4), 555–559 (1998)
50. Keller, H.: Nonlinear observer design by transformations into a generalized observer canonical form. Int. J. Control **46**, 1915–1930 (1987)
51. Khalil, H.K.: Nonlinear Systems, 3rd edn. Prentice-Hall, New York (2002)
52. Kokotović, P., Arcak, M.: Constructive nonlinear control: a historical perspective. Automatica **37**, 637–662 (2001)
53. Krener, A.J., Isidori, A.: Linearization by output injection and nonlinear observers. Syst. Control Lett. **3**, 47–52 (1983)
54. Krener, A.J., Respondek, W.: Nonlinear observers with linearizable error dynamics. SIAM J. Control Optim. **23**(2), 197–216 (1985)
55. Krstić, M., Kanellakopoulous, I., Kokotović, V.P.: Nonlinear and Adaptive Control Design. Wiley, New York (1995)
56. Leitmann, G.: Guaranteed asymptotic stability for some linear systems with bounded uncertainties. J. Dyn. Syst. Meas. Control **101**(3), 212–216 (1979)
57. Limpert, R.: Brake Design and Safety. SAE, Warrendale (1992)
58. Lygeros, J., Godbole, D., Brouche, M.: A fault tolerant control architecture for automated highway systems. IEEE Trans. Control Syst. Technol. **8**(2), 205–219 (2000)
59. Maciuca, D.B.: Nonlinear robust and adaptive control with application to brake control for automated highway systems. Ph.D. thesis, University of California at Berkeley (1997)
60. Marhefka, D.W., Orin, D.: Simulation of contact using a nonlinear damping model. In: Proceedings of IEEE International Conference on Robotics and Automation, Minneapolis, MN, pp. 1662–1668 (1996)
61. McMahon, D.H.: Robust nonlinear control of uncertain systems: an application to intelligent vehicle highway systems. Ph.D. thesis, University of California at Berkeley (1994)
62. Misawa, E.A., Hedrick, J.K.: Nonlinear observers: a state-of-the-art survey. J. Dyn. Syst. Meas. Control **11**, 344–352 (1989)
63. Mu, X., Wu, Q.: Development of a complete dynamic model of a planar five-link biped and sliding mode control of its locomotion during the double support phase. Int. J. Control **77**(8), 789–799 (2004)
64. Nijmeijer, H., van der Schaft, A.: Nonlinear Dynamical Control Systems. Springer, New York (1990)
65. Noura, H., Fonte, C., Robert, M.: Fault tolerant control using simultaneous stabilization. In: Proceedings of IEEE International Conference on Systems, Man and Cybernetics, New York, NY, vol. 3, pp. 605–610 (1993)
66. Noura, H., Sauter, D., Hamelin, F., Theilliol, D.: Fault-tolerant control in dynamic systems: application to a winding machine. IEEE Control Syst. Mag. **20**(1), 33–49 (2000)
67. Oh, S., Khalil, H.K.: Nonlinear output-feedback tracking using high-gain observer and variable structure control. Automatica **33**(10), 1845–1856 (1997)
68. Ortega, R., Loría, A., Nicklasson, P.J., Sira-Ramirez, H.: Passivity-based Control of Euler-Lagrange Systems: Mechanical, Electrical and Eletromechanical Applications. Communications and Control Engineering. Springer, Berlin (1998)
69. Park, J.H., Kim, K.D.: Biped robot walking using gravity-compensated inverted pendulum mode and computed torque control. In: Proceedings of IEEE International Conference on Robotics and Automation, Leuven, Belgium, pp. 3528–3533 (1998)
70. Petersen, I.R.: Quadratic stabilizability of uncertain linear systems: existence of a nonlinear stabilizing control does not imply existence of a linear stabilizing control. IEEE Trans. Autom. Control **30**(3), 291–293 (1985)

71. Petersen, I.R.: Notions of stability and controllability for a class of uncertain linear systems. Int. J. Control **46**(2), 409–422 (1987)
72. Plestan, F., Grizzle, J.W., Westervelt, E.R., Abba, G.: Stable walking of a 7-dof biped robot. IEEE Trans. Robot. Autom. **19**(4), 653–668 (2003)
73. Raghavan, S., Hedrick, J.K.: Observer design for a class of nonlinear systems. Int. J. Control **59**(2), 515–528 (1994)
74. Rajamani, R., Cho, Y.M.: Existence and design of observers for nonlinear systems: relation to distance to unobservability. Int. J. Control **69**(5), 717–731 (1998)
75. Rajamani, R., Howell, A.S., Chen, C., Hedrick, J.K.: A complete fault diagnostic system for the longitudinal control of automated vehicles. In: Proceedings of ASME Winter Conference, Dallas, Texas, November 1997
76. Rajamani, R., Howell, A., Chen, C., Hedrick, J.K., Tomizuka, M.: A complete fault diagnostic system for automated vehicles operating in a platoon. IEEE Trans. Control Syst. Technol. **9**(4), 553–564 (2001)
77. SAE International: SAE truck and bus control and communications network standards manual, 2001 edn. (2001)
78. Sandell, N., Varaiya, P., Athans, M., Safonov, M.: Survey of decentralized control methods for large scale systems. IEEE Trans. Autom. Control **23**(2), 108–128 (1978)
79. Slotine, J.J., Hedrick, J.K.: Robust input-output feedback linearization. Int. J. Control **57**(5), 1133–1139 (1993)
80. Slotine, J.J., Li, W.P.: Applied Nonlinear Control. Prentice-Hall, New York (1991)
81. Song, B.: Decentralized dynamic surface control for a class of interconnected nonlinear systems. In: Proceedings of the American Control Conference, pp. 130–135 (2006)
82. Song, B., Hedrick, J.K.: Design and experimental implementation of longitudinal control for automated transit buses. In: Proceedings of the American Control Conference, pp. 2751–2756 (2004)
83. Song, B., Hedrick, J.K.: Observer-based dynamic surface control for a class of nonlinear systems: an lmi approach. IEEE Trans. Autom. Control **49**(11), 1995–2001 (2004)
84. Song, B., Hedrick, J.K.: Simultaneous quadratic stabilization for a class of nonlinear systems with input saturation using dynamic surface control. Int. J. Control **77**(1), 19–26 (2004)
85. Song, B., Hedrick, J.K.: Fault tolerant nonlinear control with applications to an automated transit bus. Veh. Syst. Dyn. **43**(5), 331–350 (2005)
86. Song, B., Hedrick, J.K.: Design of dynamic surface control for fully-actuated mechanical systems. In: IFAC World Congress (2011)
87. Song, B., Hedrick, J.K.: Nonlinear observer design for Lipschitz nonlinear systems. In: Proceedings of the American Control Conference (2011)
88. Song, B., Hedrick, J.K., Howell, A.: Robust stabilization and ultimate boundedness of dynamic surface control systems via convex optimization. Int. J. Control **75**(12), 870–881 (2002)
89. Song, B., Hedrick, J.K., Howell, A.: Fault tolerant control and classification for longitudinal vehicle control. J. Dyn. Syst. Meas. Control **125**, 320–329 (2003)
90. Spong, M.W.: Modeling and control of elastic joint robots. J. Dyn. Syst. Meas. Control **109**, 310–319 (1987)
91. Spong, M.W., Vidyasagar, M.: Robot Dynamics and Control. Wiley, New York (1991)
92. Spooner, J., Passino, K.: Fault-tolerant control for automated highway systems. IEEE Trans. Veh. Technol. **46**(3), 770–785 (1997)
93. Sturm, J.F.: Using SeDuMi 1.02, a matlab toolbox for optimization over symmetric cones
94. Swaroop, D., Hedrick, J.K., Yip, P.P., Gerdes, J.C.: Dynamic surface control for a class of nonlinear systems. IEEE Trans. Autom. Control **45**(10), 1893–1899 (2000)
95. Šiljak, D.D.: Reliable control using multiple control systems. Int. J. Control **31**(2), 303–329 (1980)
96. Šiljak, D.D.: Decentralized Control of Complex Systems. Academic Press, San Diego (1991)
97. Thau, F.E.: Observing the state of non-linear dynamic systems. Int. J. Control **17**(3), 471–479 (1973)

98. The Mathworks: Symbolic math toolbox 5 user's guide. Available in http://www.mathworks.com/products/symbolic/ (2010)
99. Therrien, C.: Decision, Estimation, and Classification. Wiley, New York (1989)
100. Toh, K., Tütüncü, R., Todd, M.: SDPT3 4.0 (beta) (software package) (2006)
101. Tzafestas, S., Raibert, M., Tzafestas, C.: Robust sliding-mode control applied to a 5-link biped robot. J. Intell. Robot. Syst. **15**, 67–133 (1996)
102. Utkin, V.I., Chen, D.-S., Chang, H.-C.: Block control principle for mechanical systems. J. Dyn. Syst. Meas. Control **122**, 1–10 (2000)
103. Vidyasagar, M.: Control System Synthesis: A Factorization Approach. The MIT Press Series in Signal Processing, Optimization, and Control, vol. 7. MIT Press, Cambridge (1985)
104. Vukobratovic, M., Borovac, B., Surla, D., Stokic, D.: Biped Locomotion. Springer, Berlin (1990)
105. Wen, C., Soh, Y.C.: Decentralized adaptive control using integrator backstepping. Automatica **33**(9), 1719–1724 (1997)
106. Willsky, A.S.: A survey of design methods for failure detection in dynamic systems. Automatica **12**(6), 601–611 (1976)
107. Won, M., Hedrick, J.K.: Multiple surface sliding control for a class of nonlinear systems. Int. J. Control **64**(4), 693–706 (1996)
108. Yang, S., Song, B., Um, J.: Heterogeneous sensor fusion for primary vehicle detection. In: Proceedings of International Conference on Control and Applications (2010)
109. Yi, J., Alvarez, L., Howell, A., Horowitz, R., Hedrick, J.K.: A fault management system for longitudinal vehicle control in ahs. In: Proceedings of American Control Conference, vol. 3, pp. 1514–1518 (2000)
110. Yip, P.P.: Robust and adaptive nonlinear control using dynamic surface controller with applications to intelligent vehicle highway systems. Ph.D. dissertation, University of California, Berkeley (1997)
111. Yip, P.P., Hedrick, J.K.: Adaptive dynamic surface control: a simplified algorithm for adaptive backstepping control of nonlinear systems. Int. J. Control **71**(5), 959–979 (1998)
112. You, B.-J., Oh, Y.-H., Choi, Y.-J.: Survey on humanoid researches. J. Korean Soc. Precis. Eng. **21**(7), 15–21 (2004)
113. Zhang, Y., Jiang, J.: Integrated design of reconfigurable fault-tolerant control systems. J. Guid. Control Dyn. **24**(1), 133–136 (2001)
114. Zheng, D.Z.: Decentralized output feedback stabilization of a class of nonlinear interconnected systems. IEEE Trans. Autom. Control **34**(12), 1297–1300 (1989)

Index

B. Song, J.K. Hedrick, *Dynamic Surface Control of Uncertain Nonlinear Systems*,
Communications and Control Engineering,
DOI 10.1007/978-0-85729-632-0,